HOME THEATER HACKS™

Other resources from O'Reilly

Related titles

Smart Home Hacks™
Gaming Hacks™
TiVo Hacks™
Building the Perfect PC
Hardware Hacking Projects for Geeks
Home Hacking Projects for Geeks

Hacks Series Home

hacks.oreilly.com is a community site for developers and power users of all stripes. Readers learn from each other as they share their favorite tips and tools for Mac OS X, Linux, Google, Windows XP, and more.

oreilly.com

oreilly.com is more than a complete catalog of O'Reilly books. You'll also find links to news, events, articles, weblogs, sample chapters, and code examples.

oreillynet.com is the essential portal for developers interested in open and emerging technologies, including new platforms, programming languages, and operating systems.

Conferences

O'Reilly brings diverse innovators together to nurture the ideas that spark revolutionary industries. We specialize in documenting the latest tools and systems, translating the innovator's knowledge into useful skills for those in the trenches. Visit *conferences.oreilly.com* for our upcoming events.

Safari Bookshelf (*safari.oreilly.com*) is the premier online reference library for programmers and IT professionals. Conduct searches across more than 1,000 books. Subscribers can zero in on answers to time-critical questions in a matter of seconds. Read the books on your Bookshelf from cover to cover or simply flip to the page you need. Try it today with a free trial.

HOME THEATER HACKS™

Brett McLaughlin

O'REILLY®

Beijing • Cambridge • Farnham • Köln • Paris • Sebastopol • Taipei • Tokyo

Home Theater Hacks™
by Brett McLaughlin

Published by O'Reilly Media, Inc., 1005 Gravenstein Highway North, Sebastopol, CA 95472.

O'Reilly books may be purchased for educational, business, or sales promotional use. Online editions are also available for most titles (*safari.oreilly.com*). For more information, contact our corporate/institutional sales department: (800) 998-9938 or *corporate@oreilly.com*.

Editor: Brian Jepson
Series Editor: Rael Dornfest
Executive Editor: Dale Dougherty
Production Editor: Mary Anne Weeks Mayo
Cover Designer: Hanna Dyer
Interior Designer: Melanie Wang

Printing History:

November 2004: First Edition.

This book uses Otabind™, a durable and flexible lay-flat binding.

ISBN: 0-596-00704-3
[I]

Contents

Credits

About the Author

Brett McLaughlin is best known for his writing on Java (as in the programming language rather than the coffee). However, those who are closest to him realize Java is largely a means to an end—something to finance his more expensive habits, such as home theater.

Hooked on movies early, Brett has put together numerous complete home theater setups, including a total remodel of an attached garage, complete with columns, wood trim, and a 7.1-channel speaker bonanza. In the process, he's spent hundreds of hours running cables through walls, connecting components, and generally cursing George Lucas, Steven Spielberg, and anyone else who created a market for such a complicated hobby.

More often, though, you'll find Brett online, writing and editing for O'Reilly Media, Inc. You can find most of his books at *http://java.oreilly.com*, and when he's not neck deep in Java, he's probably watching the Dallas Stars or marveling at the marvelous pennant run of the Texas Rangers in 2004. Most of all, he's into what's interesting, which ranges from his two children to long discussions about absolutism and relativism. You've been forewarned...

Contributors

Lots of people still make my head spin when it comes to talking about home theater. I take great pride that almost all of them took time out of their ridiculously busy schedules to work on this book. Some contributed a single hack, and others went to town and practically wrote entire chapters. It turned this book into a monster, and I'm proud to run down the lineup:

- Dustin Bartlett is a graduate of the University of Saskatchewan with a degree in computer science and biology (focused on genetics). He is currently working as a programmer/analyst with Point2 Technologies, Inc.

He entered the world of home theater during his second year of university knowing next to nothing about the subject. Seven years and three major upgrades later his addiction to the hobby has only grown stronger. His personal home theater web site is *http://dustin.bunnyhug.net* (you have to be from Saskatchewan to get it ;).

- G. Alan Brown entered the custom home theater industry in 1997. He started CinemaQuest, Inc. (*http://www.cinemaquestinc.com*) in 1998, offering products and services perfecting home theater. CinemaQuest, Inc. designs and installs custom home theaters and integrated electronic residential systems, and also provides electronic display calibration and multichannel audio system calibration services. Alan also recognized a need for providing ideal viewing environment technologies and solutions for consumers and professionals (*http://www.ideal-lume.com*). CinemaQuest, Inc. is the world leader in electronic display viewing environment solutions.
- Michael Chen has been an enthusiast of home video ever since 1980, when his parents bought the family's first Betamax VCR. The hobby grew from there to constant upgrading of equipment and the addition of poor-man surround sound in those early days. A calibrationist for Lion A/V (*http://www.lionav.com*), he is unaligned with any company, so is free to criticize any type of display device on the market. As Michael says, "The man with the brutal truth—as I am not there to tell you how great your equipment purchases are. I only care about the image and I won't stop until the job is done to my satisfaction. (My Curse.)"
- Dr. Robert A. Fowkes has been a computer coordinator, a chemistry teacher, a staff member on National Science Foundation Summer Institutes, a consultant, and a moderator on the Home Theater Forum. He's been into computers since 1957 (when he built one from scratch!) and home theater since there was such a thing.
- David Gibbons is an engineering technician whose experience spans radio, telecommunications, early computer technology, high-end test and measurement, and (more recently) environmental simulation testing technology. (OK, he's a jack of all trades.) His web site (*http://www.sonic.net/~dgibbons*) has annoyed some "high-end" folks already, and probably will continue to do so.
- Joseph Jenkins is the site owner and operator of HDTVoice.com (*http://www.hdtvoice.com*), a discussion forum dedicated to all things high-def! When not stuck to his PC, he is either attending a Philadelphia Eagles game (diehard fan), or spending time with his wife, Kim, and their cats, Indiana and Ripley.

- Robert Jones (*http://www.imageperfection.com*) is a professional ISF calibrator—the final and most visually crucial stage in the long process leading to the completion of the video phenomenon we have come to know as home theater. Already in the audio-video business for more than 20 years in the San Francisco Bay area, he left all that behind and went calibrator nearly exclusively. He has worked on and repaired countless units, from small Walkmans to high-precision, ceiling-hung video projectors. He also has calibrated many of the different types of venue currently available, which presently include CRT, DLP, LCD, plasma, LCOS, and DILA. He is fascinated by any calibration venue, even those that are not yet in existence, and works on most of the brands available.
- Gregg Loewen is the proprietor of Lion Audio/Video Consultants (*http://www.lionav.com*). He is a long-time home theater enthusiast and has been a professional calibrator since the summer of 2001. His consulting work includes Atlantic Technology, Outlaw Audio, Optoma Projectors, 20th Century Fox, Widescreen Review, and Milori. He is also a Runco Certified Installer.
- Vince Maskeeper-Tennant (superhero alter ego of the mild-mannered Vince Tennant) is a graduate of Kent State University, currently working as an audio engineer for film and TV in Hollywood. A lifetime member of the A/V Club and self-proclaimed uber-nerd, Vince spends the majority of his time obsessing endlessly over the aural spectrum. When he finds spare time, he serves as a moderator and guru on the Home Theater Forum (*http://www.hometheaterforum.com*). You should stop by and say hello.
- Robert McElfresh knows how to deal with the simpler things in life. An administrator on the Home Theater Forum (*http://www.hometheaterforum.com*), Bob specializes in making complex problems easy (such as choosing from several-hundred-dollar sets of speaker cable).
- Kenneth L. Nist is the author of the *HDTV Primer* (*http://www.hdtvprimer.com*). He holds a BS and MS from Carnegie Mellon in electrical engineering, and is an avid FCC nut. Catch him at KQ6QV.
- Tim Procuniar always has been one to experiment with unusual audio setups. His last great two-channel stereo system included two EV-30w 30-inch subwoofers, University 12-inch woofers, EV horns for the midrange, and Heil ESS Air Motion Transformers for the highs. Because no crossovers were available for this unique setup, he had to made his own—hand-winding the coils. No stranger to home theater, he's even started and owned a speaker shop, complete with his own original prototype designs.

- Mark D. Rejhon of *http://www.marky.com* was the moderator of the first Home Theater Computers forum at *http://www.avsforum.com* from 1999 to 2001, and made major contributions to DScaler (*http://www.dscaler.com*) as well as several magazine articles. He now runs a contract software engineering business for the home theater industry, Rejhon Technologies Inc., online at *http://www.rejtech.com*.

You can't do any research with audio cables without eventually finding references to a gentleman named Chris VenHaus. Mr. VenHaus has become known as one of the pioneers of the high-end DIY (do-it-yourself) cable world. His highly regarded cable and interconnect designs have been made and used by thousands of audiophiles and videophiles worldwide. The company he founded (VH Audio, at *http://vhaudio.com*) offers finished cables/interconnects, DIY supplies, and "tweaks" for audio and home theater systems. You could spend hours visiting his audio and photography web sites, all accessible from *http://www.venhaus1.com*.

BetterCables.com (*http://www.bettercables.com*) is an *eight*-time inductee into AudioReview.com's cable Hall of Fame and winner of the prestigious ConsumerReview.com's 2000 CHOICE Award from AudioREVIEW.com.

Acknowledgments

I'll keep it brief; I've said most of what I need to say in other books by now. Thanks first to Brian Jepson, the editor who took on this project with little understanding of what a pain I am to work with. Without Brian, I'd still be piddling away on this book, and probably be very frustrated. Rael Dornfest turned out to be a Really Cool Guy™ and someone who is great fun to argue with. He also made a remarkable offer to help me out halfway through this book in some very unique ways, and I'm grateful for that.

The O'Reilly gang was monumental on this one; I really do love where I work. No Limit Texas Hold 'Em, and everything that came out of those fateful first nights in Sebastopol, made so many of the long nights possible. Thanks, all. You know who you are.

Preface

It would be hard to argue that *any* invention has affected more people in the 20th century than the television. That might sound ridiculous when you think of achievements such as the polio vaccine, or mapping the human genome, but then again, you don't see a lot of DNA models sitting in the armoire in people's dens and living rooms, do you?

Once TV hit it big, movies quickly followed. Soon, Friday night at the movies was as much a national pastime as baseball! At some point, though, some lazy couch potato figured out that if he (and we all know it *had* to be a man with a death grip on the remote, right?) could get these movies to show on his television set, potato chips and bathroom breaks were no longer a problem. And thus, home theater was born.

Now, some 15 years later, a 13-inch version of *The Matrix* just doesn't cut it. If you don't duck when guns go off, you're laughed at and ridiculed. The good news is that big TVs, expensive speakers, and gadgets galore are available in three of four stores (at least) in every town in the world. The bad news is that even if you drop six months worth of cash, you might have just a really big, really loud version of that 13-inch setup. *Home Theater Hacks* tackles these problems head-on, finally providing answers to all the questions you can imagine—and a whole lot you can't!

This collection of advice, tips, tricks, warnings, and screaming admonitions reflects the best wisdom in the rather significant home theater community. Just getting started and want to know what TV to buy? We cover it. Comfortable with computers and want to use your home PC for playing DVDs? We'll show you how to get resolutions better than $1,000 commercial DVD players. Seasoned pro looking for a new challenge? Learn to wire up your own super-high-end power and speaker cables. Whatever your experience and whatever your interest level, you'll find something to pique your interest and push the envelope of even the highest-grade home theaters. So, what are you waiting for? Let's get to it.

Why Home Theater Hacks?

The term *hacking* has a bad reputation in the press. They use it to refer to someone who breaks into systems or wreaks havoc with computers as their weapon. Among people who write code, though, the term *hack* refers to a "quick-and-dirty" solution to a problem, or a clever way to get something done. And the term *hacker* is taken very much as a compliment, referring to someone as being *creative*, having the technical chops to get things done. The Hacks series is an attempt to reclaim the word, document the good ways people are hacking, and pass the hacker ethic of creative participation on to the uninitiated. Seeing how others approach systems and problems is often the quickest way to learn about a new technology.

There is hardly an area where this is truer than in home theater. For years, there have been two options: spend a little money for a mediocre system at your local chain store, or pay a professional tens of thousands of dollars to do things you can't even pronounce, let alone understand. As long as there have been home theaters, though, there have been those few who have cracked open their TVs, pulled apart power cables, and been determined to *figure things out*. Finally, the best advice and wisdom have been brought together in one place, and only a Hacks book can get that information to you without all the mumbo jumbo that created this problem in the first place.

How to Use This Book

You can read this book from cover to cover if you want, but each hack stands on its own, so feel free to browse and jump to the different sections that interest you most. If there's a prerequisite you need to know about, a cross-reference will guide you to the right hack.

How This Book Is Organized

The book is divided into several chapters, organized by subject.

Chapter 1, *Buying Gear*
: It's easy to buy cheap gear, and it's hard to find quality components. The hacks in this chapter pull back the curtain on chain stores, show you the wonders of home theater boutiques, and reveal the tools of the trade for buying online. There are even hacks to help you locate hard-to-find items such as component racks that *won't* break your budget, theater lighting, movie posters, and fancy molding for your walls.

Chapter 2, *Video Components*

When you sit down to watch a movie, the first thing you notice is invariably...the movie! This section is chock full of tricks to get the most out of your video components. From choosing a television set to defeating the ridiculous VHS/SVHS/DVHS scam, from removing glare on your TV to dealing with letterboxing and those pesky black bars, you'll have your video sources cranking out killer picture in no time.

Chapter 3, *Audio Components*

The best picture in the world won't make a silent movie interesting. This section deals with video's sister, audio. You'll learn how to choose the units that drive your speakers, as well as amplification tricks for getting the most out of your main speakers. You'll even learn how to listen to those big action movies when everyone else is trying to sleep.

Chapter 4, *High Definition*

It's everywhere you go now—high-definition TV, high-definition VCRs, ESPN HD, HBO in lifelike color. High definition is much more complicated than Best Buy makes it seem, though, and this section breaks it all down. The right TV, the accessories that play nice with HD, getting your local networks in stunning resolution: these are all covered, and then some.

Chapter 5, *Speakers and Wiring*

Here's where things begin to get plain nasty. Electronic components take a few paragraphs to deal with, but speakers are an order of magnitude more complex. These hacks teach you when to choose speakers from the same manufacturer, or even the same line; how to balance the super-sleek, small speaker with big movie sound; getting a literal rumble in your house when the bass kicks in; and much, much more. This is where the big boys start to play.

Chapter 6, *Subwoofers*

Did someone say bass? Without a subwoofer (or two), a home theater is just a game room. This section teaches you how to move from seeing a plane take off, to *feeling* its engines wash back across the pavement. The right sub, as well as careful placement and proper connections, make even a mediocre theater sound professional.

Chapter 7, *Connectivity*

As unsexy a topic as it sounds, keeping all your gear connected and running is crucial. The relatively simple hacks in this section will ensure that you're not wading through cables with a flashlight gripped between your teeth, all while your friends are waiting to see the last disc of *Band of Brothers*.

Chapter 8, *Calibration*

If home theater is nothing more than a whimsical hobby, this is the section that will turn you into a bona fide audio- and videophile. Calibration, simply put, makes a decent picture great, and a great picture unbelievable; it turns a good set of speakers into a smooth soundscape; it provides $10,000 quality via your $2,000 TV. This section is tricky, tedious, and time-consuming—but oh, the benefits! You'll blow your neighbors away before you're halfway through the chapter.

Chapter 9, *Do It Yourself*

If you've got a hankering to break out your power saw, the hacks in this section are the excuse you've been waiting for. From speaker stands to 100-inch movie screens to power cables, you'll learn how to put together much of your home theater from scratch.

Chapter 10, *Remote Controls*

No system is complete without a remote control; unfortunately, most theaters have three, four, or even more! Consolidate into one killer remote, and even learn how to make it look better than an Atari 2600 in the process.

Chapter 11, *HTPC*

No hardcore home theater book would be complete without a discussion on adding computers into the mix. This section details everything from home theater PC display resolutions to cranking up *DOOM 3* in glorious HD resolution.

Chapter 12, *TiVo*

Once you're all set up, all that's left is to make sure you don't miss anything good. These TiVo hacks will help you catch every show you ever thought about watching, and do it all from your local Starbucks via a web browser!

Conventions

The following is a list of the typographical conventions used in this book.

Italics

Used to indicate URLs, filenames, filename extensions, and directory/folder names. For example, a path in the filesystem appears as */Developer/Applications*.

Bold

Used to indicate a user-supplied value.

`Constant width`
: Used to show code examples, the contents of files, console output, as well as the names of variables, commands, and other code excerpts.

Color
: The second color indicates a cross reference within the text.

You should pay special attention to notes set apart from the text with the following icons:

This is a tip, suggestion, or general note. It contains useful supplementary information about the topic at hand.

This is a warning or note of caution, often indicating that your money or your privacy might be at risk.

The thermometer icons, found next to each hack, indicate the relative complexity of the hack:

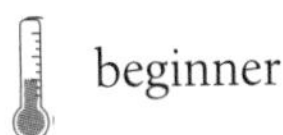

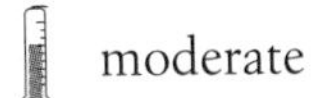

Using Code Examples

This book is here to help you get your job done. In general, you may use the code in this book in your programs and documentation. You do not need to contact us for permission unless you're reproducing a significant portion of the code. For example, writing a program that uses several chunks of code from this book does not require permission. Selling or distributing a CD-ROM of examples from O'Reilly books *does* require permission. Answering a question by citing this book and quoting example code does not require permission. Incorporating a significant amount of example code from this book into your product's documentation *does* require permission.

We appreciate, but do not require, attribution. An attribution usually includes the title, author, publisher, and ISBN. For example: "*Home Theater Hacks*, by Brett McLaughlin. Copyright 2005 O'Reilly Media, Inc., 0-596-00704-3."

If you feel your use of code examples falls outside fair use or the permission given above, feel free to contact us at *permissions@oreilly.com*.

How to Contact Us

We have tested and verified the information in this book to the best of our ability, but you may find that features have changed (or even that we have made mistakes!). As a reader of this book, you can help us to improve future editions by sending us your feedback. Please let us know about any errors, inaccuracies, bugs, misleading or confusing statements, and typos that you find anywhere in this book.

Please also let us know what we can do to make this book more useful to you. We take your comments seriously and will try to incorporate reasonable suggestions into future editions. You can write to us at:

O'Reilly Media, Inc.
1005 Gravenstein Highway North
Sebastopol, CA 95472
(800) 998-9938 (in the United States or Canada)
(707) 829-0515 (international/local)
(707) 829-0104 (fax)

To ask technical questions or to comment on the book, send email to:

bookquestions@oreilly.com

The web site for *Home Theater Hacks* lists examples, errata, and plans for future editions. You can find this page at:

http://www.oreilly.com/catalog/htheaterhks

For more information about this book and others, see the O'Reilly web site:

http://www.oreilly.com

Got a Hack?

To explore Hacks books online or to contribute a hack for future titles, visit:

http://hacks.oreilly.com

Disclaimer

Much of the information contained in this book is based on personal knowledge and experience. Although I believe that the information contained herein is correct, I accept no responsibility for its validity. The hardware designs and descriptive text contained herein are provided for educational purposes only. It is the responsibility of the reader to independently verify all information. Original manufacturer's data should be used at all times when implementing a design.

The author (Brett McLaughlin), contributors, and O'Reilly Media, Inc., make no warranty, representation, or guarantee regarding the suitability of any hardware or software described herein for any particular purpose, nor do they assume any liability arising out of the application or use of any product, system, or software, and specifically disclaim any and all liability, including, without limitation, consequential or incidental damages. The hardware and software described herein are not designed, intended, nor authorized for use in any application intended to support or sustain life or any other application in which the failure of a system could create a situation in which personal injury, death, loss of data or information, or damages to property may occur. Should the reader implement any design described herein for any application, the reader shall indemnify and hold the author, contributors, O'Reilly Media, Inc., and their respective shareholders, officers, employees, and distributors harmless against all claims, costs, damages and expenses, and reasonable solicitor fees arising out of, directly or indirectly, any claim of personal injury, death, loss of data or information, or damages to property associated with such unintended or unauthorized use.

CHAPTER ONE

Buying Gear

Hacks 1–8

Although there are literally thousands of options and dials you can tweak and spin on most home theater components, the components themselves are the most important part of any solid home theater system. All the adjustments in the world won't make up for having the wrong pieces of equipment in your theater.

In fact, before you even start to turn dials and pull up menus, you might want to see what you've got sitting in your cabinet. Sometimes making some small changes in your equipment list can have a huge effect on your theater's audio and video, and even its usability. This is especially true in today's world of combination devices. Merging VCRs, DVDs, TVs, and CD players might make for less wiring to have to hide, but it doesn't necessarily make for a better movie-watching and music-listening experience, and you can forget swapping out or trading up to newer and better technology.

Of course, all of this rides on finding the right equipment, preferably at the best possible price. Buy the right components at the best prices, and you'll stretch your dollar further. You'll end up with a bigger and better system if you do your homework. In this chapter, you'll learn the basic language of the trade, and then explore the common (and not-so-common) places to pick up equipment, oftentimes at way below retail prices.

Master Theater-Speak

You're not going to get very far in the world of home theater if you don't understand how the techies talk. Learn the lingo, and you won't get bullied into buying something you don't want; you'll also understand how different components interact, and you'll end up with a much better setup.

If you've ever walked into a home theater boutique, you've probably been quickly overwhelmed by the strange language that's coming out of some well-intentioned salesperson's mouth. It's sort of like showing up at a Ferengi yard sale and not knowing the difference between a warp coil and a plasma conduit. What's worse, it's easy to be convinced that you need something you don't, or that what you intended to buy isn't really the right component for your system.

The basic definitions you'll need to be familiar with are listed here. The following sections go into further detail on each item, and explain other important acronyms and terms related to each.

Televisions
: This is a pretty obvious one: the television, of course, is what you actually watch video on. However, TVs have become increasingly complex these days, and some TVs are self-contained theater systems; you can buy a TV that includes its own VCR, DVD player, satellite receiver, and virtual surround sound system. For the purposes of this book, I'll include computer monitors and other video sources in this category, except when they are specifically called out in the text.

DVD players
: For those of you not stuck in the '80s, DVD is the medium of choice for watching movies and, now, even television series. DVD stands for digital video disc, and these discs look just like CDs, although they hold a lot more data. Players can be as simple as a deck that does nothing more than play your disc, or complex enough to enhance the sound and picture of a disc, and even make copies of a disc.

VCRs
: VCR stands for video cassette recorder. The predecessor to DVDs, the VCR still is an important part of most home theaters. For those of you who have cases of VCR tapes with all the episodes of *The X-Files* on them, it's still the best-understood means to capture your favorite television show; however, DVD recorders and personal video recorders such as TiVo and ReplayTV are changing that in this century.

Satellite and cable receivers
: Satellite receivers provide you an audio and video signal, generally of television/cable channels, from a satellite dish. DISH Network and DirecTV are the most common providers, and both use roof- or pole-mounted minidishes pointed up into the sky. Cable receivers provide the same basic functionality, but they use underground cable and obtain signals from providers such as Time Warner Cable or other local outfits. Although there are some differences between satellite and cable receivers, they generally are interchangeable in terms of basic operation.

Receivers

A receiver is a unit that essentially consolidates and redistributes signals. Receivers usually get video and audio from devices such as a VCR, DVD player, or satellite receiver, and play those signals through TVs (for video) and speakers (for audio). They also make switching between input and output sources simple, and they are the cornerstones of any decent home theater system. Receivers, then, provide preamplification and signal distribution as well as speaker amplification.

Separates

The term *separate* doesn't refer to a specific component; instead, it indicates that the tasks that are typically rolled into a single receiver unit are broken up among several components. A simple separate-based setup might include a preamplifier and a single amplifier; more complex setups could involve preamplifiers, signal enhancers, equalizers, and an amplifier for each speaker in the system.

This certainly isn't an exhaustive list, and it actually leaves out several common components such as speakers, cabling, and equalizers. However, it's enough to get you past the tech-speak of the typical salesperson, and help you know what you're actually buying. The next several subsections provide you further detail on each category.

Televisions

Although televisions were the very first entry into what has now become the home theater market, they remain the most important. A receiver is the basis of all your audio, but it's the television that ensures you get a killer picture; it's also what most people notice first.

Although there's not much complexity involved in choosing a television [Hack #9], there are a few terms you need to be clear on.

HDTV

HDTV stands for high-definition television. In a nutshell, HDTV is all about trying to reproduce the picture you get on 35mm film, the gold standard in picture quality. A true high-definition (HD) picture will have 1080 lines of data, interlaced, or 720 lines, progressively scanned. *Progressively scanned* simply means the picture is drawn one line at a time, line by line. The alternative, *interlacing* a picture, means half the lines are drawn on the screen, and then the other half (in other words, the first, third, fifth, etc., lines are drawn, and then the second, fourth, sixth, etc., lines are drawn).

Writing or saying "1080 lines of data, interlaced" and "720 lines, progressively scanned" is a pain. It's more common to see these expressions abbreviated by stating the number of lines, and then either "i" for interlaced or "p" for progressive. So, 1080i and 720p are the HD formats. You also will see other formats use the same notation: 480i, 480p, 720p, and so forth. For the time being, though, the highest available resolutions are 1080i (there is no 1080p, although it's coming), and 720p (there's no 720i).

As for comparing HDTV to 35mm film, the dividing line is getting smaller every day. Many experts say that direct replacement of 35mm film will occur when 1080p (progressive, not interlaced) pictures are being shown at 24 frames per second. Because most broadcasters are gearing up for HDTV broadcasting (if they're not already there), 24 frames per second is very achievable; all that's left is to get resolutions running at 1080p, and that's not far away either. In short, HDTV is already nearly a direct replacement for the film you see in theaters, and in coming years (months!?) it will be an exact replacement, as far as visual information goes.

DTV

DTV stands for digital television, and you see this in reference to a lot of first-generation HDTVs, as well as a few of today's higher-end televisions. More often than not, DTV simply means the television accepts a digital signal as opposed to an analog one. As a result, a DTV doesn't really offer you anything special at all, other than ensuring your TV was made sometime after about 1990! However, on a more technical level, DTV really refers to any signal other than the true HD signals, 1080i and 720p. When most people refer to HDTV, they actually are referring to a DTV signal because very few signal sources are being shown in true HD today. In fact, even DVDs can't currently display a true HD signal.

Aspect ratio

The only other term you need to be clear on when dealing with TVs is their aspect ratio. An aspect ratio is the ratio between the horizontal width of the picture and the vertical width. The two ratios you need to be particularly comfortable with are 16:9 (pronounced "sixteen by nine") and 4:3 ("four by three"). 16:9 pictures are widescreen, and 4:3 pictures are the standard TV format you see so commonly today. That said, it's important to understand that the aspect ratio of a picture can be different from the aspect ratio of a television. For example, you can watch 16:9 (widescreen) movies on your 4:3 TV; that results in the black bars you see on the top and bottom of your screen. In the same fashion, a 4:3 picture has extra space on the right and left sides when viewed on a 16:9 television. I'll talk more about the TV aspect ratio you want [Hack #13] later.

If this seems confusing, don't worry: TV manufacturers are going to plaster these terms all over televisions because they are the primary selling points. If you get a widescreen HDTV, you get all the bells and whistles; a 4:3 HDTV is the next step down; from there, you're into plain-vanilla TVs. That's really all there is to it.

DVD Players

DVD players are a lot more complex than the average TV, at least from the standpoint of the consumer. They have been made even more difficult because DVDs now generally double as CD players, and must handle multiple video formats as well as some of the newer audio formats. Here are the highlights of what you need to know.

DVD

You already know that DVD stands for digital video disc. Against almost all odds, DVDs are a mostly standardized format, and you don't need to spend lots of time figuring out if such and such DVD will play in this or that DVD player. Given technology and the insurgence in HDTV, though, this is beginning to change. Over the next several years, expect DVDs capable of displaying 1080i and 720p pictures to appear and to be incompatible with older DVD players. With this future exception, though, DVDs as a rule play in any DVD-compatible player, from the cheapest to the most expensive.

The one notable exception was in consumer-burnt DVDs, such as from a computer. These discs often won't play on very low-end DVD players.

Progressive scan

You already know what progressive scan is from the discussion on HDTVs. However, it bears mentioning again in this section. Although most DVD players provide progressive scanning, some older or less expensive ones don't; avoid these if at all possible. What good is a DVD if you're not getting a flicker-free picture?

DVD-Audio

This is a term related to audio more than DVDs, but due to DVDs serving as CD players these days, it fits well in this category of components. DVD-Audio is a format that allows a musical track to play in more than a simple stereo format (left and right channels); instead, DVD-Audio discs play in 5.1-, 6.1-, or 7.1-channel formats (see Chapter 4 for details on these formats). The intention is to provide a live-like listening experience, with sound coming from every direction. You can find DVD-Audio and SACD sections in chain stores such as Best Buy and Circuit City these days, so these formats have become both popular and easily obtainable.

SACD
SACD stands for Super-Audio CD, and is somewhat in competition with DVD-Audio. SACD is focused on audio quality in general, while DVD-Audio is focused specifically on multichannel sound. As a result, you will find some SACDs that are multichannel and some that are simply stereo; however, the sound quality of even the stereo discs is far superior to that of a standard CD.

The bad news is that, at least for now, you'd be hard pressed to find a progressive scan DVD player that supports both DVD-Audio and SACD without dropping at least $1,000. Most DVD players are progressive scan these days, and many also support DVD-Audio formats. If you're looking for a SACD player, you're probably going to have to add it to your system as a separate component, in addition to your existing DVD/DVD-Audio player.

Before spending lots of time looking for a killer SACD player, or even a DVD-Audio player, make sure you actually want one. The offerings in these formats still are relatively sparse; additionally, SACD discs don't play in DVD-Audio players (other than at normal CD quality), and vice versa. If you're not a music student or a serious music aficionado, you're probably better off just picking up a solid DVD/DVD-Audio player and leaving SACD alone. However, if you want to listen to Rachmaninov in C# minor, SACD might be for you (if none of that made sense, don't worry about SACD too much).

VCRs

What is there to say about VCRs anymore? They really do seem to be yesterday's news, and the only high point in sight is the new wave of high-definition VCRs (HD-VCRs). There aren't even enough VCR terms to warrant their own list! There was a day when you had to worry about two heads, four heads, high fidelity, and even Betamax (remember those?). Today, anything that costs you more than a twenty will provide what you need for easy recording; nobody expects VCR tapes to look like DVDs, so picture quality is not much of an issue.

The one exception to this rule is the HD-VCR. This component is exactly what it sounds like: a VCR that plays tapes in HD format. These tapes come in true HD as well, meaning the picture you get is 1080i or 720p. For those of you paying attention, this does indeed mean that an HD-VCR playing a correctly formatted tape will provide a better viewing picture than a DVD (remember that DVDs don't play in HD formats yet). HD-VCRs also can record your favorite high-definition TV program for repeated viewing. Does this mean the VCR is coming back? Probably not. These units are quite expensive, and with a new DVD format looming I don't expect these units to be anything more than the next fad.

Satellite and Cable Receivers

Satellite and cable receivers also are well-understood devices, so I won't spend much time here. Satellite receivers bring in a signal from a satellite dish, usually from either DISH Network (*http://www.dishnetwork.com*) or DirecTV (*http://www.directv.com*). Both providers offer similar packages and comparable prices. If you're trying to decide between the two, there are a few important things to consider.

Local channels
: Getting local channels through your satellite is a major coup; nobody wants to switch to the VCR just to pick up these channels. Although most major cities are covered, smaller cities are hit-or-miss. Pay attention to this before making a decision.

HD programming
: If you're into home theater, you should be into high definition. The two satellite providers are not equal in the offerings here, and frequently seesaw in the balance of power. I switched to my current provider, DirecTV, to get ESPN in high definition; I also wanted HDNet so that I can watch the Dallas Stars in glorious high def. However, by the time you read this, DISH might be the better choice for HDTV. Check with your provider before signing up.

Package deals
: Last but not least, you might want to just look at what the coolest deals are. A lot of times you can get a certain number of rooms hooked up for free, or a complimentary TiVo (see Chapter 12), or free installation. All of these are nice bonuses, and you shouldn't ignore them.

If you aren't interested in putting a dish up on your roof, most likely cable is for you. Generally, cable setups are simpler to maintain, although they're often slower to provide innovations such as high-definition channels and digital sound (both are now available, by the way).

One thing I don't see as a distinguishing feature is susceptibility to poor weather. My DirecTV system rarely goes out, even in major storms, and the few times it has, my mother-in-law's cable system has been out as well. Go figure.

If you do go the cable route, Time Warner Cable is the dominant provider, although there are others, such as Comcast and the like. Take your pick: if there aren't major differences in price, most cable providers provide similar equipment with similar capabilities, and usually even similar channel lineups.

Receivers

Receivers are the most important part of your home theater setup; at the same time, they often are the last thing the typical consumer thinks about, as they don't seem to "do anything" on their own. A good receiver simply distributes and amplifies signals—video from input sources to a TV, and audio from those same input sources to speakers. However, because all of your audio and much of your video pass through this device, it's critical to get this component right.

Dolby Digital
: Dolby Digital (*http://www.dolby.com*) is arguably the major force in sound formats today. This term is usually representative of several sound formats: Dolby Surround Sound, Dolby Pro Logic (and Pro Logic II), and Dolby Digital itself. This grouping, taken as a whole, allows for two-channel, six-channel (5.1), seven-channel (6.1), and eight-channel (7.1) sound, from almost any type of input source. Systems that don't support Dolby might as well give up competing; it remains the gold standard for audio formats.

The (.1) in these formats refers to the low-frequency effects channel, most commonly just referred to as the *sub* (as in subwoofer). Because it's not providing directional material (bass is omnidirectional), it is relegated to a decimal place.

DTS
: DTS (*http://www.dtstech.com*) stands for Digital Theater Systems, Inc., a company that produced an audio alternative (and now several alternatives) to Dolby's suite of formats. Initially the format was just called DTS, and was most commonly used for listening to music, or to movies with sweeping musical scores (*Gladiator*, for example, sounds majestic in DTS, as compared to Dolby Digital). Today, DTS offers a number of additional musical and theater formats, such as DTS-ES (for 6.1-channel

sound), Neo:6 (for surround playback of stereo music), and DTS Virtual (for down-converting surround tracks to stereo). DTS is just as prevalent as Dolby Digital these days, and it remains the killer suite of formats for highly musical audio, as well as being great for movies.

THX

THX (*http://www.thx.com*) is George Lucas's set of specifications for sound reproduction. THX really isn't as much a sound format as a set of specifications for listening. When listening to a THX system, you'll still have to choose a Dolby Digital or DTS audio format. THX systems provide audio correction and equalization in an attempt to provide the home theater audience an experience as close as possible to what you can get in movie theaters. The downside is that because THX makes changes on the fly, you might not hear the exact DTS or Dolby track that is encoded on your favorite DVD. For this reason, some audio experts consider THX a nuisance at best, and a real problem at worst; the THX brand sells components, but it's not worth breaking the bank over.

Most receivers will support, at a minimum, Dolby Digital in 5.1 channels and DTS in 5.1 channels. Spend a little more money and you'll be able to add 6.1- and 7.1-channel sound, as well as THX and THX-EX (the EX flavor supports 6.1- and 7.1-channel sound) capabilities; you also can easily get the various DTS music formats, such as Neo:6. Still, there's more to a receiver than the acronyms that can be splashed across its face; Figure 1-1 shows the front of a higher-end audio receiver.

Despite the acronym frenzy found on the front of a receiver, one of the key features of a great receiver turns out to be what is on its back. As a general rule, you want as many connections, for as many devices, as possible. Figure 1-2 shows the back of a fairly high-end receiver, and you can see the wealth of places to hook up cool toys.

Picking the right receiver is a lot like choosing a car; even if you can narrow down exactly what you want, there are plenty of options in your functionality spread, and budget and brand preference often become the prime considerations. Discussion of exactly what to buy is left for the hacks in the rest of this chapter, as it largely depends on what sort of theater you're trying to put together.

Separates

The final category to look at is separates. As mentioned earlier, this doesn't refer to a specific type of component, but rather, to the idea of splitting out functionality from the receiver into individual pieces. As a rule, the more

Figure 1-1. Front of a Pioneer Elite receiver

Figure 1-2. Back of a Pioneer Elite receiver

focused a component can be on one job, the better it will be at that job. The typical receiver has to perform preamplification tasks (preparing a signal for distribution to speakers), as well as amplifying the signal itself. In many cases this involves some form of equalization, with the ability for consumers to add more equalization as they desire. This doesn't take into account the need to route signals properly, either, allowing seamless switching between VCRs, DVDs, your trusty TiVo, and your favorite video game system.

A more effective approach to home theater, albeit an expensive one, is to split up these tasks. The most common approach to separates is to break the receiver's job into two tasks: preamplification and amplification. This involves one component for signal routing, connection to your DVD, cable receiver, and other input sources, and equalization, and another component for speaker amplification. Of course, that one amplifier must support as many as seven (or, in some cases, nine or more) speakers, and can be broken up into several separate amplifiers. Not surprisingly, the more amplifiers you buy, the more expensive the overall system cost becomes. Figure 1-3 shows a preamplifier, and Figure 1-4 shows an amplifier.

Figure 1-3. Outlaw processor/preamplifier

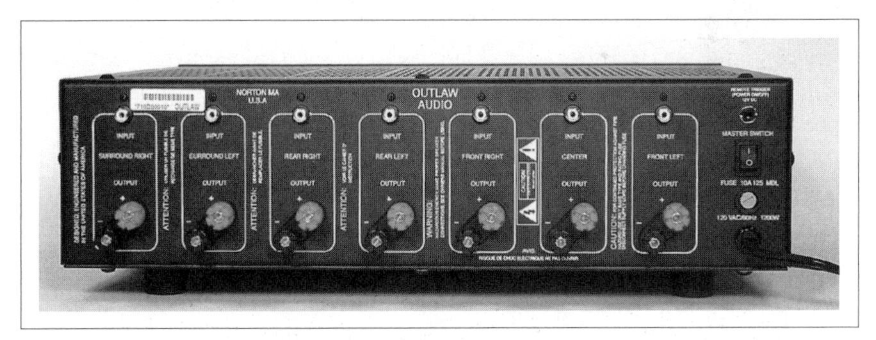

Figure 1-4. Rear of a seven-channel Outlaw amplifier

You can take this scenario as far as your budget allows. Many of the best home theaters you will find will have a preamplifier, an equalizer, an amplifier for each speaker in the theater, and one or more line doublers and scalars, both of which are used to increase picture clarity and density. At some point, though, you're getting only a fraction of a percent improvement at a cost of thousands of dollars (yes, you read that correctly. In other words, know when to say "when!"

With these basics down, you're ready to actually look at getting your own theater started, and the hacks in the rest of this chapter will help you know what you need. First, you'll learn some basic tricks to ensure you get the best sound and video components; then, you'll get to see the various options available for buying equipment.

HACK #2 Audition Before You Buy

There's no substitute for listening and watching equipment in action before buying. However, there's as much science as there is art to choosing a good home theater. Preparation and a few tricks will help you pick the best system for you.

There's a lot of pressure when a salesperson is hovering over you, waiting to see if you're going to buy a particular piece of equipment. In fact, this is the number-one reason people walk out of chain stores [Hack #3] or boutiques [Hack #4] with equipment they're ultimately not happy with. But this pressure is largely due to a lack of preparation and method. If you have specific criteria in mind and a particular method you always follow, you'll feel less pressure, have no trouble telling a salesperson "I'd like to take a little more time," and will usually be a lot happier with your purchase.

Prepare Your Ears with AM Radio

Auditioning speakers and audio components is a lot like critiquing food. Professional critics know that their taste buds become saturated after a few bites of something so they use a palate cleanser such as raspberry ice to reset their sense of taste between dishes. Although there isn't a raspberry ice for the ear, you can reset your hearing by not listening to loud music, bubbled up in your front seat, on the drive over. This will prevent you from building up any preconceptions or expectations about how something should sound.

One weekend I went to a chain store and listened to a system that failed to impress me. The following weekend I was back at the same store and I listened to the same system, and it sounded much better! I realized that on my second trip I took a car that had AM radio, so I listened to news on the way over, and probably had my window rolled down. The lower-quality sound in the car conditioned my ears so that the equipment in the demo room sounded better than I first thought. Some might think this means you would buy a system that isn't as good. On the contrary, loud music, especially when it's coming out of several speakers within a few feet of your ears, dulls your hearing. You lose a sense of dynamics, subtlety, and all the other intangibles that make good music "good." The system I didn't like at first was providing those nuances, but by the time I got to the store on my first trip, all I could hear were screaming guitar breaks. The second time, I noticed the sound had more texture, the dynamics were terrific, and the smallest background sounds were present. I missed all this the first time because of the loud music in my car.

In the same vein, be willing to take breaks between speaker auditions. Over time, sounds can begin to blur together; just like the food critic, your senses have been saturated. Take 15 or 20 minutes to walk around the store or stroll outside, or even consider coming back the next day when your ears have had a chance to relax. Taking time to make a good purchase is always a great idea.

Bring a Favorite DVD and Audio CD for Auditioning

No two people are completely the same, and this is certainly true when it comes to movies and music. When you're trying out home theater components, you'll often find that salespeople play the same disc (usually a DVD) over and over again. Although that might be their favorite movie, it's more often a disc that's perfectly suited to the system they are trying to sell (this is particularly true in larger chain stores). However, unless the disc being played is your favorite movie as well, you might not like the sound you get when you take those speakers home. Suddenly your favorites sound boomy or brittle. Instead of relying on the store to select auditioning material, bring your own.

It's important to bring material you like and are familiar with. You might even want to watch or listen to a disc a few times before going into the store, just to refresh your memory. When trying out speakers, you'll find that some speakers are laid-back and work great for classical or jazz and dialog-heavy movies. Other speakers are a bit harsher, forward, and in-your-face. These work great for rock, rap, and action movies. By bringing your own styles of music and movies, and your favorites within those styles, you're getting a great system for your particular tastes.

Boutique stores **[Hack #4]** actually expect you to bring your own media because they know it makes a difference. There is no reason you should not bring your own disks to a chain store as well. Some great demo DVDs are listed here, along with a good track to try out on each:

- *Star Trek: First Contact:* Chapter 2 (the Borg battle scene)
- *Toy Story 2:* Chapter 1 (Buzz versus Zoltron's robots)
- *The Fifth Element:* Chapter 24 (Diva solo)

If you're a seasoned home theater guru, you might want to consider bringing a sound pressure level (SPL) meter **[Hack #61]** and some calibration DVDs

[Hack #62]. Although your friends might laugh at you for this, a boutique salesperson will not mind in the least (and probably will appreciate how serious you are about making a good purchase).

Avoid Switch Boxes

Lots of stores, both chain and boutique, will have multiple speaker systems connected to a receiver or amplifier through a switch box. Refuse to make a decision based on these setups! Most of these switches have a 4-ohm resistor to prevent a receiver or amplifier from overdriving a speaker and potentially damaging it. This means that signal is getting filtered between the source and the speaker. Unless you plan to use the same switch box at home, you're not getting an accurate sonic picture. Insist that any speakers you listen to are connected directly to a receiver or amplifier.

The same principle is true for video devices. Just as a switch often is used to connect multiple speakers to the same receiver or amp, multiple display units (TVs, in particular) often are connected to the same DVD player. Again, you're not getting an accurate picture of what's going on. Even if the switch uses only component video for connectivity, there is still a tremendous variety in the available switches' qualities [Hack #59]. Again, a salesperson interested in your desires rather than making a buck won't mind taking the extra time to connect your sources directly to a display unit.

Evaluate the Whole, Not the Parts

One of the biggest mistakes you can make in trying out several components is choosing them independently. This might sound strange; shouldn't you base your decision about a component on its own merits? However, components and speakers don't operate in a vacuum. A set of speakers might sound amazing with a high-end preamplifier and amp, but if you're using a receiver, those same speakers might not be as clear, or as focused.

It's great to narrow down your choices to a few specific pieces of gear. However, don't buy anything until you test everything together! I recommend taking down the brand of your existing components, as well as the model, and seeing if the store has the same models to test with. If you're auditioning equipment at a boutique, you even can bring in your existing components and try them out. Again, this takes a lot of extra time and effort. However, the end result is a system that sounds like what you expect, and that's worth some additional work.

Always Look at the Manual

Even the best salesperson, in the smallest boutique, can't know everything about every model from every manufacturer. This is where manuals come in: you never should buy anything without taking a look at the owner's manual. Whether it's a television, speaker, or DVD player, the manual is going to tell you things a salesperson won't remember (or doesn't want you to know, in some cases). If a salesperson isn't willing to crack open a box and let you see a manual, politely tell him you're not interested in shopping at his store. Just be sure you're willing to back up this assertion, or you'll end up looking pretty foolish (and lose any ability to negotiate down the line).

Buy from Chain Stores with Skepticism

Buying from a local electronics store has the advantage of letting you check out gear in person, but you'll have to endure crowds, obnoxious music, and frequent interruption from salespeople along the way.

Buying from a chain store requires a lot of forethought; you need to dress appropriately, prepare your ears, and have demo discs in hand (all discussed in detail in this hack). These preparations might make it seem that it's better to avoid people altogether and just shop online. However, the advantages of a brick-and-mortar store are significant:

- You can watch and listen [Hack #2] to the specific gear you take home.
- You can visually inspect components to ensure new or like-new condition.
- You avoid the often-dangerous shipping process (and the possibility of having to return equipment, often at your own cost).
- You can see your favorite movie or hear your favorite CD on a variety of equipment.

All of these are great reasons to at least look at your local electronics stores. However, each advantage comes with its own set of drawbacks.

The Pitfalls of Chain Stores

Although it's nice to listen to a specific set of speakers or audio components, chain stores often set up systems to favor the systems at the expense of the listeners. You won't be sitting (or standing) in the exact same area, in the same size room, in your own home theater. In fact, most chain stores have huge rooms, with at least one glass wall, and you're standing up—about as different from a typical home theater as you can get. So, although you will get an idea of the sound you would obtain at home, realize that it's not a perfect picture of what to expect. If you want a closer-to-home experience, boutiques [Hack #4] often have viewing and listening rooms that closely simulate a home theater, with regular walls, only one door, low lighting, and even theater seating.

It's also nice to be able to make sure the components you are buying are in perfect condition (even new equipment can easily be smudged or scratched in transit from the manufacturer). To really gain this advantage, though, be sure you insist on getting the specific model you are viewing rather than another unit already boxed up. You'll also need to ensure that you don't get stuck with a display unit that doesn't have a remote, instructions, or other materials that should be included.

Sometimes you can negotiate a lower price on a unit that has been demoed in the store; don't be afraid to dicker, even at a chain store.

You will avoid shipping charges, but these often are offset by sales tax, especially on larger purchases. If you're buying more than $1,000 worth of gear, you'll often break even in comparing local sales tax with shipping on non-taxed items. The only notable exception to this rule is larger TV sets, which typically cost a bundle to ship.

There's an ugly rumor going around that the days of buying tax-free across state lines are coming to an end. Although it might still be years before this comes into play nationwide, you might want to do some reading on the subject. The term for this is *use tax*—a tax on goods bought out of your state of residence, but intended to be used or consumed there. As an example, visit *http://www.tax.ri.gov/info/whats.pdf* to see what the State of Rhode Island expects from its citizens.

It is also appealing to see your choices in movie and audio on a specific system. However, remain skeptical: as mentioned here and in Chapter 5, chain stores often ensure that systems are perfectly tuned with thousands of dollars of equalization, causing you to hear something more than just a receiver, speaker, or CD player. Assume that what you hear is the best possible case, not the average case.

Dress for the Best Service

Everyone pays more attention to people when they dress up, and no serious job applicant walks into an interview in old jeans and a ragged T-shirt. It's obvious, at least in those cases, that although dress might not define you, at least it plays a part in identifying you. Dress can say you're serious about a job, a person, even a purchase. It's in this last category that a lot of ignorance lies. Taking the time to put on an outfit a few notches just below what you'd wear to a job interview will make a huge difference in the attention you get from qualified salespeople at home theater stores.

Dress might not make as much of a difference in chain stores, but particularly in boutiques and other small shops, a well-dressed consumer will almost always get more attention, better information, and generally, a better deal. Ten extra minutes in the closet can result in hundreds of dollars saved, and a system that pleases you for years rather than months.

A Limited Inventory

Of course, it should go without saying that the biggest problem in chain stores is the lack of selection. You won't see Marantz, Pioneer Elite, Onkyo, Meridian, Lexicon, B&K, and most other high-end component manufacturers in even the largest chain stores. This means you're effectively limiting yourself to lower- to medium-ended manufacturers. If you're on a shoestring budget, this might be acceptable, but if you're looking for the best, you'll want to investigate boutiques [Hack #4] and higher-end online sites [Hack #6].

When you shop in chain stores, realize this going in. If you're looking for the very best in home theater, you're probably going to leave frustrated and upset. If you're on a tighter budget, though, you often can find some good buys. However, realize that any audiophile friends you have might turn up their noses at you. Don't get in a tiff about this; you bought within your budget! Ask them how their credit card bills look, and then you can do the laughing.

Buy in Person from Electronics Boutiques

An electronics boutique offers the advantage of in-person tryouts without many of the downsides of a larger chain store.

Most home theater enthusiasts realize that chain stores [Hack #3] rarely cut it, for either quality or selection. If you want more than a "home theater in a box," you're going to have to spend some money, and in that case, you deserve more than a too-bright, too-crowded chain store. In these cases, an electronics boutique is ideal.

Boutique stores are smaller stores, often on a side road, that specialize in home theater. You'll have to look for them; they often don't advertise as much, or have as much visible presence, as the Best Buys and Circuit Cities of the world. However, if you find a good boutique, you can easily become a customer for life.

Understand the Pricing Model

Before you walk into your first boutique, you should realize you are probably going to pay anywhere from 10% to 15% more on items than you would if they were in a chain store. Boutiques have higher overhead and less revenue, and they pass those (lack of) savings on to you. However, you have to realize many of the components in a boutique aren't available in chain stores, so this theoretical price increase becomes just that—theoretical.

If you want to pay the absolute lowest price for equipment, you're going to have to shop online [Hack #6]. But on the Internet, if you don't know exactly what you want and what to watch out for, it's easy to pay less for something that doesn't serve your needs. Some will urge a potential consumer to figure out what he wants at a boutique, write the items down, and then shop online. I have a real distaste for this; if you're going to spend several hours at a boutique (and you will, if you're serious about choosing the best gear), you've taken up valuable time for the owner and salesperson. I think it borders on outright dishonesty to then devalue the information and assistance they provided, buying the gear they helped you select from an anonymous Internet dealer. Realize that 10% of a purchase is more than warranted if you have a great salesperson who helped you find what you wanted, was patient with you, and gave you plenty of options.

Prepare First, Shop Second

Another important tip when shopping in boutiques is to go in prepared. You should have, at a minimum, the following items written down.

Room dimensions
: This is one of the single most important preparation tasks. If you don't know the room size, including ceiling height, you really handcuff even the best salesperson. Speakers perform completely differently based on the size, shape, and structure of a room. This information is invaluable in a good boutique providing you a tailored system.

Room construction
: Detail whether your room is an interior room or an exterior room. Know the difference between a wood-framed house and manufactured housing. You also should note the location and size of windows, doors, eaves, overhangs, and anything else that could affect where you place speakers.

At a minimum, I need...
This is your list of desired applications. Don't worry about detailing that you want a receiver, DVD player, CD player, and so forth. Instead, list the uses you want your theater to serve—movie playback, watching cable TV, listening to CDs. If you want to listen to surround-sound music, let your salesperson know; on the other hand, if that's really not a big deal, know that going in. Additionally, let your salesperson suggest the equipment, unless you're really sure of what you're doing.

Budget
Although this might sound pedantic, it's really frustrating to spend hours with a customer showing him incredible systems, and then as you talk price, he mentions he wants to spend only $1,500 on a complete system. Know what you can spend, and let your salesperson know up front. She can pick better systems, and ensure you can afford what you hear.

Shop for the Complete Package

One of the biggest advantages of a boutique is that they usually offer installation services and even equipment calibration. Although this is going to cost money (doesn't everything?), it can be a huge advantage to have the people who sold you your equipment install it. They know the gear inside and out, and they might even offer you a discount if you buy equipment and let them install it. Additionally, if something goes wrong, you're not on the line; the people who sold the equipment are the ones who have to deal with it.

Another real advantage to this approach is that you often can swap out equipment if something goes wrong. If your boutique installers get to your house and realize something isn't going to work, they'll likely allow you to exchange the gear for something else that works. Finally, if you forgot to get that one component video cable, or if you run out of speaker cable, a boutique can take care of these little details easily.

Shop Intelligently at eBay

Although ebay.com isn't the best place to buy all your gear, it's a great source for certain types of gear, often at lower-than-normal prices.

No book on buying anything would be complete without mentioning eBay. One of the largest and most diverse marketplaces around (online or otherwise), eBay is a good source of a lot of electronics, including home theater components. You'll often find slightly used or B-stock gear at killer prices that aren't available anywhere else. With all that said, you'll need to be just as careful with eBay (if not more so) as you are when shopping at a chain **[Hack #3]**.

What Should I Get Online?

One of the keys to using eBay to your advantage is determining what you can and can't buy online. More accurately, this could be phrased as figuring out what you should and shouldn't buy online.

The basic rule of thumb is that if an item is essentially a *stock item*, it's fair game for buying online. Stock items are anything that is manufactured and is generally the same across the board. For example, receivers are a stock item; as long as you have the same manufacturer and model, they're going to sound more or less the same, unit after unit, as long as they're in good condition. The same goes for DVD players, VCRs, CD players, and most other audio/video components.

Where you get into trouble is with items that, even when manufactured, don't always come out exactly the same. Speakers fall into this category. I've listened to two "identical" speakers that don't sound at all alike. You also need to be very cautious about items that have great potential for damage in transit. Although you can pack a DVD player in a ton of foam, it's very hard to ship speakers, as well as larger items such as TVs, without causing some shakeup. At a minimum, you're probably going to have to calibrate **[Hack #61]** these devices (which might mean paying a professional, if things are really out of whack). This also is true, although somewhat less so, with receivers and preamplifiers. These can be shipped safely, but be prepared to spend a lot of time explaining what you want to the seller, and ensuring that they follow through.

In general, stick to your more basic components, such as DVD players, CD players, and other video units, and you'll have far fewer problems.

Avoid Scams

One of the greatest boons of eBay, and yet one of its potential downfalls, is the ability for literally anyone to sign on, in complete anonymity. Like any good thing, it hasn't taken long for people who are out to make a fast buck to figure out ways to exploit the system. As a result, you'll have to be careful to avoid getting ripped off; home theater, along with musical instruments, is a primary target for these scam artists. You can do several things to minimize your risk if you do shop online, though.

Insist on pictures. A great way to separate scammers from real sellers is through pictures of the item for sale. Often, these are part of the listing, and you can form an initial impression. One giveaway of a potential scam is to see a picture pulled off of a product web site. For example, Figure 1-5 shows a Lexicon MC-12 from the Lexicon web site (*http://www.lexicon.com*). It's as stock as it gets, and no more represents a real unit than a hand drawing.

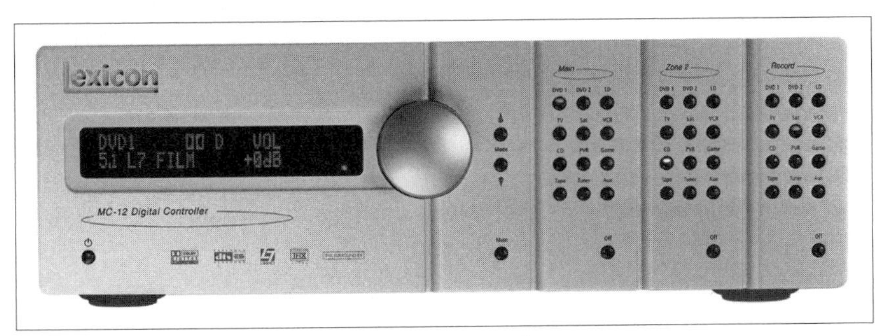

Figure 1-5. Lexicon MC-12 picture from Lexicon web site

Figure 1-6, on the other hand, shows a picture that obviously was taken in someone's home.

Figure 1-6. "Real-life" picture of Lexicon MC-12B

If you don't see any realistic pictures, email the seller and ask about it; a real seller won't be offended in the least. If you're still concerned, ask for additional pictures of the equipment in an unusual (and therefore inaccessible from a vendor web site) position; pictures of the top, bottom, or sides of the unit are always great. Again, a legitimate seller won't be upset in the least by this.

The same principles apply for "new, in the box" auctions, which sometimes show up. Although the seller might not want to take the unit out of the box, you can ask for different shots of the box, such as close-ups. In one auction, I even asked the seller to make a small mark on the box and photograph it; I had no doubts about the seller once he did what I requested. Remember, nothing says a buyer has to accept things just the way they are; only a fool buys from an anonymous, seemingly illegitimate seller.

Pay attention to feedback. This should be obvious, but you need to take a close look at feedback on eBay; it's there for a reason. Buying a high-end piece of equipment from someone who's sold only two items before is a risky venture, at best. Also be on the alert for users who have changed their username or are brand-new (signified by the eBay sunglasses), as there's another potential for problems.

In addition to looking at the number of transactions, always click through to the actual feedback itself. Reading through these comments often alerts you to how the seller handles payment, how quickly he ships, and how easy he is to work with. Of equal importance is to check out how many times the seller has actually sold before. Figure 1-7 shows my profile, where you see a good mix of buying and selling. It's also OK if someone sells all the time, but don't buy high-end gear from someone who's never sold before!

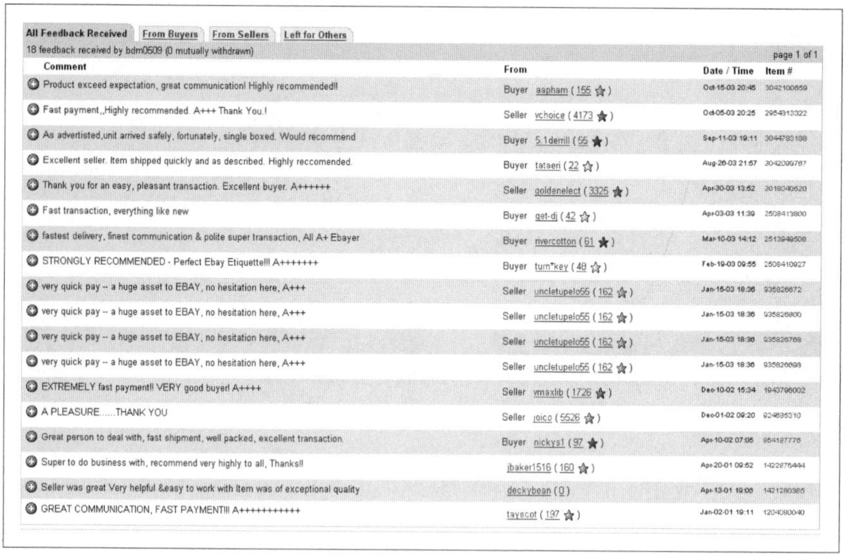

All Feedback Received | From Buyers | From Sellers | Left for Others

18 feedback received by bdm0509 (0 mutually withdrawn) — page 1 of 1

Comment	From		Date / Time	Item #
Product exceed expectation, great communication! Highly recommended!!	Buyer	aapham (155)	Oct-15-03 20:46	3042100659
Fast payment,,Highly recommended. A+++ Thank You.!	Seller	vchoice (4173)	Oct-05-03 20:25	2954913322
As advertisted,unit arrived safely, fortunately, single boxed. Would recommend	Buyer	5.1demll (55)	Sep-11-03 19:11	3044783108
Excellent seller. Item shipped quickly and as described. Highly reccomended.	Buyer	tataeri (22)	Aug-26-03 21:57	3042099767
Thank you for an easy, pleasant transaction. Excellent buyer. A++++++	Seller	goldenelect (3325)	Apr-30-03 13:52	3018040620
Fast transaction, everything like new	Buyer	get-dj (42)	Apr-03-03 11:39	2508413800
fastest delivery, finest communication & polite super transaction, All A+ Ebayer	Buyer	rivercotton (61)	Mar-10-03 14:12	2513949508
STRONGLY RECOMMENDED - Perfect Ebay Etiquette!!! A+++++++	Buyer	turn*key (48)	Feb-19-03 09:55	2508410927
very quick pay -- a huge asset to EBAY, no hesitation here, A+++	Seller	uncletupelo55 (162)	Jan-15-03 18:36	935826872
very quick pay -- a huge asset to EBAY, no hesitation here, A+++	Seller	uncletupelo55 (162)	Jan-15-03 18:36	935826800
very quick pay -- a huge asset to EBAY, no hesitation here, A+++	Seller	uncletupelo55 (162)	Jan-15-03 18:36	935826768
very quick pay -- a huge asset to EBAY, no hesitation here, A+++	Seller	uncletupelo55 (162)	Jan-15-03 18:36	935826698
EXTREMELY fast payment!! VERY good buyer! A++++	Seller	vmaxlib (1726)	Dec-10-02 15:34	1943796002
A PLEASURE......THANK YOU	Seller	joico (5526)	Dec-01-02 09:20	924895310
Great person to deal with, fast shipment, well packed, excellent transaction.	Buyer	nickys1 (97)	Apr-10-02 07:06	854187778
Super to do business with, recommend very highly to all, Thanks!!		jbaker1516 (160)	Apr-20-01 09:52	1422876444
Seller was great Very helpful &easy to work with Item was of exceptional quality		deckybean (0)	Apr-13-01 19:06	1421280385
GREAT COMMUNICATION, FAST PAYMENT!!! A+++++++++++		tayscot (197)	Jan-02-01 19:11	1204080040

Figure 1-7. eBay feedback

Email, email, email. A great way to feel out a seller is to simply email him—repeatedly. Email certainly isn't as personal as a phone call or a face-to-face meeting, but you often can get a feel for someone through a few emails. Ask about the equipment, and make sure you get detailed responses. If you've exhausted your equipment questions, ask about the shipping policies; and then ask if he's sold similar items before; then ask about his payment policies. Often, you don't need to pay attention to the specifics of his replies as much as to the tone and feel. Are you dealing with an open person who seems to care about the transaction? Or is the seller testy, impatient, and prone to be slow and unresponsive? The latter could indicate that you're not going to have a pleasant experience.

If you're buying a high-dollar item, it's perfectly reasonable to ask for a brief phone discussion. Any seller not willing to spend 10 minutes on the phone for the sake of selling a multithousand-dollar item isn't worth your time.

HACK #6 Find High-End Equipment Online

Online sites such as audiogon.com and videogon.com provide high-end gear, both new and used, at low prices.

So, here's the basic problem: you don't have a good boutique [Hack #4] nearby; you hate the chain stores (and their selection is terrible) [Hack #3]; and eBay [Hack #5] doesn't offer the really high-end gear you're looking for. This is actually a common situation (I'm in it myself, for what it's worth), and thankfully there's a solution. There are a couple of high-end online sites that offer even the most unusual gear, at less-than-factory prices.

Audiogon
: Housed at *http://www.audiogon.com*, this is the original site for killer equipment online. Audiogon has lots of home theater gear, but caters to the stereo (two-channel music) crowd as well.

Videogon
: Videogon is a follow-on to Audiogon, and is online at *http://www.videogon.com*. Videogon is more squarely focused at home theater, and is probably where you want to start if you're into...well...what this book is about.

Just in passing, there's also a high-end camera marketplace at *http://www.photogon.com*.

You'll find everything from the standard fare seen on eBay to $20,000 components on these sites; needless to say, you won't be limited in selection.

There's also a nice network of reviews that will help you find just the right gear for your particular system.

Both of these sites have essentially the same navigation. The main page (shown in Figure 1-8) has a listing of categories down the left. You also have subcategories to choose from. However, I often find things miscategorized (or not placed in a subcategory at all), so stick with the main headings and work from there.

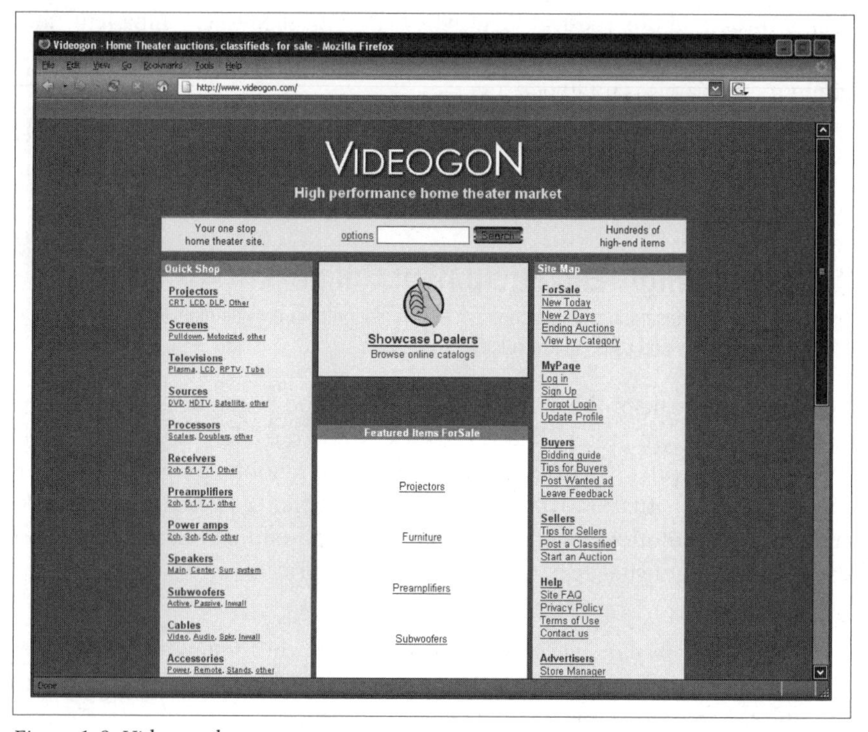

Figure 1-8. Videogon home page

Once you drill down into a category, you'll see a list of featured ads (see Figure 1-9), auctions (seen in Figure 1-10), and finally, dealer demo listings (Figure 1-11). Each area is worth some investigation, as long as you know what to watch out for.

Be Careful of Auctions

The high-end sites have auctions, just like eBay **[Hack #5]**. However—unlike eBay—you don't have the same robust set of protections on your purchases. There is not a well-established recourse for filing grievances, there is no safe harbor, no escrow, no feedback…you get the idea. Although these sites

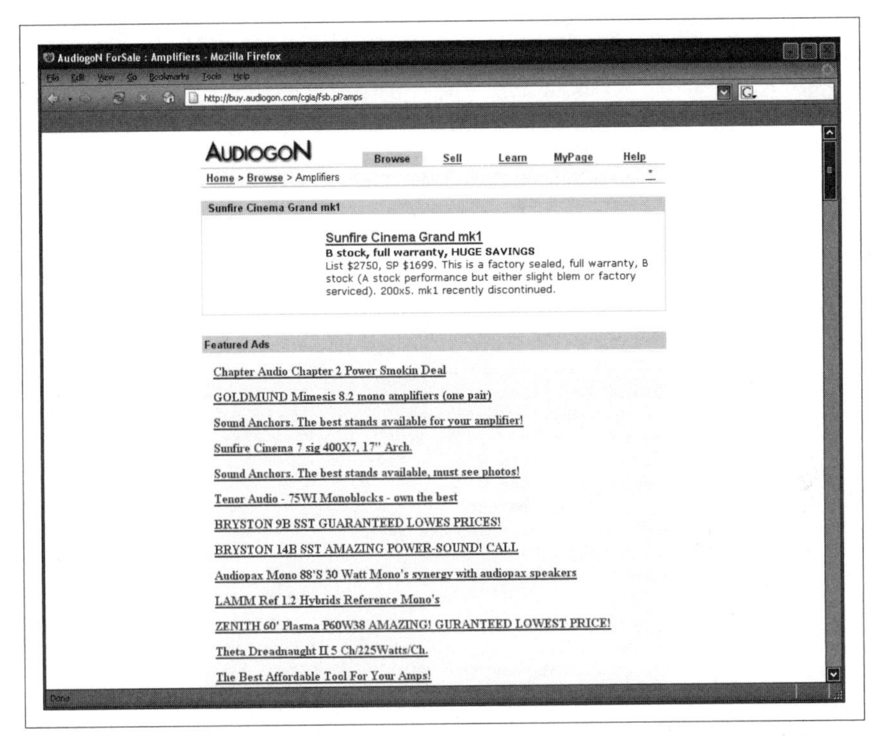

Figure 1-9. Audiogon featured ads

include many items that you simply can't find on eBay, you should exercise real caution. You also should start by reading the terms of use: *http://cgi.audiogon.com/auction/rulesauc.html*.

Realize that I'm not ragging on the auction system at these sites; I'm just saying you might want to take a few extra steps as their infrastructure develops.

The best way to reduce risk, though, is not to simply bail on any gear up for auction. In fact, that would defeat much of the value of these sites. Instead, go back to "old school" eBay tactics. In the early days of eBay, it was common to trade 5 or 10 emails, and often to have a phone call, with a seller during a transaction (at least, that's the way I did things). Although this won't guarantee you will avoid any charlatans, it certainly will help you avoid a lot of bad situations.

Finally, follow all the tips outlined in working with an eBay seller [Hack #5]. Request pictures, get details, go the whole nine yards. Additionally, you can ask the seller if he has done any business on eBay. If he has, get a user ID

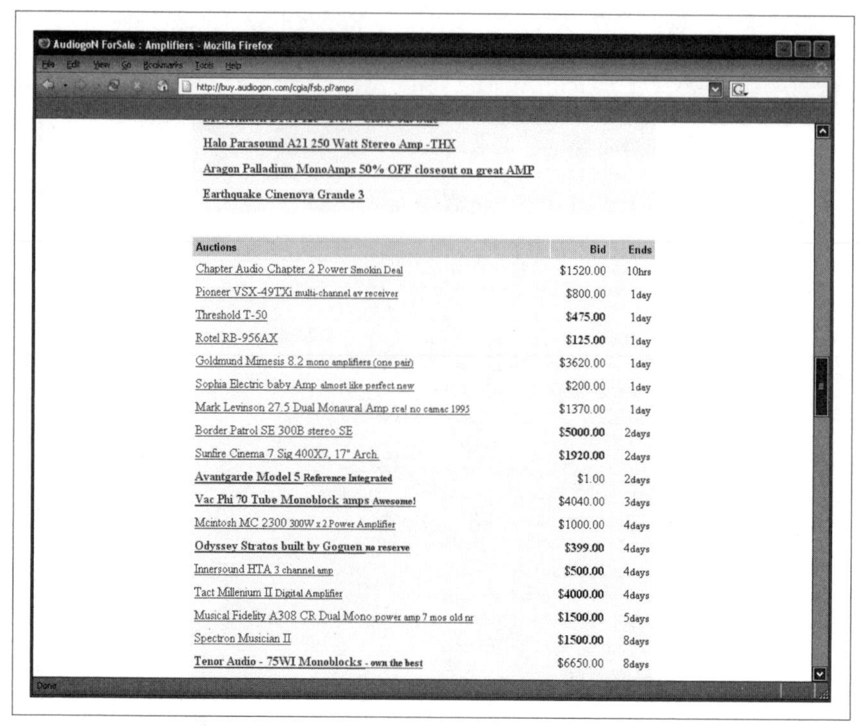

Figure 1-10. Audiogon auction listings

and check his feedback. This is just one more step in making sure you get the best gear, at the best price, in the best condition.

Retail Prices Are Useless

Many of the listings on Audiogon and Videogon have a "sale" price and a retail price. This is pretty typical fare for sites such as these; the retail price always is much higher than the sale price, which makes you feel like you're getting some sort of deal if you buy. That said, you should realize retail prices are absolutely useless. Even the manufacturer inflates its retail prices; certainly you realize that even dealers don't sell gear at retail, don't you? So, never get too excited just because a sale price is some huge percentage off "retail."

The best way to figure out if you're getting a good deal is to hunt around. Most popular items have multiple listings, and you can compare listings to see what the street price is. Another good idea is to see if you can find the item being sold in a brick-and-mortar store, and call that store for a price quote. Even if you can't find a local store, this technique will help you determine if the price online is reasonable. As a general rule, buying direct from

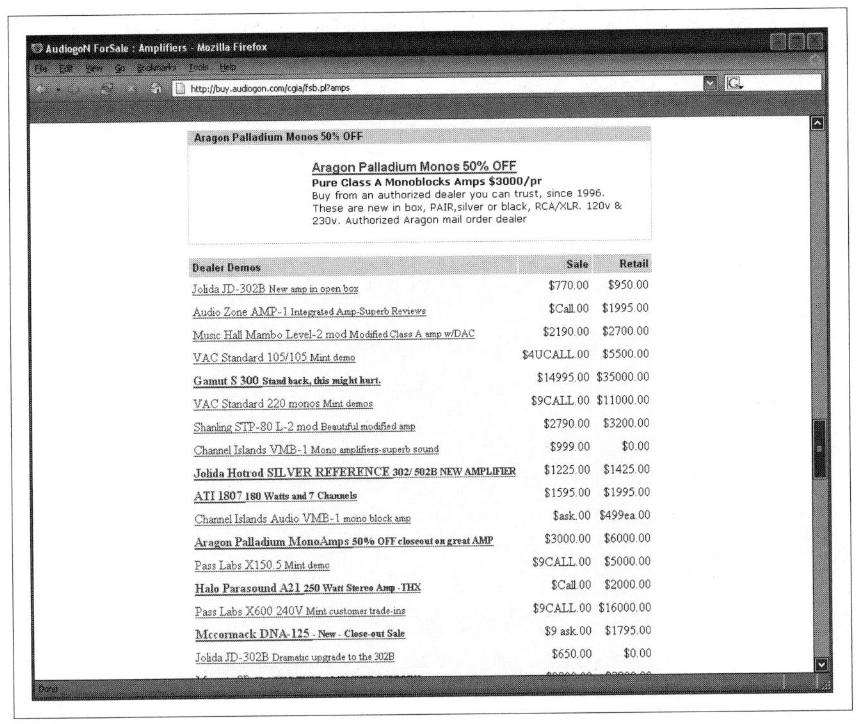

Figure 1-11. Dealer demos

someone online should save you at least 10% to 15% off of buying it in person, and you avoid taxes in most cases.

Always Call Dealers

One of the real steals at sites such as Audiogon are the dealer listings. This might come as a surprise; I actually ignored these for the first several months of shopping online. Typically, dealer demos are like web site ads; you just want to click through them and make them go away. This couldn't be further from the case, though, at these high-end sites. First, you'll often find dealer *demo units*. These are units that have been shown on the floor, or perhaps have been opened up or are surplus stock. In almost all cases, though, these units are new and are in great condition, but have below-market prices. You'll also find some dealers just advertise regular stock on these sites to attract business.

Whatever the situation, you need to call these dealers and ask them for their pricing. Equipment manufacturers (especially the high-end ones) have very restrictive agreements with their dealers as to what prices can be listed.

However, a phone call might reveal that a dealer is willing to go far below what he can legally list a unit at.

For the ultraparanoid, this isn't illegal. It's just a technicality you find in areas like this. The same applies, incidentally, for many high-end guitar builders, including Martin, Taylor, and Collings.

As a point of reference, I was shopping for a Lexicon MC-12. These listed at about $10,000 at the time, and the best you could buy them used on eBay was for around six grand. I called up a dealer I found on Audiogon, and he offered me a new unit at just over $4,500! This wasn't even B-stock or a refurb; it was a new, never-been-opened unit, and I got it at well below used prices. Needless to say, I've been back to that dealer many times, and I've found the same paradigm (albeit not always at such a drastic savings) is in effect all over the place. Five minutes on the phone have saved me thousands and thousands of dollars over the years.

Buy Cabinets for Your Gear

Purchase the best cabinets you can afford for your gear to ensure that it runs as cool as possible, and looks good doing it.

Let's face it: gear is expensive. You easily can spend $20,000 on a good home theater over the course of a couple of years, and that's still leaving plenty of room to upgrade. But that gear isn't worth much, if you never let anyone listen to it. So, here's the common scenario: you invite your friends over for the "ultimate" movie experience. But instead of being in awe of the sound and picture, the first comment you hear is "Gee...why's it all just sitting out on the floor?" Not a pretty picture, is it? One of the most important purchases you'll make, in going from cool gear to cool theater, are cabinets. You'll be able to organize your gear, tuck it away neatly, keep the cable situation under control, and look high-end, all for a fraction of what you've already dropped into the black hole which is your home theater expense sheet.

Starting Out "On the Cheap"

If you've squeezed out every penny to get what you could in terms of equipment, you might have to begin with a lower-end solution than nice wood cabinets. With a decent set of tools and some effort, you actually can put together a nice set of component shelves. Unless you're Bob Vila, you probably won't have hinged doors, custom-fit racks, and the like, but it's still a lot better than a $5,000 receiver sitting on a cardboard box.

Purchase some medium-density fiberboard (MDF) from Home Depot or Lowe's, and cut it to size. For a nicer look, bevel the edges (this will make your shelves look much more expensive!). Then drill holes and run rods through the holes to create a support structure. Connect the whole thing with your basic bolts and washers, and you're done! For extra support, especially on carpets, add some casters to the bottom shelves to level things out. As an example of this sort of setup, check out Figure 1-12; it should give you some ideas.

Figure 1-12. Home-built component rack

Be sure you heavily prime the MDF, especially if you're painting the shelves black, to ensure a nice matte finish. I've also heard of folks lining the shelf tops with truck bed liner for extra traction—go figure! In any case, I've not gone into detail because most of you will want to jump to the higher-end cabinets as quickly as possible, especially if you've got bigger TV sets.

The Real Deal

When you've decided to move up from the homemade equipment, you have two basic options available to you: custom shelving and store-bought home theater racks.

If you happen to be into woodworking, that obviously is a third option, but you'll find it takes many hours to put together a top-notch home theater cabinet. This is more than just shelves and dowel rods; you need to take into account cooling, some sort of door or enclosure, rear openings for cables, and a lot more. Go for it if you like, but take plenty of time to plan ahead.

The most custom solution is to find a local carpenter, and have her put in shelving. This is a killer option, as you can tailor the shelving and cabinetry to your specific needs. You'll also be able to mount the shelving in a closet or on a wall, if you've got the space, which provides a really clean, integrated-looking system. The downside, though, is that you'll often get a carpenter who doesn't know a thing about audio and video gear. You'll have to walk her through proper ventilation and your cable system. Keep in mind that you need access to the rear of your equipment for cable checks, calibration, and the inevitable upgrade. If you can't find someone who has already done a few similar installations, you should exercise caution, lest you end up with really expensive bookshelves!

The second option—the road most taken—is to buy cabinets. Just skip right over eBay and chain stores for this, though, as you'll be dissatisfied. You should check out Videogon first; it has a section just for this category: *http://cgi.videogon.com/cgi-bin/fs.pl?furncabi*. Unfortunately, it's still somewhat undersized; nonetheless, you might find just what you need there.

If Videogon fails, search Google for "home theater cabinets." You'll get a slew of results to wade through and check out.

I've purposefully omitted specific sites, as I've found that the most reliable locations seem to change every few months. What worked for me might not work for you six or nine months down the line.

Personally, I recommend wood cabinets (see Figure 1-13), as metal just doesn't appeal to my eye. You'll also save a bundle in shipping wood as opposed to metal. I also urge you to overbuy; if you have 10 components, look for a unit with at least 12 shelves; you never know when you'll find the next piece of must-have equipment, and you want room to add it into your system. Also, ensure the shelves are adjustable. All components are not created equally!

In the same vein, please, please measure your components! I can't tell you how many horror stories I've heard of someone buying shelves that their

Figure 1-13. Home theater cabinet

gear didn't fit into, and having to spend more than a hundred bucks to ship back the cabinets. Get the widest component you have, and record its width. Also get the height of your tallest component, and the depth of your deepest component. Then, ensure you get the interior measurements of the cabinets you're looking at, and ensure things will fit. I always add at least two inches to my requirements to ensure a comfy fit and room for cables and the like.

As a last piece of advice, I recommend you find home theater cabinets at a furniture-style shop rather than the other way around. I find that the stores that major in cabinetry have better and sturdier cabinets than the home theater stores. This is a big enough market these days that this shouldn't be an issue, but watch out anyway.

Cable Management?

One of the big draws for cabinets these days is *cable management*. This is usually marketing jargon for a set of conduits through which you can run your cables, supposedly hiding them from view. The problem is that, even in the high-end gear, these conduits are small. You'll spend hours threading cables through one end, hoping (praying!) that the cables will make it out the other end. And, once you get even a few cables threaded through, things get really complicated. I've yet to see a unit with larger cable runs, and honestly, if the conduit were big enough to be useful, it probably would look obnoxious anyway!

As if that's not bad enough, when you have a problem, you'll be ready to murder someone once you realize you've got to mess with cables, so carefully run through those little conduits (I'm getting frustrated just thinking about it). The best cable management involves a little Velcro [Hack #36], not expensive routing.

Avoid Glass Like the Plague

A final word on racks before you spend your hard-earned money. Under no circumstances should you ever buy glass shelves. I say that with great reluctance because some of the coolest setups you'll ever see are built out of glass; in fact, I once spent almost $900 on two glass component racks. The problems, though, are many.

Dust
: I mean, it gets on everything, you can't hide it, and it's constantly blowing up off of the (often static electricity-charged) glass shelves.

Clarity
: You might think clear glass is cool, but it doesn't hide anything. Your one silver component suddenly jumps out of the rack, while all of its black-finished neighbors hide. This gets really annoying after the fifth time someone asks you "Now, why didn't you get a black DVD player?"

Cable management
: What a mess: if you can't hide a silver component, imagine the quite-visible mess of component, coaxial, RGB, and other cables that litter your racks.

I've yet to find a single rack with glass shelves that addresses even one of these issues, let alone all three. And, to add insult to injury, the glass racks are usually more expensive! So, save your money (and sanity) and stick with wooden shelves (or, if you prefer, metal).

Plan Your Room Décor

Take your theater to the next level by adding creative lighting, wall sconces, seating, and decorations.

So, now you've got all the gear—televisions, receivers, speakers, DVD players, and the rest—and you've even spent a bundle on enclosures **[Hack #7]** for your components. However, your theater still doesn't feel like...well...a theater. What gives?

Most likely, the answer is that you've simply got a room with great audio and video. And that, my friend, does not a theater make. You'll be amazed at what the accessories in your room can do; you'll soon find that carpet, paint, lighting, posters, and seating add up to a better experience than even some of your components! Ultimately, components are subtle things, and sound and video are in the eye (and ear) of the beholder; but everyone knows what a theater looks like!

In my experience, the bar-none best place to get room décor is the Home Theater Market (HT Market), located online at *http://www.htmarket.com/index.html*. These guys have the best selection, and are the most responsive, that I've found. Make no mistake: I've probably spent a few bucks more here than if I'd hunted for each individual piece online. That said, I've received stellar service, and they've even thrown in free shipping from time to time as a thanks for repeat business.

Paint and Carpet

There's a more complete discussion on paint **[Hack #18]** in the section on video components, as you'll need to understand how lighting affects the color palette before selecting paint. Still, begin to consider how theaters are painted. You probably can't think of a single one with light colors; they're usually done in black, midnight blue, or some other deeply tinted dark color.

Also think about the ceilings of these rooms. They are almost always painted the same dark color. This is often a common mistake in home theaters;

enthusiasts get the side walls right but forget about the ceiling. You also should consider removing any ceiling fans. For those of you up north, this is no big deal; but I'm in Texas, and pulling the ceiling fan was a serious decision, albeit one I think I made for the best. I probably pay a little more to air condition my theater, but it looks like a theater (who ever got up from watching a movie and got distracted by the fan over their head?).

Finally, think carefully about carpet. At a minimum, choose a dark color to go with the rest of the room. If you've got room in your budget, though, consider a themed carpet. The Home Theater Market has some killer designs available at *http://www.htmarket.com/homtheatspec1.html*. These aren't super cheap, but if you have someone local provide installation and the carpet pad, it's actually not all that bad.

I've received more comments about my carpeting than any other single piece of home theater gear or décor I have!

Lighting

Another major addition to any good theater is lighting. Again, the gold standard is the real movie theater. Normal in-ceiling lights just don't provide that movie atmosphere. Decorative sconces, though, are just the ticket. Once again, HT Market is the answer: *http://www.htmarket.com/homtheatligw.html*. You might want to brace yourself for sticker shock here, though; this stuff is very expensive. I seem to recall spending about $800 on lighting, but it was well worth it. Adding four sconces throughout the room, and placing them on a dimmer, makes for a killer setup.

Sconces rarely will provide enough lighting for your entire room, though, and often serve as effect more than anything else. For the main lighting, you might want to consider some in-ceiling pot lights (see Figure 1-14). These provide good light sources, and yet they are focused and can be precisely pointed. I personally like the areas of light that they produce, as opposed to a more general "light the whole room" approach. A good electrician can purchase and install these for a fairly low cost.

Posters

I'm a nut for posters. I have three movie frames and I'm constantly changing out movie posters in each. I'll have the *Lord of the Rings* trilogy one month, the *Matrix* trilogy the next, *Star Wars* the next (I'm into trilogies, can you tell?)...and I'm always getting compliments on the frames. For a

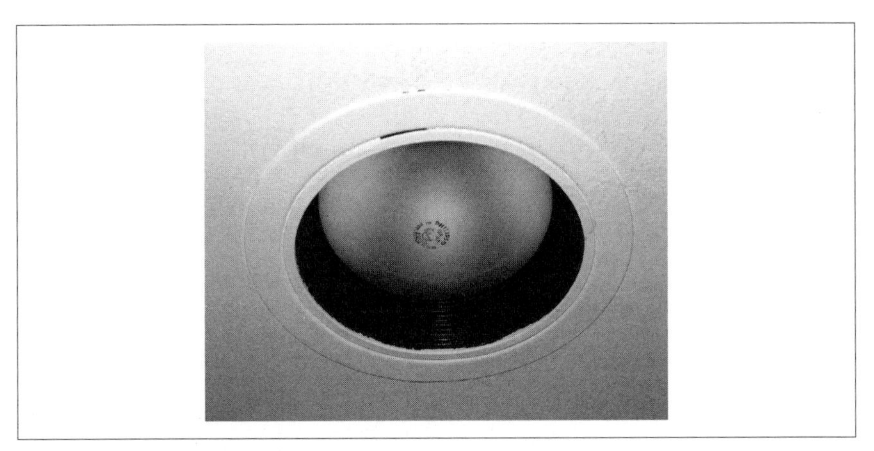

Figure 1-14. In-ceiling pot lights

small investment, you can pick up nice frames at HT Market: *http://www.htmarket.com/posframandca.html*.

The only real "gotcha" here is to ensure that you get a frame that mounts solidly to the wall, and that can be changed out easily. I prefer the brass Loc frames for a classy, yet manageably priced, theater. You can mount them to drywall or to studs in your walls, and it's trivial to pop posters in and out.

You can buy posters almost anywhere, so don't feel limited by HT Market's selection. *http://www.allposters.com/gallery.asp?aid=754680* is another good source for posters, and of course Google is the ultimate resource for finding anything you want.

Seating

Home theater seating is one area where people can get a bit contentious. Some feel that the only way to really outfit a theater is with true "home theater seating." In other words, you want the bucket seats that rock back, just like you find in a theater. However, these are both fabulously expensive (more than $500 a pop in most places) and not that comfortable.

Personally, I'm a big fan of the La-Z-Boy recliner, and I think you can get a few comfy models from them, save big money, and be more comfortable. You can buy models with cup holders, so you really don't lose anything (if you're especially adventurous, add a massage chair, and you're in heaven!). So, don't get too hung up here; the main thing is to have comfortable seating that doesn't look out of place with the rest of your room.

The Little Things

Once you've got all of this covered, there are still plenty of little things you can add to take your theater to the very top of the elite. One of my favorites is decorative woodwork. Most hardware stores offer all sorts of decorative chair rails and molding. If you're so inclined, this sort of thing can really class up a room. Figures 1-15 and 1-16 show some of the woodwork in my theater; notice the fancy-looking joints. This is pretty easy to do; just cut a small rectangle and attach it to the wall, and then join the molding at the wood block. Add a little bevel to the rectangle, and you add yet another level of sophistication.

Figure 1-15. Ceiling molding

Another small, but easy, touch is ensuring that everything matches. Wood finish is so cheap these days that you can easily ensure that your end tables, shelves, cabinets, and woodwork are all the same color. This brings the room together, and helps make things look as though they belong. You'll have lots of other ideas, as well: experiment, and go with what works.

Figure 1-16. Corner molding

CHAPTER TWO

Video Components

Hacks 9–20

No matter how you cut it, home theaters are about great pictures. You can have killer audio, amazing décor, and even an old-fashioned popcorn machine, but nobody will care if your picture is fuzzy and dull. That means you need to spend lots of time getting the right components, and making sure they perform at their very best. This chapter will help you make the right selections, and get them performing at their peak levels.

Get the Right TV

Getting the right television to provide that picture isn't trivial, but here are plenty of good tips to help you pick the right unit.

Figuring out which TV to buy, even with a fixed budget, is worth some explanation. The first decision you need to make relates to the aspect ratio [Hack #13]. If you've got less than $700 or so in your TV budget, this might be a moot point; you're going to have to spend at least that much to get into the widescreen TV market. You might consider saving up a little longer and making sure you can afford a widescreen TV, for several reasons:

- You'll get to watch movies in the closest possible format to how they were filmed, without losing much of the picture to the screen edges.
- Most HD broadcasts [Hack #28] are coming in 16:9 these days, so your favorite episode of *CSI* will look right on your TV.
- Xbox, PlayStation 2, and GameCube all support widescreen TVs; you simply can't beat *Halo* on a 16:9 screen.
- Widescreen TVs invariably come with the highest selection of inputs for video sources.

These are all worth considering, and if you can't afford a widescreen TV on your budget, you might want to wait an extra month or two and save up.

Once you've determined the aspect ratio of the TV, you should simply look for the biggest TV with the best picture available. Although you can buy most components for great prices online [Hack #6] these days, it's still best to buy a television in person. You'll pay a little more in many cases, and often get hit with sales tax, but you save the stress of hoping someone with a death wish isn't driving your TV's delivery truck. Judging picture quality is also highly subjective. So, browse at a local store—I prefer boutiques [Hack #4] when possible—and see what you like. There are several good tips to ensure the set you like in the store looks great when you get it home.

Look for Dim Lighting

When you visit a chain store [Hack #3] that sells HDTVs, notice that the televisions are all on brightly lit shelves. Then go visit a middle- to high-end store [Hack #4]; the lighting is almost always dim. They also usually have demo rooms with one or two sets in a fairly dark room. These stores are trying to re-create your living room environment. Not only do they turn down the television brightness, but also they might have hired a calibration specialist to come in and correctly set the color temperature, focus, and so forth [Hack #62]. The stores that dim their lights in the television area are doing a more honest job of showing you how the units will look in your home. You might want to reward these stores with your business.

Consider the Source

The television signal we all grew up with was designed in the 1940s, for a 9-inch screen. Today we magnify this image to 50 inches (or more), and display it on a high-resolution HDTV monitor [Hack #29]. The result often looks like junk. I see people look at a highly rated HDTV in a store with a nice, dim environment and remark "I don't like that one, let's look over here." People also often get their nice, new HDTV home, plug it into their cable box, and suffer buyer's remorse because the picture doesn't look as good as it did in the store. It's not the television's fault!

If your family watches a lot of cable or satellite television, you need to see how it works with these sources. Some television manufacturers spend extra time and money to make low-quality signals look good, but you should ask to see how the television looks with a DVD (480p) and a true HD signal (either 720p or 1080i). You might be shocked at how the same television will look poor with one source and stunning with another.

Connectivity

The final step is to ensure you get the right sort of connectivity on the back, so you'll be able to connect all the components you'll eventually buy. The key here is to ensure that you have *component video* inputs. Component video cables are the top of the line in carrying an analog video signal. After that comes S-video and then RCA cables. The TV box you're looking at should mention something along the lines of "component video inputs" or sometimes (especially on Toshiba TVs) "colorstream inputs." For reference, these connectors look like those shown in Figure 2-1.

Figure 2-1. Component video connectors

If you went with a widescreen or otherwise HD-capable set, this should be a given, but it's still good to make sure. Now, make your purchase! You should have the biggest and best picture that all but a small bit of your budget can afford.

Size Is Everything...

I realize that one of the primary considerations in buying a TV is size. Unfortunately, what's not often paired up with size is your house layout. Although that might sound silly, it's not a trivial thing to take a monstrous 58-, 60-, or even 72-inch rear projection TV and get it into your viewing room, unless you've got an exterior door in that room.

A word to the wise: you can do a lot more with a theater room that is in a converted garage, or has direct access to the outdoors. With only an exterior door to limit equipment entry, you're virtually unlimited in what you can get into the room.

If you do have to get a big TV around some corners, you might want to spend a little time before purchasing, just to make sure the TV will fit. With an hour of free time and some scrap wood, build a rough mockup of the TV. You can get the height, width, and depth of the potential TV off the box (or the manufacturer's web site) and a simple mockup shouldn't cost more than 10 bucks to make. Be sure to put some boards in the middle of the front, back, and sides, so you simulate the entire TV rather than just its edges (see Figure 2-2).

Figure 2-2. TV mockup

With the mockup built, take it outside, and ensure you can navigate through the house. If you've got a smooth trip, you're all set. Also keep in mind that packaging often adds an inch or so to the dimensions, so if it's really tight with the mockup, you might be out of luck. Still, you just saved the hassle of trying to return an opened TV to your local store **[Hack #4]** (and heaven forbid you had it delivered after buying out of state or online **[Hack #6]**).

Which Brand?

Questions get asked all the time on which brand is good or better when choosing an HD-capable set. This very question was asked recently, and Michael Chen gave his opinions on the big players.

Hitachi
: Nice units, but not too much tweaking knowledge out there. These are also not considered to be major players in the TV market.

Mitsubishi
: A very nice set, but only after a significant amount of work inside the TV's service menus had been done. In some cases, you'll have to go even beyond these menus and get a professional calibrator to work the set over.

Panasonic
: Panasonic has come into the TV market from obscurity this year with a very nice product that is very tweakable, and an image that looks as good as any in the group.

Pioneer
: Pioneers always have a great image, but sometimes are a bit on the high side in terms of price. These units have a very friendly service menu, and are very tweakable.

Samsung
: Samsung isn't a major player, and there is too little information out there on image quality.

Sony
: Sony has a very nice image and a very good price point this year. Of course, you get great name recognition, and these units are easy to calibrate, although somewhat limited in terms of fine-tuning options.

Toshiba
: Toshiba sets took a step backward in 2004 in the image quality department. They are still at a good price point, and the images are pleasing, though slightly worse than in 2003. The good news is that unless you know what to look for, the image is still pretty good.

Zenith
: Zenith makes a good product, but they leave most people with a very poor impression of quality. Even though I think this is an incorrect perception, the impression is still there. Zenith just isn't a major player in the rear projection TV (RPTV) field.

Your TV's First Steps

After the credit card has taken a hit, there are several tasks you need to take care of right away on a new television set. This will allow you to get the TV in the house and make sure it's working, and will guard against early burn-in.

The most difficult time in buying a new TV is the period between checking out at the register (or entering your credit card information online) and actually having the TV in your theater room. You'll be impatient, grumpy, and probably a real pain to be around; who likes to wait? The bad news is that this is exactly the time when you need to be careful and cautious, as many problems can occur in shipping and you're the only one who cares about catching them.

While the Delivery Van Is in the Drive

When the TV arrives on a big delivery truck, you should test the TV—say, in your garage—before you let the shippers leave. You might want to plug it in and connect it to a DVD player just to see if it turns on and if an image appears. If not, you have the option of refusing the shipment.

Some also recommend that you look through the vent holes behind the TV with a flashlight to check whether anything is out of place. If you want, you can remove the whole bottom back panel to closely inspect the innards. Many folks might not even know what to look for, but if you ensure that everything looks clean and in place, you've already prevented a lot of major problems.

Sometimes delivery men get cranky and impatient while you do this. Frankly, that's just too bad; you've put too much money into a good TV at this point to allow something silly to happen in transit.

As an example, the heat sink tray, like those in Toshiba units, looks like a little table that sits on a circuit board at the bottom left side of the TV. If it is not level, the tray probably broke off due to mishandling during shipment. You can then look into having the shippers take it back. Some, however, suggest that if the TV still works, simply work with Toshiba in getting the tray resoldered onto the circuit board. That way, you can enjoy the TV now rather than have to wait for a replacement, which might take another few months! Can you really stand to wait that long again?

Prevent Early Burn-In

After you've gotten the TV into your house, there is one very crucial thing you must do to prevent damage to your TV.

This applies to any projection HDTV.

Bring up the screen menu, select Picture settings, and go to Contrast. You'll notice that the factory setting is 100! Some call this *torch mode*. You'll remember from the section on selecting a TV [Hack #9] that many stores crank this up to make the picture look brighter in the store.

Back it off to less than 50 to prevent early burn-in. This is one of the best things you can do to ensure a longer life for your TV screen.

Move Your TV Safely

Even if you have a professional deliver your TV, you might eventually need to move it—to a new home, to a new city, to a new state. Avoid losing thousands of dollars by following a few simple instructions.

This might not apply to a short move, but it definitely applies to a cross-country move, and everything in between. Whenever you are considering a move—near or far—that involves a big-screen TV, you should go down to U-Haul and buy half a dozen unbuilt cardboard boxes, and don't build them.

Instead, lay the boxes flat in your moving truck. Place them where the wheels or flat bottom of your TV or projector is going to go. Use two sets—one for each side of the TV—if it is an extremely long RPTV, such as a 73-inch unit. As you roll or lift the TV into the truck, set it down on these boxes.

Then tie down the TV very carefully (you've just raised the center of gravity for the unit) and allow the 12 layers of corrugated cardboard per corner to absorb the road shocks. You also should pad the large contact areas at the sides of the truck; hopefully you can place the TV in a corner of the moving truck. Just padding that area should do it there.

Be extra careful of the screen, of course, by keeping padding away from the screen's center. Try to ensure that nothing touches the TV screen, and if you can't avoid the screen entirely, just use the edges. Facing the TV up against the side of the truck should present the screen face with a safe parallel surface, and protect it from something falling against it during frisky cornering on the part of the driver.

Keep in mind that if your TV is a Sony, or some other brand with very little fortification at the plastic, slanted/curved, outside upper-rear surface that encases the mirror, the mirror is paralleled up against that surface and anything that hits it will take out the mirror also. You'll never even have evidence that anything untoward has happened; at least, until you turn on the unit and find a very unusual picture, if you still have one at all. The Sony 41-inch table model T15 is very susceptible to this; I've seen two of them need new mirrors after a move.

Feel free to also add moving blankets over the cardboard, but keep in mind that carpets and blankets bunch up and compress, whereas corrugated cardboard does not. Remember, not mechanically transferring the road shocks is the goal here.

In tying off your cabinetry, for center-of-gravity reasons, be sure and pad the contact points well, so the rope or binding doesn't rub into the corner's finish. Again, corrugated cardboard on the outside of the padding, at the rope, helps keep the physical pressure spread out rather than centered.

It also might be a good idea to glue up your focus block (these are knobs, dials, and buttons commonly found on projectors near the lenses) before the move, as Sony and Runco do before their units leave the factory. Don't use Super Glue, or anything else that would be highly penetrative. Just use a light-duty, thick glue, such as caulking or a light strip of Elmer's. Apply the glue to just one edge of each control; after all, you don't want to make them permanently unmovable. The hold should just be temporary and should suffice until and unless the focus block settings need to be altered again.

This will immobilize all electrostatic focus controls, and your grayscale's screen controls, if they are located in a focus block. Some Mitsubishi table model TS units have only three controls present on the focus block, while older Mitsubishis have both sets of controls located elsewhere; some are on the convergence board, and some have no focus block at all.

I think this will keep your convergence, mechanical focus, and the settings on your focus block the best they can be under those circumstances, unless someone out there has thought of something even better.

Of course, all of this is especially important if you've just had your unit finely and professionally calibrated to the tune of several hundred dollars.

—*Robert Jones, Image Perfection*

Fix Panasonic's Picture Glitch

Many of Panasonic's 16 × 9 rear projection TVs have a nasty picture glitch: if power is lost, the picture looks odd, with the top and bottom cut off, and the convergence goes screwy on you. A little service menu magic will get you back to normal in no time.

The PT47/56/65 and WX51/49 series of Panasonic rear projection TVs have a glitch that results in what appears to be a hardware problem with the sets. If power is lost (a fuse shorts out, or even if your surge protector trips) and is then restored, the TV picture suddenly looks strange. No matter what you try, you won't be able to get that picture looking good.

The easiest way to verify that this is indeed the glitch discussed here is to examine the picture on the screen (with the TV on, of course). The top and the bottom of the picture will be cut off if you've got this bug. If you're not sure how to check for this, tune to either a sports channel such as ESPN or a news channel such as FOX News. These types of channels usually have a ticker running at the bottom of the screen. If it's cut off, you're missing part of your picture. Additionally, you'll often see a fishbowl effect, where the picture almost appears to be curved on the sides; convergence won't help this, either.

Entering the Service Menu

The first thing you need to do is enter the Panasonic service menu. This requires your remote (often, programmable remotes **[Hack #84]** are a pain for this; the original remote is recommended). Then, follow these steps:

1. In the setup menu, set the antenna to "cable."
2. In the timer menu, set the sleep time to 30.
3. Exit all menus and tune to channel 124.
4. Adjust the volume (on the TV, not your receiver) to 0.
5. Press the Volume Down button on the television set panel until a red CHK appears on the screen.

Once you're in this menu, you should know how a few buttons on your remote function, until you exit this menu:

- The TV/Video button changes video inputs.
- The Power button (on the remote) toggles menus.
- The up, down, left, and right arrows are used for navigation.

Once at this point, you can actually fix this power glitch once and for all.

Fixing the Picture

Here's what you need to do:

1. Move the cursor to VER and press the Enter button on your remote.
2. Select 16:9 (you are able to choose either 4:3 or 16:9), and press Enter.
3. Unplug the TV set.
4. Wait 60 seconds.
5. Plug the television back in.

When you plug the TV back in, it should still be in service mode.

6. Save your changes and exit the service menu.

Things should work normally now. The key thing to look for is that curved image, and check out the top and bottom of the picture. If you can see those news tickers, you're all set.

What Happened?

Panasonic's 4:3 and 16:9 TV sets share some common parts. In fact, when Panasonic started producing 16:9 TVs, it used the majority of the guts of its existing 4:3 TVs. Essentially, a 51-inch 4:3 set is the same as a 47-inch 16:9 set; you're getting the same basic display unit, with a different cabinet and aspect ratio (not to mention, you're paying a good bit more). However, some of these widescreen TVs don't "realize" they are actually 16:9s; the VER parameter in the service menu takes care of this. In particular, when power is lost, these sets often will revert to the 4:3 setting, causing this glitch.

What's even more interesting is that you can make your 51-inch 4:3 TV behave just like the 16:9 set, using the same procedure! The image area on the 4:3 TV is reduced to—surprise, surprise—a 47-inch 16:9 image! Of course, you'll need to perform a lot of calibration [Hack #62] to get the picture to look right, but that's still a lot cheaper than buying a widescreen set!

Keep in mind that if your TV is hit with a power shock again you might have to repeat these steps. Also, understand that none of this has to do with the Aspect Ratio button on the remote; this problem affects all pictures, in all formats.

—Michael Chen

Figure Out Aspect Ratios

This hack explains the difference between the 4:3, 16:9, and 1.85:1 aspect ratios. You'll learn what TV is best for you, why black bars are good, and how to make sure you get the "theater" in home theater.

Many first-time DVD buyers notice that DVDs are mostly in the widescreen or letterboxed format, which have black bars at the top and bottom, and they wonder why. The next time you go to your local movie theater, take a close look at the movie screen. You will find that modern movie theater screens are actually rectangular in shape (they are much wider than they are tall), as Figure 2-3 illustrates.

Directors shoot their movies in such a way that the shape of the finished picture is a rectangle. While you watch a movie in the theater, take note of how wide the movie picture is.

Now, when you get home, look closely at your TV set. Your TV set (if it's a standard TV) is basically a square (see Figure 2-4).

Figure 2-3. Typical movie screen aspect ratio

Figure 2-4. Typical TV screen aspect ratio

From the Big Screen to the Small Screen

To make the rectangular picture you saw at the movie theater fit in your square TV the studio uses one of three processes in preparing the DVD release.

Pan and scan

The studio cuts off the sides of the rectangular picture and makes it a square (obviously losing picture area from the original film). In many cases the missing information is important, so a technician must pan the center of focus back and forth digitally to fit the important action from the rectangular frame onto the square TV screen (thus the name "pan and scan"). This is probably what you have seen if you watch movies on VHS or on cable TV. You might have seen the warning "This film has been modified to fit your screen." This warning lets you know that someone has cut off big pieces of the original rectangle to make it fit inside your square.

Open matte

Some film cameras actually shoot on a square negative, but the director composes his shots for the rectangular movie screen and plans to throw away the extra material at the top and bottom. When the film is shown in theaters, it is matted to eliminate this empty space above and below, and shows only the director's framed action. Sometimes for home video, to create a square TV transfer, the square negative is transferred without the mattes. This results in the appearance of "more image" above and below the action than what was seen in theaters, but still differs from the director's intended theatrical framing. In fact, production elements such as boom microphones and lighting equipment often are revealed when the mattes are removed for home video. Like pan and scan, this is modified from the theatrical shape.

Letterboxing

Alternately, the image can be zoomed out a little. This allows your square set to fit the full width of the rectangle inside the square TV shape, but leaves unused areas at the top and bottom. You see black bars because you are seeing the full width of the rectangle, which leaves no picture for the top and bottom of your square-shaped TV.

To create a truly theatrical experience in your home, it is important to present the film as it was originally intended. To respect the film and present it as the artist intended, you always should display it without cutting off parts of the picture. Thus, we have widescreen DVDs (and now widescreen TVs). This respect for the original shape of the film is called *original aspect ratio* (OAR). The term *aspect ratio* simply means the ratio of the width of the image to its height, and is usually represented by numbers—*width:height* or *width×height*, as in 16:9 or 4×3. This number is often divided and expressed as a relation to the number 1, like 1.78:1 (which is the same thing as 16:9, if you do the math).

Prevalent Aspect Ratios

Five aspect ratios are in use today. The next sections expound on each.

1.33:1 (4:3). The 1.33:1 aspect ratio was the dominant shape for television programming, from its creation up until the advent of HDTV. In the earliest days of cinema, films also were produced in approximately this same 4:3 shape (in the film world, this was sometimes called the *Academy Ratio*), shown in Figure 2-5.

Figure 2-5. The 1.33:1 (4:3) aspect ratio

As a side note, it is important to realize that many classic films such as *Citizen Kane* and *The Wizard of Oz* were produced in this almost-square 4:3 ratio. When these films are released on DVD you will find they are not letterboxed, but rather, are presented in their original 4:3 aspect ratio.

About 50 years ago, the movie industry was losing a large portion of its audience to television, so they created a wider aspect presentation to give audiences something they couldn't see in their homes. Widescreen aspect ratios were born.

1.66:1. This is a less-popular widescreen aspect ratio seen more often in Europe and in independent films. The image is only slightly wider than the standard 4:3 frame, as Figure 2-6 illustrates.

Figure 2-6. The 1.66:1 aspect ratio

1.78:1 (16:9). This is the official aspect ratio of HDTV and the aspect used on widescreen TV sets. The majority of high-definition television programs are created in this aspect, and some consumer camcorders are beginning to offer this aspect ratio as an option as well. Figure 2-7 shows an image in this aspect ratio.

Figure 2-7. 1.78:1 (16:9) aspect ratio

1.85:1. This is the first of two common widescreen formats used in the theatrical presentation of film. It is the narrower of the two shapes used in theatrical exhibition but is still considerably wider than the standard 4:3 television picture. It is important to note that the 1.85:1 ratio is nearly identical to the 1.78:1 HDTV ratio, and the two sometimes are used interchangeably (note how close Figure 2-8 is to Figure 2-7 in appearance).

Figure 2-8. 1.78:1 aspect ratio

2.35:1. This is the second of two common widescreen formats used in the theatrical presentation of film. This aspect is very wide, bordering on panoramic—providing a very pleasing theatrical experience. This aspect is considerably wider than the standard 4:3 TV and is also wider than the 1.78:1 screen of a widescreen television set, as evidenced in Figure 2-9. In some cases modern films are presented at a slightly wider 2.40:1, and some older films went as wide as 2.76:1, but for the most part common usage of this widest aspect ratio falls right around 2.35:1.

Figure 2-9. 2.35:1 aspect ratio

What Does This Mean to Me?

With the advent of widescreen DVD and widescreen TV sets, TV salespeople have been quick to tout the purchase of a brand-new widescreen HDTV set as a cure to those dreaded black bars. However, it's important to realize that a widescreen set is not a cure-all for black bars. The shape of a widescreen TV is 16×9. If you attempt to play any material that isn't 16×9 (but rather, is wider or narrower), without stretching or cutting off parts of the picture, some of the area still will be unused and will appear as black bars.

The advantage to a 16×9 set is that it contains 1.78:1 aspect ratio or 1.85:1 widescreen films perfectly, and you will see them without bars, which means that all HDTV broadcasts, and many movies, appear perfectly on your TV. However, films with the wider 2.35:1 aspect ratio (such as *Gladiator*, *Armageddon*, *T2*, *Harry Potter*, etc.) are larger than the 1.78:1 (16×9) TV frame. To present the film in the proper shape without cropping, a small amount of unused space will still exist above and below the picture area. Still, most home theater gurus vastly prefer this to 4×3 sets.

In summary: watch the pictures, not the bars!

—Vince Maskeeper-Tennant and Brett McLaughlin

Avoid Cheap Projectors

One of the more difficult decisions you'll encounter in the evolution of your home theater is whether you should add a projection system to replace your television. However, this isn't always a matter of adding quality; sometimes you're trading picture depth for picture size. Make sure you understand the difference between a better picture and a bigger one.

Although it's true that throwing a movie picture onto a huge screen in your theater makes you look cool, the picture might sometimes suffer as a result. Before buying any old projector just to brag to your buddies, make sure you're getting something that's going to perform better than the TV you've already got.

The Players

One of the first things to take note of is the differences between display technologies. Here's the basic rundown.

Cathode-ray tube (CRT). One of the earliest types of front projection technology, CRT uses the same *scanning electron beam* technology used in old-fashioned picture tubes to create the video image. In the case of CRT projection, the projector increases the light output to project an image across a distance onto a screen. CRT display is an analog process, meaning it is not locked into a specific display resolution like a digital display, which is based on a fixed number of pixels. The only display limits on a CRT display concern how fast the electron beam can draw (called the *sync* or *refresh rate*).

You probably have seen a CRT projector in a conference room or your favorite sports bar. The majority of CRT projectors feature three lenses—one green, one red, and one blue—that are specially configured to project overlaying colors to create an image on a reflective screen. Many of the big-screen rear projection sets also use CRT technology, but have the projection elements mounted inside the case, projecting from behind the screen.

The main advantages of CRT projection are the smoothness of picture, the solid black levels, and the variety of sync rates it can display. The main disadvantages are the low level of light output, the overall size, and the complexity to install and maintain. CRT gives an excellent picture, but the projectors tend to be large and unwieldy, they require absolute darkness, and they often take hours to install and configure properly.

Liquid crystal display (LCD). The first major technology in digital projection, LCD uses a light projected through three grids of colored liquid crystal pixels to create the image. The red, green, and blue panels illuminate as needed, coloring the bright light passing through them accordingly, creating a projected video image. The LCD display process is digital, meaning that it has a native resolution tied to the number of pixels in its display panels.

If a projector panel has a native resolution of 800 pixels by 600 pixels, it can properly display only an 800×600 resolution signal, and will process and convert incoming video (such as DVD) to be compatible with this native resolution. This processing of incoming video signal is called *scaling*. You might have experienced this with an LCD computer monitor; using any resolution other than the suggested resolution results in a processed, less-than-perfect image.

The main advantages of LCD technology are the size and high light output. LCD projectors tend to be smaller and brighter than their CRT projection counterparts. The main disadvantages include the resolution issues outlined earlier, and the dreaded "screen door effect" where pictures suffer from a visible grid structure due to the nature of the panel. Occasionally, LCD will suffer from "stuck pixels," resulting in the grid having a few single pixel spots as a result of a pixel getting stuck in an on or off position.

Digital light processing (DLP). DLP has become the common catch name for what was previously and more broadly referred to as Digital Micromirror Devices (DMD). A technology developed by Texas Instruments, these devices consist of a tiny chip with anywhere from hundreds to millions of tiny mirrors, all arranged in a grid. These mirrors are attached to a hinge of sorts, with an electrical support post that allows it to be moved slightly when electricity is applied. By tilting each mirror a few degrees in either direction—through the use of electricity—and shining light onto the grid, you can manipulate the reflection of on/off pixels to create images.

This on/off pixel grid is only half the equation; there is also a color component. To add color to the reflected light, the white reflected light from the mirror grid is sent through a spinning wheel consisting of the primary colors: red, green and blue. The brain of the DLP engine divides the image into these primary colors. The spin of the wheel is synchronized to the grid of mirrors, and as each color passes in front of the reflected beam of light, the grid triggers the needed pixels for that color. So, as the light is passing through the green section of the wheel, the pixels that need green added are turned on. This process repeats for each color, for each frame, thousands of times per second. Because it all happens so fast, your eye is able to integrate the overlaying colors into a single, full-color image.

Texas Instruments has heavily promoted this technology as the replacement to film projection in cinemas, and most major cities now have at least one theater fitted with a DLP projection system. This technology also appears in some of the more popular small-format projectors and now is emerging in rear projection televisions.

The main advantages of this technology are a further reduction in size and a finer-quality projected image. Because the mirrors on the DLP chip can be very close together, it doesn't cause the "screen door" effect present in LCD. The main disadvantage of this technology is a "rainbow effect." Because of the nature of the spinning color wheel, some viewers see rainbow bands or halos around images projected using the DLP system. Although increasing the speed of the color wheel spin has reduced the rainbow effect for many viewers, some with more sensitive eyes still complain and can even get headaches from viewing DLP images for extended periods.

Liquid Crystal on Silicon (LCOS). A conceptual offshoot of traditional LCD technology, LCOS takes the liquid crystals off the glass plates and moves them to the surface of a silicon chip. The control circuits for the LCD are etched onto the chip and the chip surface is made reflective. Although on the surface this is very similar to a traditional LCD chip, it opens a whole new world of options in LCD technology. LCOS chips can be smaller and of higher resolution than traditional LCD chips, and they are cheaper and easier to manufacture as well!

LCOS display technology has been integrated into a few projector products; however, its primary use has been in microdisplays (small LCD monitors that can be integrated into everything from wearable computers to greeting cards). The main advantages of LCOS are the improved resolution, the reduction of the "screen door" effect, and the potential for even smaller sizes than traditional LCD. The main disadvantage of LCOS is that it is not very common, so prices remain higher than they would be if they were manufactured on a large scale and selection of projectors is very limited.

Black Is Beautiful

Tweaking most LCD projectors produces acceptable black levels in an ambient light-controlled room, but most DLP projectors can produce black levels that please the most critical viewer. And not only are the blacks "blacker," but also, the detail in dark scenes takes viewing to another level entirely. Of course, this is the nature of DLP versus LCD technology to some extent, but it still has to be seen to be appreciated. I'm reminded of those experiences where you don't miss something until you have it, and then you can't seem to live without it (power windows in cars is just such an example).

One of the key indicators of a good picture is black: how "black" does the color "black" actually appear? Poorer pictures have blacks that tend toward just being dark rather than being a true absence of color. This has the effect of making the picture appear less sharp and less focused.

But if black is beautiful, it still doesn't tell the whole story. Color rendition with higher-end projectors isn't the same as it is with lower-end ones. Reds, in particular, jump right out in the nicer Runco (*http://www.runco.com/*) projectors; all the other colors seem richer and more three-dimensional as well. To put this in perspective, an old, but still trusty, Pioneer Elite Pro-75 Rear Projection TV (45-inch 4:3 unit) outperformed some of the mid-level Sony projectors! Running tests comparing the Sony projector with the Pioneer TV, it was obvious that the blacks were blacker and the colors were more vibrant on the TV. However, when comparing a Runco CL-710 to the Pioneer television, the advantage shifted. The picture on the CL-710 matched the color and black level of the Pioneer, even allowing for the fact that the Runco is producing a 110-inch diagonal picture compared to the more concentrated image on the Pioneer. Where the Sony made up for what it lacked in picture quality by sheer size when compared to the Pioneer, the Runco doesn't need to take a back seat to the Pioneer in any aspect (no pun intended.)

All that said, picture quality sometimes isn't the only concern; you might be dead-set on having a 110-inch screen, and in those cases, television stops being an option, and fast. You also might have space concerns, where a large TV, and its accompanying depth, just don't fit. In all these cases, keep in mind that ultimately, seeing is believing and assuming that any projector is better than any TV can leave you with a very expensive mistake. Move into projectors only if you've got a nice-size budget.

High-End Projectors

Once you decide to take the plunge into projectors, or at least take a gander at them, you're going to have yet another episode of sticker shock. This is especially true if you go to the high end and, in particular, to Runco projectors (*http://www.runco.com*). That said, there are good reasons to lay out money in these areas, even if you are paying for a name to a certain extent. Although Runco isn't the only name in high-end projectors (Sanyo, Sony, and Philips have offerings in this space), the comments here are indicative of the entire class of high-end, high-dollar projection systems.

Runco, in particular, has long been associated with high-quality (and high-priced) video products. Some have implied that *elitist* might be an appropriate word to use. I've read the remarks by some on the Internet and have heard comments such as "Grossly overpriced," "You get what you pay for," and everything in between. So, who's right? I can talk only from personal experience and really don't have any axe to grind. Nor do I owe any particular allegiance to the company. All the disclaimers taken care of, here's my take.

There are many, many projectors to choose from and quite a number of manufacturers out there peddling their wares. You can spend less than $1,000 for a projector or far more than $100,000 for one. Usually a law of diminishing returns is in effect. I saw this within the Sony line, when I first purchased my VW10HT for $5,800 in late 1999. Sony also sold the excellent G90 CRT projector for about seven times the cost of the VW10HT. And I just couldn't see a seven-times-better picture (neither could at least one of the reviewers of the VW10HT, who canceled his order for a G90 and bought the VW10HT instead). My guiding principle was a solid ratio between price and performance.

In the intervening three years since my initial front projector purchase, technology has advanced and the prices have come down; so much so that a true "home theater" experience is well within the means of most people. However, there is no substitute for a big picture, especially when it's a great picture. You turn down the lights and you are at the theater! Because of my good fortune I have become part of the Runco "family" and now see things in a different light. True, you can get DLP projectors that apparently have similar specs for less money. However, you must look beyond the specifications and consider all the aspects of your purchase. Look at the company involved. Look at the support structure, both technical and informational. Look at the workmanship involved.

The first difference I noticed between Runco and my older Sony VW10HT (outside of the superior build quality of the Runco projector) was the support structure for initial setup and technical assistance. Sony manufactures projectors, but it also manufactures clock radios and everything in between. Trying to get in touch with the Sony projection support team is like pulling teeth. If you get lucky, your local Sony Service Center might be able to help. On the other hand, they might not know a VW10HT from a VW Beetle. And the out-of-box experience can be a real adventure; it literally took months to go from an acceptable picture on my VW10HT to an excellent picture.

The Runco unit was a completely different experience. The projector came fully adjusted and looked great right out of the box. I was provided with a technical support 800 number and was told that if I had any problems with installation a person would be sent out to make sure everything was properly adjusted. Although I had no need for such a visit, it was good to know that this resource was available if problems came up. Runco sells video equipment. I have yet to find any catalog with a Runco clock radio—or a rotisserie oven and BBQ. That's Runco! I think you see what I mean.

Bottom line: sometimes you get what you pay for.

—*Dr. Robert A. Fowkes*

DVHS on a Budget

The high-definition craze is beginning to affect every type of video component, and your old VCR is no exception. DVHS allows you to tape HD material, but it comes at a high price. With a little trickery, you can get all the benefits of DVHS at a fraction of the cost.

Just as TiVo enthusiasts are getting used to having the ability to record all their favorite programs, HD has come along and changed things. A standard TiVo set can't record HD content; you'll have to shell out at least $500 for an HD-recording unit. Additionally, not everyone has TiVo, and many that do don't get their local channels (where a lot of HD content is to be found) except through an antenna.

Amazingly enough, this has caused a renewed interest in VCRs—but VCRs that can record HD content. These HD-VCRs are great, but the tapes are expensive, and they make keeping a lot of HD content—such as an entire season of *Stargate SG-1* or *Alias*—difficult, if not impossible. But a little bit of work and a lot of clever experimentation can get you around this limitation, keep you in HD content, and impress your buddies.

A History Lesson

First, a little bit of history about the Super VHS (SVHS) format, which JVC introduced in 1987. This format was an upgrade to the existing VHS format; SVHS was designed to record at a higher horizontal resolution than VHS. The boost to detail came in the luminance domain (a.k.a. the black and white information), while the color resolution was still unchanged from VHS. The horizontal resolution specification for SVHS was greater than 400 lines, compared to VHS, which was only 240 lines. Strangely, the increased detail in the black and white information resulted in an image that looked like the color saturation was low all the time.

In practice, most consumer SVHS VCRs typically recorded only 360 to 380 lines at the best SP (standard play) speed, so they were not quite as good as the specification for the new format; still, this was a large leap in image quality. Some of the cheaper SVHS VCRs of the time were actually down in the 340-line range for recorded horizontal resolution. The current JVC higher-end SVHS units (circa 2001) record greater than 420 lines while their budget units are still in the 350-line range.

The catch with the SVHS format was that the improved format required a special tape for the higher-resolution recording, and thus the SVHS tape was born. Regular VHS tapes could not be used to record in the SVHS mode because the cassette shell design prevented it. However, the SVHS tapes are simply a higher grade of VHS tape, and nothing more. In fact, there are several different tape formats, all of varying quality. They're all listed here, from lowest to highest grade:

- Standard Grade
- High Grade
- Hi-Fi Grade
- Professional Grade
- Industrial Professional Grade (probably SVHS quality)
- SVHS
- Industrial Grade SVHS
- DVHS (Digital VHS)

Enter the Hackers

This wealth of varying formats required the higher-grade tapes to guarantee a certain recorded quality. Astute enthusiasts quickly figured out that the difference between the two types of tape, from a cassette design standpoint, was nothing more than a small *recognition hole* drilled on the bottom of the SVHS cassette! A sensor within the SVHS VCR would detect the presence of the hole and permit the recording of SVHS signals on the tape.

Of course, in these early days of the format, the SVHS tapes were simply far more expensive than regular VHS tapes. Even in the early 1990s, the SVHS tapes were $10 compared to $2.50 for a good TDK standard-grade VHS tape. For people who loved to archive their TV programs and maximize the recorded quality, SVHS was the only way to go. Unfortunately, the six-hour EP speed in SVHS, although superior in detail, was still just as unreliable when it came to archiving important tapes. So, the SVHS enthusiast was stuck with buying the expensive SVHS tapes and using the SP speed. But $10 for a two-hour video was (and still is) awfully expensive, especially if you've got several hundred movies archived.

Early in the life of the format, enthusiasts such as myself started to experiment with VHS tape formulations at SVHS quality. We found that we could drill a similar recognition hole into a standard VHS cassette and the SVHS VCR would then see this tape as an SVHS tape. People used drills and soldering irons to create these holes in the VHS cassettes and suddenly they were able to archive many more programs far more cost effectively and at far greater image quality.

Taping SVHS signals onto a typical cheap VHS tape resulted in a higher-resolution image that was viewable, but would never be mistaken for true SVHS taped images. The images were grainy. As one went up the VHS tape quality chain, the SVHS images rapidly approached the quality of the true SVHS tapes. The challenge in those days was to find the right standard-grade VHS formulation that came closest to the true SVHS quality. The best solution was found in TDK standard-grade tapes.

Modding the VCR

With this history behind us, we come to the new DVHS format for HD recording, and once more we have the dilemma of expensive DVHS tapes, and the need to keep the quality while minimizing the dollars we have to spend on premium tape.

Drilling holes into the cassette shells was considered a bad idea, even at the time the hacking community tried it (and we did!). Plastic shavings affected the tapes and the soldering iron didn't work. This led us to open up the first SVHS VCRs to look closely at the so-called *recognition pins* used to read the recognition holes on tapes. It did not take long to see that the recognition pin was merely a spring-loaded plastic device about six millimeters tall, with a diameter of two or three millimeters. When the VHS cassette depressed the spring, SVHS mode was disabled. When the spring was not disturbed, SVHS was fully capable.

So, what was to stop us from taking a small knife or pair of scissors to cut off the first 4 millimeters of this plastic pin? Snip... After this minor bit of surgery, all VHS tapes inserted into the SVHS player were now recognized by the VCR as being SVHS tapes (albeit poor-quality ones). If we still wanted to record in regular crappy VHS, we would simply manually shut off the SVHS mode—simple as that.

A DVHS Application

What makes these DVHS tapes so special that we have to use them for DVHS recordings? Why can't we use SVHS tapes instead, just to see what happens? Well, aside from the fact that DVHS tapes are pretty much just high-grade SVHS tapes, not much is special about them, except—another recognition hole.

Figure 2-10 shows the two holes on a DVHS cassette, and the single hole on the SVHS tape. The hole common to both permits DVHS tapes to handle SVHS signals; the hole unique to DVHS is what we want to focus on, as it allows the DVHS to work with HD signals.

For the pictures here, I am using the current JVC DVHS D-Theater VCR.

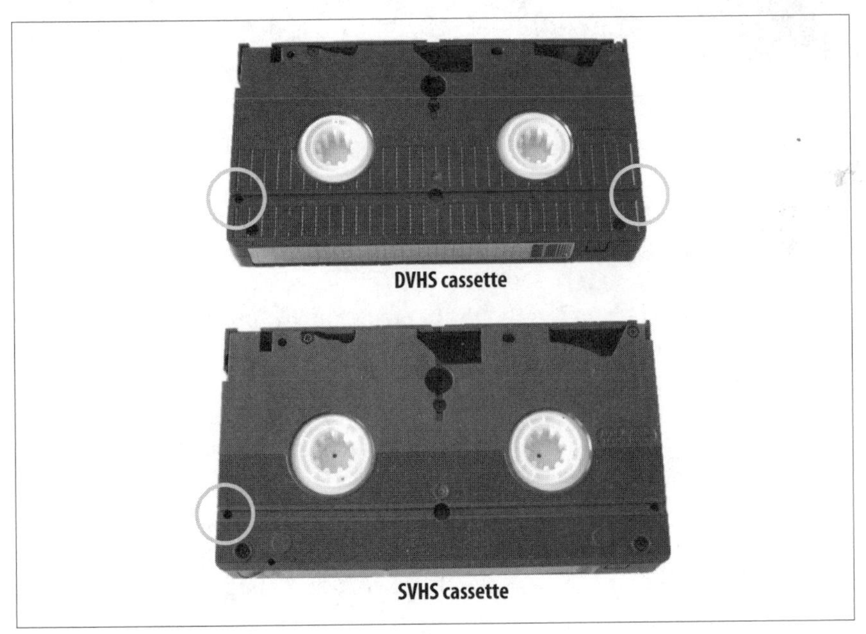

Figure 2-10. Comparing DVHS tapes to SVHS tapes

Pull the top off of your DVHS VCR, and you should see two pins, one for SVHS and one for DVHS. The inside of my VCR is shown in Figure 2-11.

Figure 2-11. Recognition pins for DVHS and SVHS

Once you remove the screws that hold the top on, you've probably voided your warranty.

Look for the pin specific to the DVHS tape, and cut off the top four to five millimeters of the pin.

Please disconnect all power to the VCR when you attempt this. If your hand slips, you could short-circuit the VCR and damage it or injure yourself.

In case you are confused about which pin is which, Figure 2-12 shows a propped-up DVHS tape, lined up with the corresponding recognition hole.

You don't need to perform this procedure to the SVHS pin on the other side because the VCR has SVHS-ET capability for recording SVHS on VHS tapes.

What a way to try to add life back into a format—disabling their own pin, and offering it to the buying public as a new improved feature! Manufacturers spent years telling us that it was bad to record SVHS on a VHS tape, and then suddenly it was OK. You gotta love SVHS-ET technology!

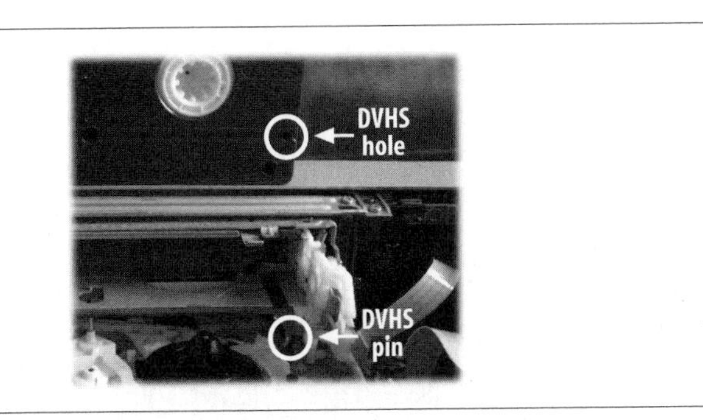

Figure 2-12. Lining up the recognition hole with the DVHS pin

Once you do this, try to fish out the small plastic piece that you cut off. Then put the unit back together. Now all tapes are DVHS tapes! Most SVHS tapes should work just fine with the DVHS signal; I've used both T-120 and T-160 SVHS tapes with no problems.

If you don't want DVHS mode recorded on the SVHS tape, simply select the SP or EP speeds instead. Playback-wise, the signal selection process of the VCR is entirely automated. The VCR will identify the signal type of the tape and play it back in the original recorded form.

—Michael Chen

HACK #16 Cover Black Bars with Letterbox Mattes

Cover the black bars that appear on the edges of your screen with mattes, creating a better-looking image and sharper picture.

The nature of the display technology inside a television set is such that it sometimes has trouble "holding black at black." The color displayed as black on your television screen is often only relative. When a scene is composed mostly of light color elements, it is said to have a *high picture level* in terms of brightness. When viewing these high-picture-level scenes, an area on the screen that is supposed to be black often appears gray. The overall picture level of the image affects the black level; light from the brighter objects often spills over and influences the darker images.

As a result, if you watch the letterbox bars at the top and bottom of your screen, they will shift in color from solid black to a much more medium gray depending on the picture level in a scene. Although you might not consciously notice this shift, it does have an effect on how you see the image.

By covering these bars with mattes you remove the distracting shift in color, allowing your eye to truly ignore this unused area. The letterbox bars become more neutral to your eye and the contrast and color level in the picture seem much more vivid. This hack is not only an excellent example of how a small investment can make a sizable difference, but also, an interesting demonstration of how our minds play a starring role in how we process visual information. You'll be amazed at the difference.

Widescreen TVs, Take Notice

It's important to note that users of standard 4:3 televisions and users of 16:9 widescreen sets can benefit from this hack; it is useful in working with almost any aspect ratio **[Hack #13]**. Standard-TV owners will find that they might need to create two sets of mattes, one for 1.78:1 and 1.85:1 content, and a second set for 2.35:1 material. Widescreen-TV (16:9) owners also will need two sets of mattes, one for 2.35:1 material and, if desired, a set for covering the bars on the sides of 1.33:1 (4:3) material.

Creating the Matte

The first step is to determine the size you will need for your mattes. You can use a calibration disc such as Avia or Digital Video Essentials **[Hack #62]**, as both feature test patterns with simulated content for each aspect ratio. If you don't own a test disc, you can simply use a few of your favorite movies to test each aspect ratio. Display the images onscreen, and measure the distance from the inside edge of the letterbox bar out to the edge of the TV set. Then measure the width of the set from edge to edge (or, if you're making mattes for 4:3 content on a widescreen TV, measure the height of the set).

Once you have determined the sizes for your mattes, you have several options for building them.

Cloth only
: Some users have found that simply using a heavy material without any backing is sufficient for mattes. This is especially true if you have a direct-view TV set with a curved screen; the cloth will attach better to your TV without a stiff backing. Some common cloth types that work well with mattes include duvetyne, felt, velvet, two-pass (for lining drapes), "Commando" black-out, velveteen, suede, velour, and sensuede.

Backing only
: Some users use a medium to heavy material such as cardboard, poster board, matte board, or wood, and paint it solid black. This type of construction results in a nice clean edge and a sturdy matte that will survive being attached and removed repeatedly. Some common backing materials that work well with mattes include poster board, foam board

(sometimes called foam core), picture board, matte board, bass wood, cardboard, Gatorboard, Styrofoam, and vinyl.

Cloth and backing
Some users combine a sturdy backing material with a nice black cloth to create a sturdy matte with a very pleasing look. By covering a backing material with fabric you get a long-lasting, professional-looking matte.

You can find most of these materials at a local framing shop, craft shop, or home improvement store. Figure 2-13 shows everything you need, all nicely laid out, ready to go.

Figure 2-13. Prep work for TV mattes

Now you just need to cut the top and bottom strips, as well as the cloth strips to go around the sides of the television. This is actually pretty trivial once you've measured things; Figure 2-14 shows the mattes cut and ready to be attached.

Attaching the Matte

Make sure when you're deciding on your matte size and type that you take into consideration how you're going to attach the mattes. If you plan to attach them on the sides of the set, you will need to allocate extra material on each side to allow for this attachment.

Using Velcro is the most popular way to attach mattes to the set. By attaching small strips or tabs of Velcro to the face, or better yet, the sides of the TV

Figure 2-14. Top and bottom mattes cut, along with cloth connectors

Figure 2-15. Letterbox mattes on television

cabinet, the mattes can be attached and removed easily. Some alternate attachment methods include poster putty and double-sided tape, but Velcro is by far the easiest. Velcro is available with adhesive backing at most hobby, craft, and fabric stores. Figures 2-15 and 2-16 show two views of a Sony WEGA TV with mattes attached.

Figure 2-16. Side view of letterbox matte attachments

—*Vince Maskeeper-Tennant*

HACK #17 Improve the Picture on Rear Projection TVs

Taking the protective reflective screen off a rear projection television increases black levels, reduces glare, and makes for a much-improved picture, especially during daylight hours.

We've all been there before: you've got the day off, and you decide to sit around the house and enjoy watching *Saving Private Ryan* in your home theater, without constraints on volume from your roommates, significant other, or spouse. Everything is great—until you realize you can't see Tom Hanks because sunlight from the rear window is glaring up your screen. What began as an enjoyable experience quickly becomes annoying, and you end up spending all day answering email and grumbling about your TV. Fortunately, the answer to this problem is simpler than you might imagine.

Rear projection televisions (RPTVs) have a protective screen placed between the actual screen elements of the television and the viewer. This protects your TV from pets, carpet dander, and the hands of two-year-olds. However, if you've got your theater in a well-controlled environment where there isn't much traffic (resulting in low carpet dander), and pets and children are either not allowed or watched fairly closely, the protective screen stops serving its purpose. In fact, it actually hurts the picture, allowing for light sources to cause glare and distortion. In these cases, removing the screen is a great idea.

First, you'll need to remove the speaker covers on the bottom of the TV. These usually are attached via thick Velcro-like attachments, and you can firmly pull them off without damage. Then you can unscrew any other coverings on the bottom portion of your TV, all of which house and protect the projection elements of the television. On my Toshiba RPTV, I had to pull off two speaker grilles, attached by Velcro, and then unscrew a center cover. The results are shown in Figure 2-17.

Figure 2-17. RPTV without bottom speaker grilles

Different models use different attachments and cover schemes, so consult your manual carefully. Pulling firmly on something that's screwed in can be futile, or in certain cases, can rip your speaker grilles.

Next, you'll need to remove the screen console itself. Before removing the screen, though, find a good clean sheet and lay it down on your floor. Make sure it's unfolded enough to allow your entire screen to lie on it. Then, remove the screen unit from your television. This usually involves removing additional screws and then lifting the entire console up and out. You might want to have a friend help, as this step is the most critical, and one during which you could damage your screen elements if you're not careful. Lay the entire assembly on your clean sheet, face down. Your RPTV should now look something like Figure 2-18.

Figure 2-18. RPTV without screen

Unscrew any restraints holding the screens in your front assembly. On my unit, there were restraints on each edge, each with three screws to remove. At this point, you should be able to lift your *display stack*, which is the set of screen elements including the protective screen, out of the screen console. This is another critical step and one that you should take with great care; you might want to simply tilt the stack up, never completely removing it from the front assembly. Carefully remove the protective screen from the stack by pulling out the clear, thick piece closest to the screen assembly. Then set the remaining stack back into the display assembly.

Now you'll need to reattach the restraints, but don't start screwing things in yet; most sets allow room for the protective screen, and you'll find that you can't tighten down the display stack. You shouldn't just grind the screw in place because this can easily damage your screen elements, or even the entire front assembly. Instead, pick up some inexpensive weather-stripping from a local hardware store, and cut it in thin strips to match your restraints. Then you can set these strips in between the display stack and the screen restraints, providing a durable cushion. With these in place, reattach the restraints. In Figure 2-19, you can just see the edges of a not-so-precisely cut bit of weather-stripping sticking out from under the restraints (I don't recommend being this sloppy—it just serves as a good illustration).

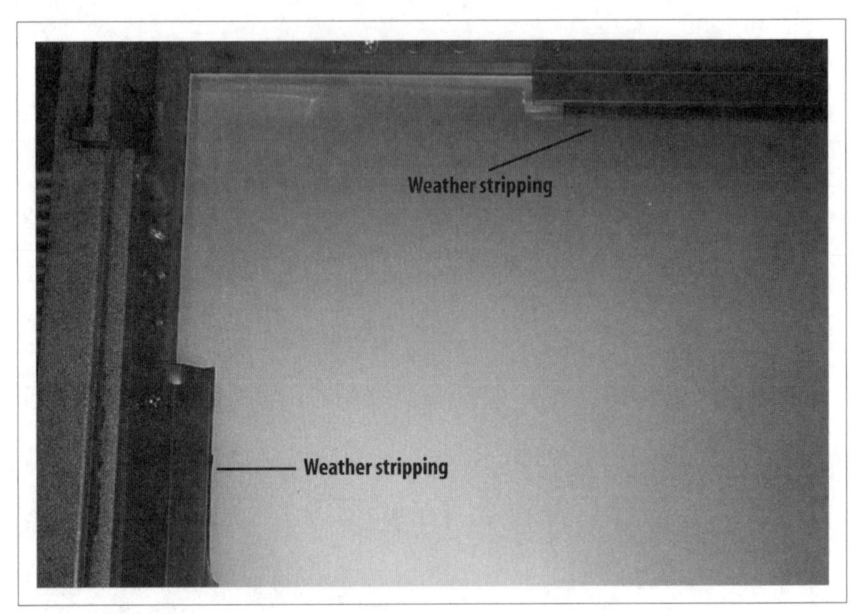

Figure 2-19. Screen with weather-stripping added

With your front assembly sans protective screen, you just need to put the unit back together. I was unable to take a before-and-after picture, as the glare on the before picture distorted everything! Of course, after these modifications, you'll have a sleek, glare-free unit that you can photograph anytime you like.

As an added bonus, I find that my blacks seem blacker, and my overall picture has a clarity that was distorted somewhat by the protective screen—a very nice improvement for the cost of a bit of weather-stripping.

HACK #18 Paint Your Theater a Neutral Color

Although your walls are certainly not a video component, they can make a dramatic difference in how your display devices are perceived. Neutral colors might not be as impressive with the lights on, but they sure make a difference with the lights off.

The Society of Motion Picture and Television Engineers (SMPTE) has recommended the use of neutral—or at least nearly neutral—colors in viewing environments for electronic displays. This specification is to preserve accurate color perception for the viewer. The way our brain processes optical information results in surrounding colors having an effect upon our perception of other colors in the same field of view. A photo of a person wearing a cyan shirt, placed in a frame with a blue matte surrounding the image, will cause the shirt to appear greener. Place the same photo against a green matte, though, and the shirt will appear bluer. It's only with a neutral gray matte that you get a true perception; the shirt will appear cyan.

The same principals apply to your perceptions of electronic images. SMPTE's research had critical applications for engineers, editors, cinematographers, and producers needing absolute accuracy of perception in the studio. SMPTE used the Munsell System of color standards to specify precisely what constitutes neutral, and nearly neutral, for professional viewing environments. Instead of ignoring these standards, you should take them into account in your own home theater.

The Munsell Neutral Value Scale

Correct color perception in a video viewing environment is ideally enhanced and preserved by providing a neutral-colored surround within the field of view of a monitor's screen. Popular home theater magazines have consistently featured stylish viewing environments that look great with all the lights on; unfortunately, these colorful designscapes tend to contaminate the image produced by the video display when it's time to actually watch a program (and the lights go down).

Rich colors near the screen skew the viewer's perception of the image on the display in a subtractive manner. Flesh tones can be perceived as slightly pink, green, yellow, blue, etc., depending on the color of the surrounding walls. Other picture elements, such as grass and sky, can take on an unnatural hue that disrupts a sense of realism. The careful efforts of program producers and cinematographers to set a mood, via the subtle use of color, can be rendered ineffective. That fancy theater décor is suddenly hurting your theater rather than helping it! It's much more important for viewers to feel that they are in the theater when the movie is playing than when it's not.

For these reasons SMPTE states it is most critical that areas in the field of view be devoid of vivid colors. Similar problems occur in front projection theaters, even though they typically are used with no lights on. Light from the screen itself during bright scenes will illuminate the entire room. Tinted light will bounce back onto the screen when room surfaces or major furnishings are vividly colored. The ruling décor used in a front projection theater should be neutral or nearly neutral, and dark. Some vividly colored accents can be used to add interest but should not dominate the room.

For all these reasons, CinemaQuest, Inc. provides some excellent tools to help you select attractive room colors, without having to just point your finger at some random shade of gray. Figure 2-20 shows the Neutral Value Scale, which is ideal for use in a paint store.

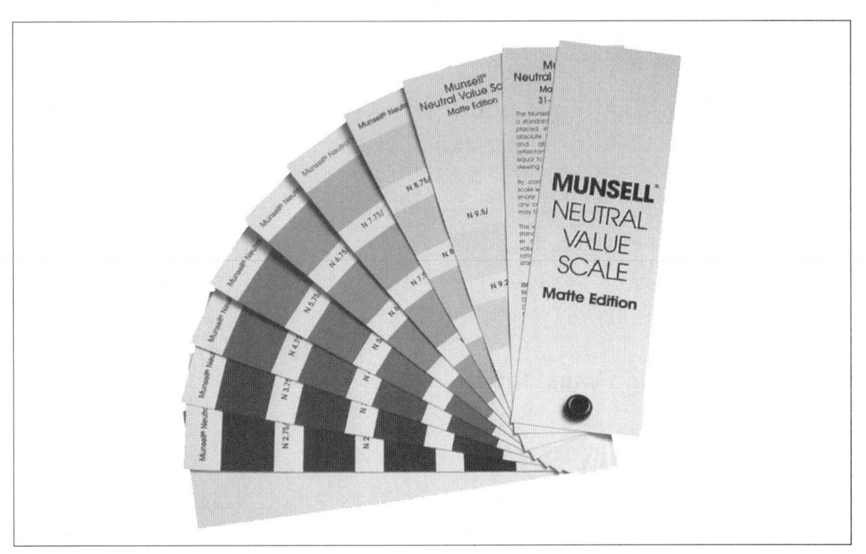

Figure 2-20. Munsell Neutral Value Scale

Take this with you when selecting colors and you can quickly determine a good color scheme that will help your theater rather than detract from it.

The Munsell Book of Color

If you want to move beyond the basics of just choosing a good color, you need to pick up the *Munsell Book of Color* (shown in Figure 2-21), available from CinemaQuest, Inc. This reference resource is a valuable tool for professional designers of video viewing environments and state-of-the-art, custom home theater interiors. SMPTE's Recommended Practice Document #166, titled "Critical Viewing Conditions for Evaluation of Color Television Pictures," which can be found at:

    http://www.smpte.org/smpte_store/standards/index.cfm?stdtype=rp&scope=0

makes reference to the Munsell "nearly neutrals" as being preferred where a change from neutral gray is desired. They specify that they are appropriate only for use in areas outside the field of view of the display. In other words, you can come up with a great color scheme for your entire theater, using neutrals for your viewing wall and other accent colors for the rest of the room.

This book also contains color chips (see Figure 2-22), which are arranged according to the Munsell color-order system. Each page represents one hue, and there are 20 pages, five hue steps apart, displaying about 1,100 chips. On each page, the chips are arranged by their Munsell value and chroma. An

Figure 2-21. The Munsell Book of Color

additional page displays a gray scale of values (dark to light) from 6 to 9.5, in steps of 0.5. Also on this page is a 20-step hue circle called Nearly Whites.

Munsell color standards are made by applying a stable coating to a paper or polymer substrate, using the most stable colorants available. Samples of each production lot are measured by spectrophotometry and are visually inspected at the time of production and periodically thereafter. The manufacturing tolerances for the colors are +0.05 in value, +0.2 in chroma, and +2/C in hue (where C is the chroma).

Putting It Together

With all these pieces and tools in hand, you can come up with a great, suitable color scheme. Once you've got these colors, you have two basic choices:

- Try to match these colors in a local paint store.
- Order the exact colors through CinemaQuest.

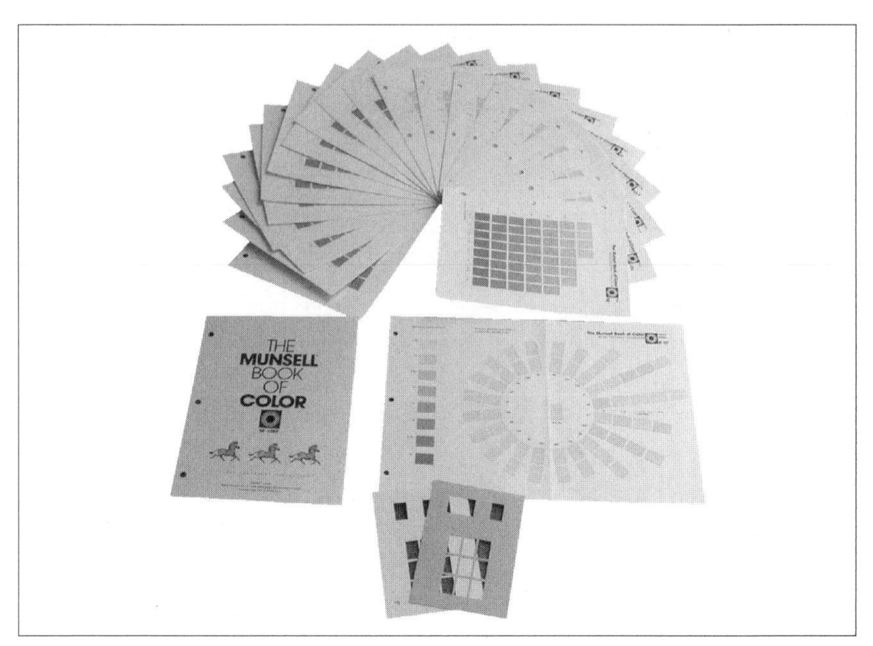

Figure 2-22. Color chips and inserts

For honesty's sake, the president of CinemaQuest wrote much of this hack. However, the following advice is authored by me, Brett McLaughlin. I feel strongly that CinemaQuest offers a terrific solution, so I want to recommend them wholeheartedly.

If it's at all possible, consider ordering these colors direct through CinemaQuest. They offer a good value, and you're guaranteed to get a perfect color match. What's more, if you are going to spend thousands of dollars on a killer TV, DVD player, speakers, and the like, why don't you spend several hundred dollars on your wall colors?

—*Alan Brown with Brett McLaughlin*

Backlight Your TV

Watching a large TV in a dark room can cause a lot of eyestrain. By placing a simple light behind your TV, you create the illusion of depth, resulting in less eyestrain and a more pleasant viewing experience.

I know you've been there—you're five hours into a *Lord of the Rings* marathon (watching the extended versions of each movie, of course), and your head begins to hurt. Your eyes get tired, and you can't understand what it is about those Uruk-Hai that make you tired. Well, it's not the movie; it's the lack of lighting in the room. More specifically, it's the lack of depth perception that the darkened room is causing.

When lights get dim, your television begins to "fade" into the back wall. Eventually, as lights go almost out (the optimal viewing environment), your eyes can no longer separate the picture on your screen from the wall behind it—but these two objects aren't the same distance away. If you're still unclear as to what I'm talking about, think about a camera trying to focus in on two objects, different distances away. Just as you seem to focus in on one object, the other starts to grow blurry. This is what's happening to your eyes in a darkened room—and eventually, it's going to begin to hurt. Your eyes need some way of distinguishing between the wall and the picture, which allows them to focus on just the picture, reducing eyestrain.

The easiest way to accomplish this is to backlight your television. Key to determining what light to use is the color-rendering index (CRI), which measures light's ability to render pigments (perceived as color) according to a prescribed standard. Most lights for your house don't even take CRI into consideration, so just grabbing a lamp and sticking it behind your TV isn't going to help much—in fact, you won't see a CRI rating on most bulbs at all. However, you can make some judgment based on the light's Kelvins (K). Ratings of 5,000K and higher are referenced to daylight, based on different times of the day; 6,500K is what you want to look for in a home theater application. A white light rated at 6,500K is going to exactly match the white on a correctly calibrated TV set.

You can pick up a 6,500K light from Home Depot, Lowe's, or Sears. Look for one of the following items, if at all possible:

- Westinghouse 18-inch 6,500K 94CRI 15w T-8
- Philips Daylight F20T12/D
- Westinghouse F15T8/FS 18-inch 6,500K 48CRI

Or, if you're really getting into home theater seriously, check out CinemaQuest, Inc. (*http://www.cinemaquestinc.com/*). They carry incredibly accurate 6,500K, 98CRI lamps, made just for this application.

Place this light directly behind your TV set. You don't want the lamp visible, and you'll find the minimal light generated is not at all bothersome—in fact, as the hours roll by, you'll find it's quite pleasant.

Some Nice Side Effects

In addition to reducing eyestrain, good backlighting often increases the perceived picture quality. Your eyes have a much more difficult time determining colors without any light source. In fact, this is why colors seem so bright when walking out of a dark room into daylight: your eyes have lost their color reference, and have to adjust. The backlighting will provide just enough light for your eyes to get their color reference, and you'll find that pictures seem a little crisper and colors more vivid. Quite a fringe benefit!

My Light Is Too Bright!

All the benefits in the world won't make you happy if your TV is smaller, and this light just seems too much. Suddenly, your lack of eyestrain is coupled with what appears to be an odd glow emanating from your TV; this isn't good. To lower the light level without losing the benefits, just place some tin foil over the edges of the light, and work your way toward the center of the lamp, until you get acceptable light levels. The foil won't melt, and you'll get the same effect.

Add Metal Plating to Support a Center Speaker

Most home theater television sets have a large center speaker sitting on top of them. This is generally the ideal placement, but over time, a heavy speaker will warp your TV set. A simple piece of metal on top can solve this problem.

The bigger the speaker, the better, right? Well, that's not always true **[Hack #37]**, but it often is. However, these big speakers also are heavy speakers, and over time they'll create an ugly indentation in your TV set. Although that might be a great excuse to buy a new set, it's too easy of a problem to correct to let lie. Besides, you can't resell a TV with a warped screen!

To solve this problem, locate (or buy at your local sheet metal shop) a good-size sheet of either 1/2- or 3/8-inch aluminum. Cut this down to the size of your television top, minus about 2 inches.

If you're not comfortable working with sheet metal or don't have the tools you need, a sheet metal shop (or even an auto body shop) can cut down metal like this for a nominal cost.

It shouldn't just be the size of your center channel's footprint; in addition to leaving you no room to upgrade, all you've done is added to the weight in the middle of your set. Instead, you want a plate large enough to distribute the weight of the speaker across the entire set, as shown in Figure 2-23. You can see that a little extra time resulted in a nice-looking stand rather than an ugly piece of metal. Paint the plate the color of your TV, place it on the TV, place the speaker on the plate, and you're all set.

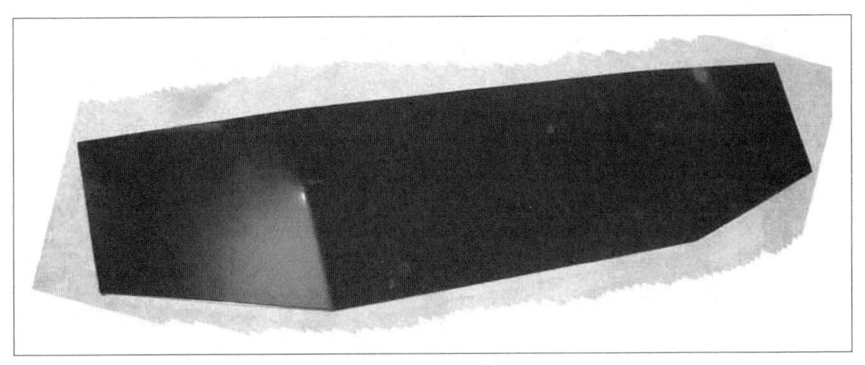

Figure 2-23. Winged center-channel speaker stand

This hack is really aimed at rear projection TVs (RPTVs). The newer CRT TV sets have some magnetic issues with plate metal, so I recommend against using a metal plate for newer CRT-based TVs. LCD and DLP televisions are not an issue, as they are rarely sturdy enough to hold a center channel in the first place.

Even better, you might want to consider taking some extra time to really make things sturdy. If you've got the tools and the inclination, bend the plate edges down. This takes all the weight off the speaker and focuses it on the edges of your TV set (which have the most reinforcement, from the sides and back of the unit). You need to test this out a bit because a heavy speaker can cause the middle of a plate bent like this to sag; you want to be sure the speaker doesn't bend the plate down so much that it ends up touching the TV set after all. Finally, you need to level off the bent ends, so you don't have aluminum edges wearing into your TV's top. You also can make the location of the plate (some have suggested using Velcro as well) so that it always rests in the same spot; check out Figure 2-24 for the completed stand, sitting on a TV, with the center channel properly placed.

—*David Gibbons*

Figure 2-24. Completed winged speaker stand

Squeeze Your TV into Your Basement

Sometimes, you'll find that your perfect TV just won't fit into your perfect room. In these cases, you might be able to take some drastic steps and disassemble enough of the TV to get it where it belongs.

In cases where building a mockup [Hack #9] actually proves you can't fit your TV into its target resting place you still might be able to make things work. This hack details taking a Sony 57HW40 apart, moving it, and putting it back together again. Your mileage may vary on different brands, but this will give you a head start on getting that TV where it just doesn't seem to want to go.

> It should go without saying that this operation is going to void your warranty. If that makes you nervous, stop reading, and find something safer to do!

First, you might want to see the set before it underwent deconstruction and moving; it's shown, still in pristine condition, in Figure 2-25.

Remove the Speaker Grill

The first thing you want to do is get a flat-head screwdriver, get down on the ground, and look for the two circles that hold the front speaker grill. Figure 2-26 shows what to look for.

Using your screwdriver, remove the two circle caps, which will reveal the two screws you want to remove.

Figure 2-25. Sony TV before moving

Figure 2-26. Screw caps for TV speaker grill

I Don't Have a Sony TV!

Plenty of folks will skim right through this hack, concerned that they don't have a Sony TV. However, the only thing Sony-specific is the location of the screws and plugs. You can easily apply these steps to all the major brands of RPTV out there. If you find yourself in a situation where you need to squeeze a big TV through a small space, these steps can serve as a guideline for taking apart your own RPTV, whatever the brand.

Wiggle the Grill Off

From the side of the grill, slowly wiggle up on one side and the grill will release. Do the same on the other side of the grill.

Don't pull this grill completely off because wires are attached to the grill from the TV.

Once you wiggle both sides free, move to the center of the grill and slowly move the grill out and up to release it. Figure 2-27 will help you understand what to do.

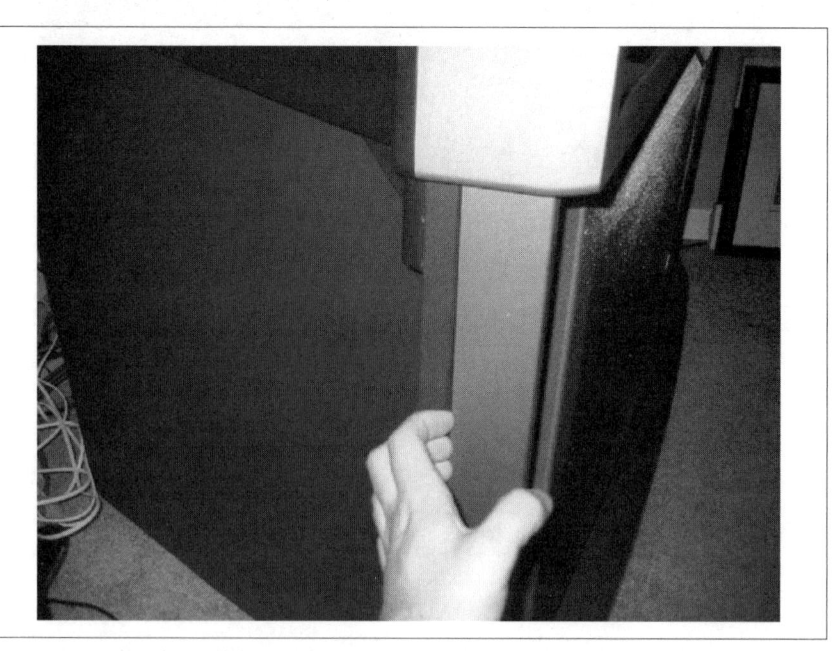

Figure 2-27. Wiggling off the speaker grill

Now you need to remove the wires from the grill that are attached to the cabinet. You need to remove five clips; I used a Sharpie marker to label them after I unplugged them, and I wrote down on a piece of paper where they went. Figures 2-28 and 2-29 show two of these plugs; you'll have no problem finding and unhooking the other three.

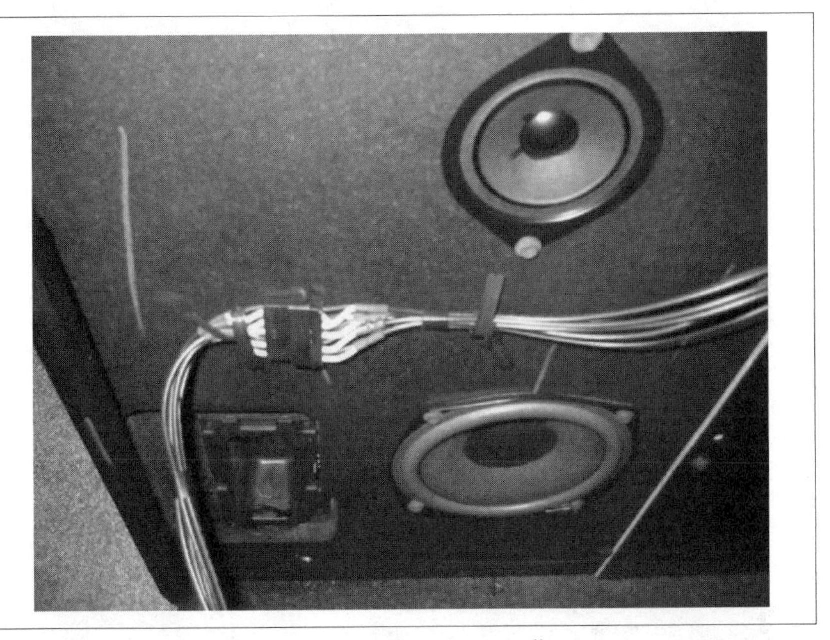

Figure 2-28. One of the plugs that connect the TV to its grill

Remove Vented Rear Cover

This is an easy step. Just go behind the TV and look for the brownish rear cover. Eleven screws should be holding this grill to the back of the set. Remove all of those screws, and you'll end up with your set looking like Figure 2-30.

Remove the Screen

The screen is pretty easy to remove as well, but you need someone to hold the screen from the front so that it doesn't fall and crash to the ground. It's really light, so don't be worried about it falling on you. Start unscrewing all the screws around the rear housing that hold the screen (see Figure 2-31).

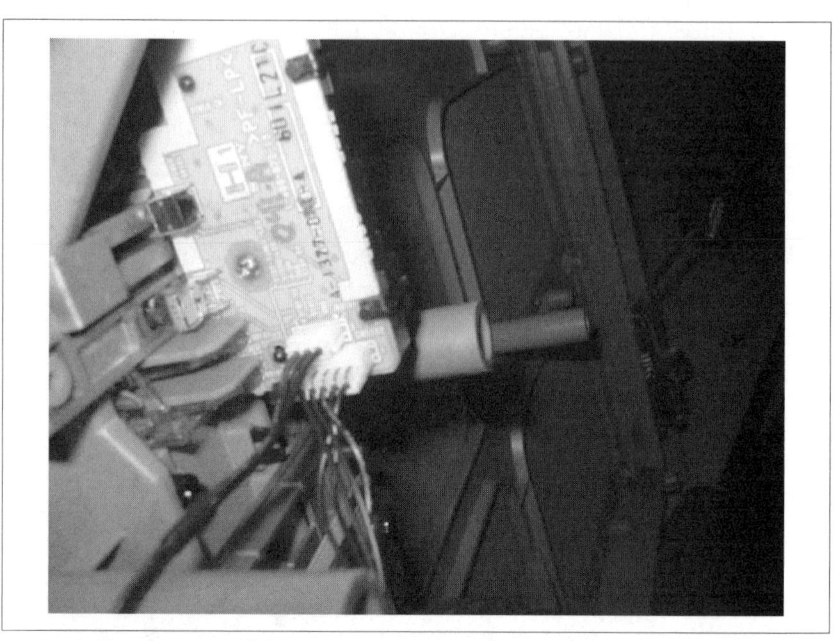

Figure 2-29. Another plug to remove

Figure 2-30. TV without rear venting attached

Figure 2-31. Detaching the screen from the TV set

The screen is attached to the base with wires, so once you get all the screws unscrewed, don't pull the screen away from the base.

While you've got the screen off, you should consider removing the protective, glare-creating front screen [Hack #17].

Detach Wiring from the Screen Assembly

This step isn't that difficult, but it is a bit of a pain to get right. Make sure someone is holding the screen while you do this. Four plugs are attached to the screen (see Figure 2-32): two red and two white. Make a note of where each plug connected to the TV; I used my trusty Sharpie marker again. Unplug each, and remove the wiring from the clips. When you've got all four plugs loose, you can pull the screen away from the TV. Set the screen aside.

Remove the TV's Top Housing

At this point your set should look something like Figure 2-33. Four remaining screws, all located on the back of the set, should be holding the top of

Figure 2-32. Plugs connecting TV screen to TV set

the TV to the bottom section. Remove each screw, and lift off the top section. Congratulations! You're ready to move things.

Move It!

Now you must use those muscles and get the heavy bottom section (see Figure 2-34) down the stairs. It's not recommended that you lift your TV at an angle, but we had to get it down through the door, so the angle we reached was about 45° (nothing calibration won't fix, right?). Then, we carted each smaller piece down to our theater room.

Reassemble

Reassembly is easy enough: just reverse the preceding steps. If you labeled things properly, this should be a piece of cake. Enjoy your too-big TV in your too-small room!

—Joe Jenkins, HDTVoice.com

Figure 2-33. TV without screen, ready for top removal

Figure 2-34. The heavy bottom part, ready to be moved

CHAPTER THREE

Audio Components

Hacks 22–27

If picture is what makes your buddies perk up, it's audio that gets the real theater geeks. Multichannel, surround sound, Dolby Digital EX, DTS-ES, Neo:6: these are the terms that make the hearts of audiophiles pound a little quicker. That's largely because you can easily buy great picture; getting killer sound is a much trickier proposition.

Although there's lots of tweaking to be done in a good home theater setup, buying good gear is still the first step. This chapter begins with that subject, and then takes you into the high-end strategies for audio.

Get the Right Receiver

The brains of your home theater are always going to be in your audio processor. Most folks at least begin with a receiver, which has to handle the demanding task of preamplification and processing, as well as pumping out power to your speakers. Getting the right receiver is crucial.

Receivers are tricky creatures, and a lot like Microsoft Office; you'll often find lots of bells and whistles, but it's not easy to figure out what you actually need. Assuming you want to impress your buddies, there are some definite requirements.

Industry-standard six- and seven-channel decoding
: Ensure that any receiver you look at has Dolby Digital EX (6.1/7.1-channel decoding) and DTS-ES (6.1-channel decoding). These are usually well represented on both the box and the face of the unit (see Figure 3-1).

Optical and coaxial audio inputs
: Ensure you have at least two coaxial audio inputs and two optical inputs. Many higher-end components will have both sets on their

Figure 3-1. Dolby Digital and DTS logos

output strip, but you'll want to use coaxial when possible, so the more the merrier.

Power
: Tim Taylor, eat your heart out. Until you break your receiver into separates, it's going to be both the controller and the amplifier in your system, so buy high. Going less than 80W per channel is a waste of time, and 100W in each channel is worth the extra cost.

Component/S-Video switching
: This is the ability to plug your video sources (DVD, VCR, PS2, satellite) into your receiver and seamlessly switch between them. Some units can even up- or down-convert different signal types (RCA, S-Video, component video) for you. These are nice features, although a good programmable remote can automate this for you **[Hack #84]** even if your receiver doesn't support it.

By the same token, there are some things to not get hung up on and certainly aren't worth wasting much of your dough on.

THX
: As much as it sounds important, as much as it gets airplay, THX just isn't that important in the typical home theater system. It often deprives your subwoofers of a lot of their primary job (to produce bass), and is rarely as effective as plain old Dolby Digital or DTS. That said, most 6.1- and 7.1-channel receivers are going to have this, but if you find a good one that doesn't, don't think twice—ignore THX.

3+ component video inputs
: Lots of folks will spend forever looking for a receiver with component inputs for all their devices: the DVD player, an HD satellite or cable receiver, their Xbox, their PS2, and so on. Most receivers have only two, and rarely three, component inputs, and it takes a lot of money to buy a receiver with more. Through clever cabling **[Hack #58]**, though, this can become a nonissue, so don't worry about it. Two component inputs are plenty for all but the most demanding setups.

Choosing a Brand

Trying to hone in on a specific brand of receiver is a tough thing to do. Part of that is because of the large variety of manufacturers out there, and part of it is because each manufacturer often has a lot of different models, from the most basic to the high-end. As a rule of thumb, though, higher-end components require specialization. In other words, I've rarely found a company that makes toasters, cameras, radios, as well as high-end receivers that can rival a company that does nothing but build audio components. For this reason, if you're serious about home theater, you ought to rule out the mainstream companies such as Sony, JVC, Pioneer, and the like.

That said, though, there are some exceptions to this rule. A few of these companies make "specialty lines" where audio is indeed the focus. For instance, Pioneer has a division, Pioneer Elite, that focuses on home theater and sells equipment that far surpasses the base Pioneer gear. These lines often are great choices for going beyond the basics, but avoiding the leap into the really expensive gear.

Next up on the rung are companies such as Onkyo (*http://www.onkyo.com/*) and Anthem (*http://www.anthemav.com/*). I've used and owned gear from both companies, and they are great units in the higher-end range.

When you finally decide to go all the way, though, I believe there's only one choice: Lexicon (*http://www.lexicon.com*). Although you're going to drop at least $5,000 or $6,000 on a new Lexicon processor, you're getting what I think is the best audio processor available for less than $20,000. I've owned Lexicon's (at the time) flagship, a Lexicon MC-12, and it simply blew away all others. Of course, at this point you're talking processor and amplifier, so this is serious audio gear.

By the way, I'm sure others will disagree with me that the Lexicon is superior to all its competitors; I'm just going with what my ears have told me.

Spending Some Money

It's been said plenty of times already, but you simply can't buy a high-end receiver or processor without either hearing it first **[Hack #2]** or getting a 48- or 72-hour satisfaction guarantee. Of course, many of these higher-end units aren't available in chain stores **[Hack #3]**, so if you want to listen, try a boutique **[Hack #4]**. If you do decide to buy from a high-end online store **[Hack #6]**, you need to insist on a tryout period. Anyone who sells you gear that costs thousands of dollars and doesn't stand behind it isn't worth your time and trust.

Be aware that you'll usually have to pay shipping, both ways, on gear if you get a tryout period. That's still a good deal, but when you're shipping heavy components, you easily could eat $200. Just be aware of what you're signing up for.

Watts Are Meaningless Without Context

One of the easiest ways to be fooled in receiver and amplifier selection is in the wattage that the component puts out. Learn how to take the raw wattage number and turn it into something meaningful.

Most newbies, when it comes to home theater, immediately want to know how much power a receiver or amplifier puts out. This is natural: most of us home theater guys love to talk statistics and flex our components' muscles. However, there's a lot more going on than meets the eye. You can have an amplifier with a ridiculous amount of power and still have a terrible-sounding setup because you didn't put that wattage into its proper context.

Getting a Handle on SPL

Instead of focusing on wattage, you need to focus on the *sound pressure level* (SPL) of the unit you're looking at. SPL is the basic measure of how loud something is, and is measured in a unit call the *decibel*. To further complicate things, decibels are measured on a logarithmic scale rather than a linear scale; in other words, if a doubling of the output level results in a 3dB increase, increasing the output by a factor of 10 results in only a 10dB increase. If decibels were measured on a linear scale, a factor-of-10 increase in output would result in a 30dB increase. The bottom line is the higher you go on the scale, the larger the difference between the levels.

Now, let's make that a little more concrete. For a person to easily perceive a difference between the intensity of two sounds, those two sounds need to differ by approximately 3dB. To perceive a doubling of the intensity, two sounds need to differ by approximately 10dB.

Relating Wattage to SPL

With this in mind, you are ready to see how a watt relates to a speaker's output. Every speaker has a *sensitivity rating*, given as dB/W/m. This number can range from the low 80s to some very efficient horn-loaded speakers with ratings higher than 100. For the purposes of this explanation, assume you have a speaker with a sensitivity rating of 91dB/W/m. This means that with 1 watt of power applied to the speaker, measured 1 meter away from the speaker, you should read an SPL level of 91dB.

Remembering that decibels are measured logarithmically; to increase the output by 3dB you have to increase the input power by a factor of 2. Table 3-1 summarizes the relation between wattage and decibels (at least, for this sample scenario).

Table 3-1. Relating watts to decibels

Wattage	Decibels
1	91
2	94
4	97
8	100
16	103
32	106
64	109
128	112
256	115

As you can see, power requirements for a meaningful increase in output go up very quickly. It also becomes obvious that the difference between an 80-watt-per-channel amplifier and a 100-watt-per-channel amplifier is not going to be all that great (so much for the Tim Taylor approach: "more power" isn't always that helpful!). Finally you also can see that a 91dB/W/m speaker will require half as much power as an 88dB/W/m speaker will to reach the same output level.

Factoring in Distance

There is one other thing you have to consider. These numbers are valid only if you are listening 1 meter away from the speaker. Of course, very few people listen to speakers, and especially subs, at a distance of 1 meter. So, now you have to apply the same sort of rules you've seen to distance. Every doubling of distance will cause a 6dB decrease in output. So, listening to that 91dB/W/m speaker at a distance of 4 meters (about 13 feet) changes things dramatically. Table 3-2 is the same sub detailed in Table 3-1, listened to at 4 meters.

Table 3-2. Listening to the sub from Table 3-1 at 4 meters

Wattage	Decibels (at 4 meters)
1	79
2	82
4	85

Table 3-2. Listening to the sub from Table 3-1 at 4 meters (continued)

Wattage	Decibels (at 4 meters)
8	88
16	91
32	94
64	97
128	100
256	103

You actually won't experience this much of a drop because the room effectively increases levels a bit, but you get the idea. The wattage your amplifier (or receiver) is pouring into your sub is meaningless without also considering the sub being driven and the room the speaker is in.

Also remember that the wattage rating on an amplifier isn't always accurate. Here are some things to watch out for:

- Amps that give their ratings at 1 kHz, instead of from 20 Hz–20 kHz.
- Amps that give THD (*total harmonic distortion*) numbers higher than 0.1 in their power ratings
- Amps that don't specify that all channels were being driven when the indicated power output rating was obtained
- Amps that rate into 6 ohms, but say not to drive 6-ohm speakers

As an example, take these two sets of specifications:

- 110W at 1 kHz, with 0.7% THD into 6 ohms. On the front of the box, the manufacturer also claims 110W×5, and the manual says not to drive 6-ohm speakers.
- 70W×5, with all channels being driven, from 20 Hz–20 kHz, with 0.08% THD into 8 ohms. The manual says driving 6-ohm speakers is no problem.

The second set of specifications is a much better (and much more honest) set of measurements. You actually should have 70W for each channel, which is what you would expect. However, the first set of specifications will fool many a potential buyer. The conditions being measured are favorable to the amp, and not to you. In fact, if measured under the same conditions as the second amp, you would in all likelihood have less than 40W into five channels! Hardly what you were looking for.

—*Dustin Bartlett*

Amplify the Front Soundstage

Take your sound to the next level by using separate amplifiers for your front soundstage—the right, left, and center channels.

In any good home theater, the dialog (and the ability to hear it clearly) is extremely important. All the surround effects in the world merely enhance the overall experience, but if you can't hear what the actors are saying, all is lost. You quickly will learn that much of the important dialog comes from center stage—the area addressed by a center speaker. The front speakers (left, center, and right) often are referred to as the *front soundstage* and represent what is happening on the screen. Unfortunately, in the early days of multichannel sound with a center channel, the center speaker often was added as an afterthought, even though it should be able to handle the bulk of the dialog. In fact, because much of the action (and the speaking) travel across the front it is important that all three front speakers either are identical (and also amplified identically) or are at least sonically equivalent. Just adding a small speaker on top of the monitor (or even worse, using the internal speakers of the monitor for the center channel) is not doing your home theater any favors.

A popular solution to ensure consistent and excellent sound quality in this area is to amplify the front soundstage, as shown in Figure 3-2, even if you are using a receiver (which normally provides speaker amplification: see Figure 3-3). It's then possible to control the wattage to each front speaker, ensuring a smooth dialog from your system. Make sure to choose amplification for the center channel that matches well in sound characteristics with your left and right front speakers. As the action pans across the front, you won't experience any dropout of sound, and it should provide a seamless listening performance.

Connecting the Amplifiers

By adding external amplifiers, you really can open up the sound. If your receiver allows you to disconnect the preamp/processor section from its internal amplifiers (usually indicated by "preamp out" jacks on the back of the unit, illustrated in Figure 3-4), all you have to do is to take signals from the L/C/R preamp outs and feed them to the inputs of the amplifiers. Then connect your front speakers directly to the speaker terminals on the amplifiers. In most cases you still can use the remaining amplifier sections of your receiver to power the surround speakers with very good results.

When considering the power for your amplifiers, err on the high side; go for at least 100 watts in each speaker, and if you can afford it, 200 watts is even better. You also don't have to worry too much about matching what your

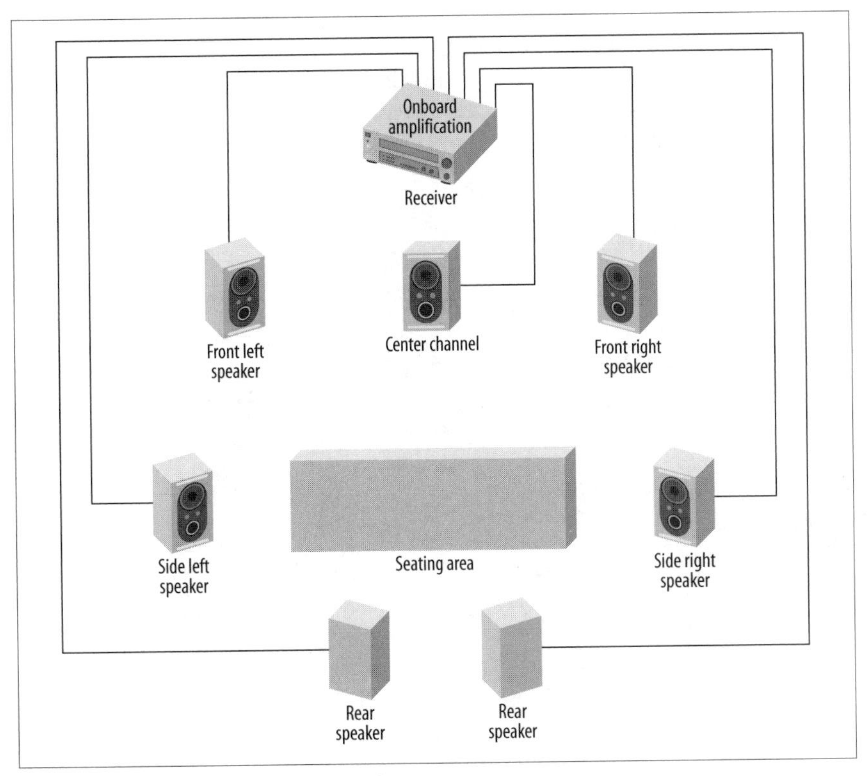

Figure 3-2. Amplifying the front soundstage

receiver is putting out to the other speakers. Generally, the power to each speaker will be identical, but in some cases, slightly less power can be used for the surround speakers as long as there is not too great a discrepancy in the speaker characteristics.

Where's the On/Off Switch?

With multiple separate amplifiers (an amplifier that powers a single channel is sometimes referred to as a *monoblock*) you need a way to turn them all on. You can control most monoblocks by 5- or 12-volt triggers from the preamplifier or receiver so that turning on your processor turns on all your amplification. Figure 3-5 shows a typical set of switched outlets on a receiver. Plug your amplifiers into these outlets, leave the amplifiers in the On position, and they will be controlled, power-wise, by your receiver.

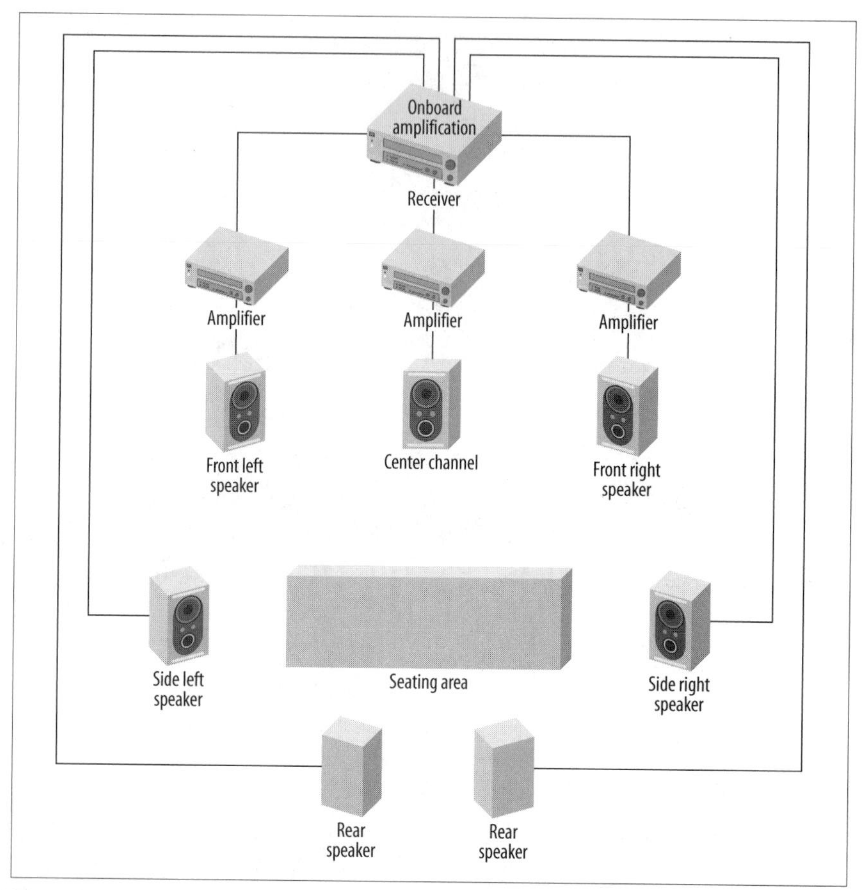

Figure 3-3. Receiver amplifying all speakers

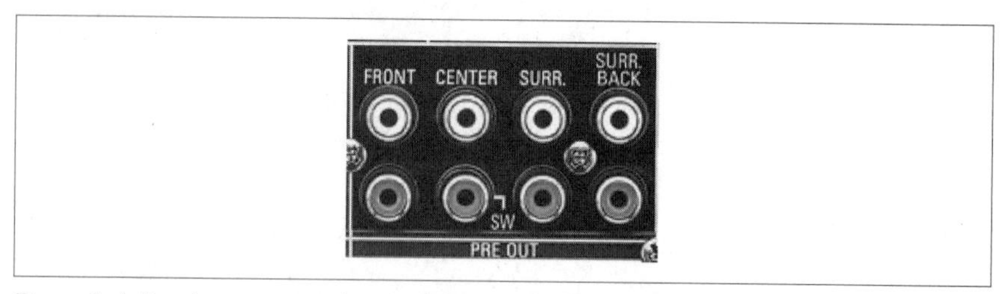

Figure 3-4. Receiver pre-outs for amplification hookup

But That's a Lot of Equipment

You might think that it would make more sense to keep everything contained in one box, but the reality is that for the best-quality sound, the amplifiers should not be compromised by having to fit into a single container that also contains the preamplification and processing circuits.

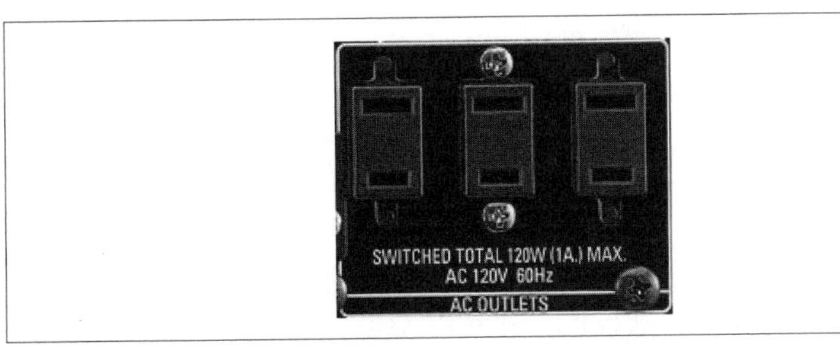

Figure 3-5. Receiver switched outlets

What might offer more convenience does so at the expense of clean amplification.

If you don't hear an appreciable difference between receivers and separates, convenience might be more important to you.

Finally, remember that the surround speakers are not always actively involved in the sound mix (unless it is a gimmicky soundtrack). It's common to hear new home theater owners complaining that there must be something wrong with their surround setups because they "can't always hear something coming from the surround speakers." Like a good subwoofer, quality surround sound should come into play only when the movie calls for it.

Experiencing the Difference

Have you ever watched something that looked fine to you until someone peeled away a fine film veneer and it looked even better? The best analogy I can conjure up is the protective coating that is on some devices such as remote controls, microwave oven touch panels, and certain trim items on a wide variety of products. Peel off the clear plastic layer and you can see all the labels and graphics much more clearly than before. Not earthshaking, but significantly better. Well, substituting three monoblock high-current amplifiers (Marantz's MA-9S1 would be a great selection: *http://www.marantz.com/p_product.cfm?id=2549&cont=u&line=amp&cat=hf*) for the front sound channel amplifiers (L/C/R) of my receiver provided the same effect in my audio framework.

When I asked for opinions about upgrading my amplifiers on the Home Theater Forum (*http://www.hometheaterforum.com/*) the replies ran the gamut from "You won't hear any difference at all" to "The difference will be astounding!" and everything in between, so I was anxious to find out where my observations fell on that scale. Now that I've had some time to listen critically to my new equipment, I can state, from experience, that the upgrade was definitely worth the investment to me. As far as I'm concerned, the overall sound is better, and in some cases, much better. Let me try to explain a bit, within the limitations of trying to describe audible sensations with the written word.

Better? How? For one thing, soft sounds are now clearer, more distinct. On those songs where the vocal fades out at the end you hear the notes longer and more distinctly into the fade—for example, the ending on Diana Krall's "All or Nothing at All" on her DTS. And the finger snapping by Ms. Krall on "My Love Is..." is eye (ear?)-awakening. In addition, when I used to turn up the volume on my receiver it made the sound louder (duh!), but the sound remained clear and clean. With the monoblocks added into the mix, now when I turn up the volume the sound gets louder by making the soundstage appear to come closer to the listener. The effect is the same as if a singer or musician were to leave the stage and approach you rather than simply singing or playing louder. And this is something that I've never experienced before—at least nothing that I ever noticed before. Piano notes sound more defined, more "anchored," for example.

Yes, there are those who will wonder what all the fuss is about and question my upgrade. And, no, I didn't do a true A/B test but did switch the wires several times to make sure I wasn't succumbing to placebo or "Hawthorne" effects.

An *A/B test* is when you have two setups that are identical (at least as much as possible), except for one component or setting. Then you can play the same audio or video on both setups, switching back and forth, and determine which you prefer.

I'd also term the upgrade not revolutionary, but evolutionary. Still, this is about as close as we get in this business to having our cake and eating it too.

—*Dr. Robert A. Fowkes, Brett McLaughlin*

HACK #25 The Mythical Burn-In Period

Running your components and cables for some arbitrary length of time when they are first purchased isn't helpful, but letting them power on and sit for a few minutes before watching movies or performing calibration is.

With even simple home theater systems now needing universal remotes [Hack #84] and calibration to perform well, a lot of well-meaning folks have passed around some home theater myths, or at best, misunderstandings. One of these is that you need to *burn-in* your equipment. Burn-in is a term used to refer to running your gear for some arbitrary length of time, usually several hours, to help it perform properly. You'll most commonly hear about burn-in when talking about speakers, which often can benefit from this sort of treatment.

Warm Your Components Up, Don't Burn Them In

Some well-intentioned folks have taken this idea and applied it to audio and video cables (discussed next), as well as audio components. In the case of cables, this is just an outright misunderstanding of electronics; in the case of components, it's more likely a misunderstanding; a confusion between burn-in and warm-up. Warming up a component is just what it sounds like—turning on a piece of equipment and letting it run for a while. Electronics need warm-up time, not burn-in time. Most modern electronics warm up in 1 to 5 minutes. In this time picture tubes and lamps begin to operate consistently, transistors perform as they should, and your system generally settles (metaphorically speaking).

Really heavy-power amplifiers might need 10–20 minutes to reach their steady-state temperature.

Giving components a few minutes to get to this optimal state will result in a better home theater experience, allowing your system to perform at its best.

Cable Burn-In Is a Waste of Time

You'll also often hear that you should burn-in your speaker cables, as well as your components. These burn-in periods have as little effect on your home theater as the length of your speaker cables (see Chapter 5).

Cables don't have some sort of "memory" that is altered in the first few hours of use. This is a slight misunderstanding of electrical fields involved. Yes, electromagnetic fields do have an effect on the *dielectric*, the white foam that surrounds the center wire of your interconnect cables. As the audio or video signal sweeps up and down (all "within" the cable), the effects of the first half of the signal are reversed by the second half. If you have an electronics background you know that the electromagnetic field depends on current moving through a wire, and it's this current that turns out to be the overriding factor in how a cable behaves over time. With interconnects, very little current is flowing through these cables, so burn-in is essentially a waste of your time.

> The one way that a signal can alter the cable is if the signal has enough current with it to heat the cable, and to melt the cable's crystalline structure. This type of burn-in is prone to bring the fire department running rather than your local home theater enthusiasts.

—*Robert McElfresh*

HACK #26 Use Gain Offset to Regulate Volume

Avoid the annoying volume switch between television programs, DVDs, and even different CDs using your receiver's gain offset feature.

Sometimes in the course of purchasing high-quality audio equipment, newcomers notice nuances in the sound of their CDs that they never heard before. One of the most common complaints stems from the perceived volume differences between compact discs. Some recordings, usually older reissues of material recorded before the mid-1980s, seem considerably quieter than more recent releases. This difference becomes even more apparent when two discs are played back to back on a carousel CD player or from your favorite MP3 jukebox.

Understanding Dynamic Range

Compact disc is a digital medium, which means it has a specific maximum sound level it is able to carry. The loudest amount of sound you can put on a CD is zero. Because zero is the maximum, all other sound works backward from that point and is expressed as a negative number (–10, –20, etc.). All CDs have this maximum; think of it as a global speed limit for the format.

Most consumers don't understand why, if all CDs have the exact same maximum level, some CDs sound so much louder than other recordings. This extra loudness is especially apparent in modern recordings. Most folks assume older recordings sound softer because they don't exploit the full volume level available on CD. They assume the music is simply recorded at lower levels, but this is rarely the case. The real reason revolves around the reduction of dynamics. The *dynamic range* of a particular piece is the range between its loudest sound and its softest sound.

Newer recordings have a much smaller dynamic range, which results in the illusion of loudness. By removing the dynamic range of a given sound using tools such as compression and limiting, more of that sound can be crammed closer to the maximum level—causing the listener to perceive the overall sound as louder. Without the wider difference between loud and soft that is present in many older recordings, everything is perceived simply as loud.

Figure 3-6 shows a visual representation of the Creedence Clearwater Revival song "Fortunate Son," recorded around 1969. Notice that the peaks of the song are at the maximum level of the disc, but that the average level of most of the material is lower: around –6.

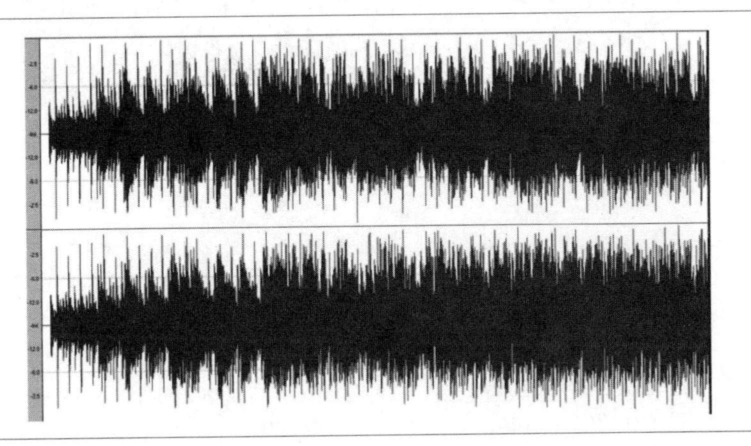

Figure 3-6. "Fortunate Son" audio graph

In contrast to this, look at Figure 3-7, the Watchman song "Stereo," recorded around 1996. The peaks of this song reach the same maximum level of "Fortunate Son," but the average level of the material under the peaks is giant: the distance that the peaks are above the body of the sound has been reduced! The dynamics of the song have been squashed, creating a loud mix that sounds better in your car or on the radio. You can see the material has just been smashed into a bigger block of sound instead of being recorded at a louder volume.

This reduction in dynamics is not a win-win situation, however. Reducing dynamics will often result in ear fatigue on the part of the listener and will often have negative effects on the perceived quality of the music. By removing dynamics, you remove one of the main elements of musical expression, often resulting in a more sterile-sounding finished product.

Dynamics Between Playback Formats

Dynamic range explains the perceived volume differences between different CDs, but it also affects the volume between different playback materials (and the components that play them). It's the reason the volume setting you

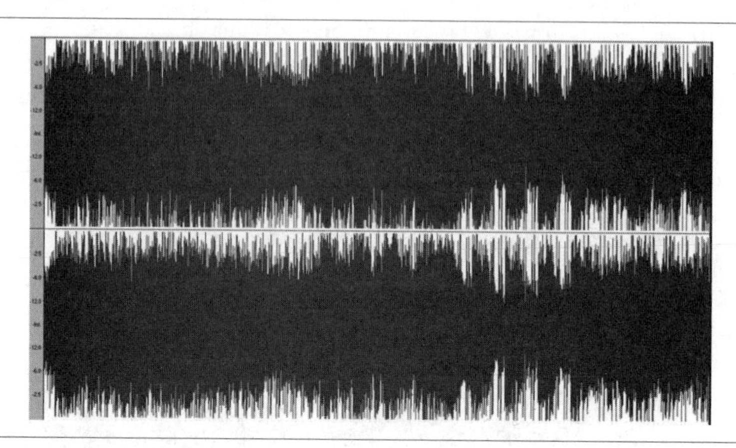

Figure 3-7. Stereo audio graph

use for DVDs seems to be much higher than the volume setting used for compact discs or television broadcasts, and why you often have to scramble for mute if you switch from watching *S.W.A.T.* on DVD to the local news on NBC.

Movie soundtracks are designed to be dynamic. They want to give you all those big, loud booms and hushed whispers. The idea of dynamic, of course, is simply that there is a big difference between the loudest sound and the quietest sound. As I said before, no matter how loud the explosion, digital material has a limit of the maximum level of sound. To have room for dynamics, and those booming explosions that shake the room, you have to make everything else softer.

As a result, movies are created to have their average overall sound level be low. If the average level is low, there is plenty of room to get loud before we hit that digital maximum level. The average dialog level for movies is about 25 or 30 steps below the maximum level available. Returning to audio graphs, Figure 3-8 shows what typical movie soundtracks look like.

Video games, television broadcasts, and CDs are usually created a different way; instead of being dynamic, they are compressed. Almost the entire signal is squashed down into a tight package, and the sound is just loud all the time. The average CD uses just the top three to six steps of available volume all the time. In other words, where movies have 30 steps of dynamics to use, CDs use only about 1/10th that much! As a result, switching from a compressed signal to a dynamic one results in a perceived loss of volume, while going the other way upsets your entire household.

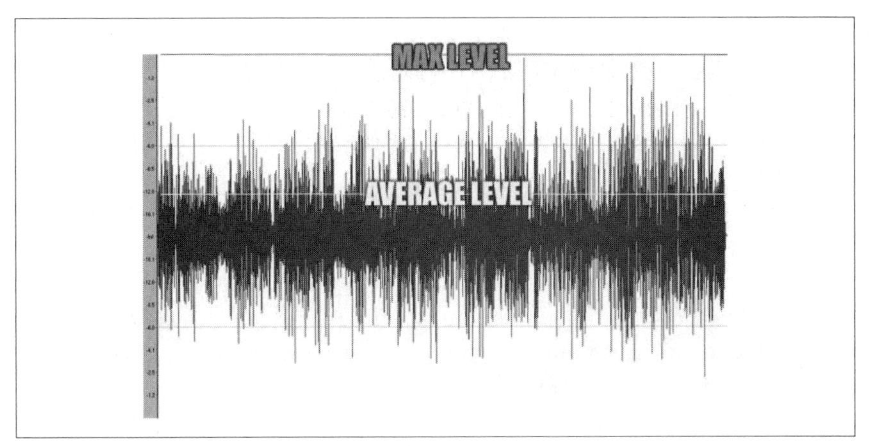

Figure 3-8. Movie soundtrack audio graph

Gain Offset

First, it is important to understand there is nothing wrong with your receiver or equipment when you experience large volume shifts. The difference in volume is normal and exactly how it should be, given the nature and goal of the different audio methods. However, that doesn't make it any more pleasant to deal with. Hope is available, though, through gain offset.

Gain offset refers to the ability to account for the change (offset) of volume (gain) in different playback materials and formats. On a receiver that offers gain offset, you can adjust the overall volume level up or down a couple of steps, specific to each input. The receiver saves these settings, ensuring that those annoying volume changes become a thing of the past. Using this feature, you can reduce the relative volume of the CD input and the TV tuner input on the receiver so that they are not so drastically different from the level of a DVD. Many newer receivers offer this feature, although it goes by a plethora of names.

Sample procedure: Lexicon MC-12. Here's the procedure for setting this feature on the Lexicon MC-12. Using this unit, the feature is simply called Volume Control Setup. Access this feature by going to the Main menu, selecting Setup, and then Volume Controls. Using the MC-12, you can set four different parameters, as Table 3-3 shows.

Table 3-3. Volume setting parameters on Lexicon MC-12

Parameter	Default value	Available values
Main Power On	–30dB	Last Level, –80dB to +12dB
Mute Level	–30dB	Full, –40dB, –30dB, –20dB, –10dB

Table 3-3. Volume setting parameters on Lexicon MC-12 (continued)

Parameter	Default value	Available values
Zone Power On	–30dB	Last Level, –80dB to +12dB
Rec Power On	–30dB	Last Level, –80dB to +12dB

For each of these, you can configure the volume that the unit will set itself to when that zone or function is accessed. For instance, Main Power On controls the volume when the unit is turned on.

The most common settings for these are either the default (–30dB) or Last Level, which sets the volume to what it was when the unit was last used in that particular zone or setting.

Zone Power On actually refers to Zone 2 on the unit; it's a rather confusing and unfortunate name.

Mute Level is actually measured as an attenuation level; in other words, if set to –30dB, hitting mute will cause the volume to be lowered by 30dB. Selecting Full will cause the unit to be completely muted when mute is pushed. Of course, this really isn't true gain offset, but instead allows some reconciliation between different zones and features of the unit. Still, it's worth including as it has direct bearing on source playback.

For more specific control, at a per-input level, you'll want to select Main Menu → Setup → Inputs → *Input Name* → Anlg In Lvl. *Input Name* will be something such as DVD or Tape. If the input is getting its source from a digital input (Digital In), gain offset isn't an issue. However, for an analog input, you can select the Anlg In Lvl option. Several values are available: Auto, which can be set to either On or Off, and Manual, which can be set to –18dB to 12dB (defaulting at 0dB). You also can set a per-channel gain offset and really tweak your sound. So, you might set an analog CD player to play at –5dB, and a VCR at +3dB.

Keep in mind that these are values added to the current volume, and not absolute volumes. So, if your listening volume is –8dB and your manual gain offset is +3dB, you'll end up listening at –5dB.

All that said, I've found the Lexicon to be more than adept at handling gain offset and I leave my personal unit on Auto, letting it do the work for me.

Sample procedure: Pioneer Elite VSX-49TXi. Some units, such as the Pioneer Elite VSX-49TXi, have a more rudimentary form of gain offset. In these units, you can't set individual gain attenuations, but you can ask the unit to handle high-gain inputs automatically. From the remote, navigate to the Receiver menu, and press the Input Att button. This will toggle the input attenuator from on to off, and back again.

With input attenuation on, analog signals are lowered when the receiver senses that they are too strong. The best way to discover the need for this is to take note of how often the OVER indicator on your receiver face lights up. If it's fairly often (more than a few times a week), you probably could benefit from input attenuation. This isn't quite as refined as the Lexicon, but it still will help out a lot.

—*Vince Maskeeper-Tennant*

HACK #27 Use Dynamic Range Compression to Regulate Volume

Dynamic Range Compression can lessen volume of Dolby Digital playback material.

All Dolby Digital–compatible receivers offer a feature called Dynamic Range Control (DRC). The DRC setting on your receiver, usually found in one of the speaker setup menus, offers three options: Off, Mid, and Max. This DRC control will reduce the dynamic range of Dolby Digital material. Dynamics are reduced significantly if you select the Max setting, reduced moderately if you select Mid, and left unchanged if you select Off. The overall effect is the squashing detailed in "Use Gain Offset to Regulate Volume" [Hack #26].

Some receivers also offer an additional Midnight mode that incorporates elements of DRC, bass reduction, and delaying the center and rear channels for better ambiance effects at lower volume. This is another, albeit similar, means of reducing the playback volume of playback material.

The most important caveat with DRC is that it operates by reading digital compression flags in the Dolby Digital bitstream; as a result, DRC usually works only with Dolby Digital signals. It is unable to affect DTS signals, or audio signals coming in via analog inputs. In addition, a Dolby Digital soundtrack must be encoded with these compression flags, and some lower-budget productions have no flags; these won't work with DRC.

Thankfully, many modern receivers have expanded the Dolby-specific DRC to apply basic dynamic reduction to analog soundtracks and DTS. Midnight mode, for example, such as the one offered on Pioneer products, works on all audio material.

It is important to remember that these compression methods alter the nature and intent of the soundtrack. If you want the theatrical experience and the best possible quality, it is best to leave these features unused. However, for late-night movie watching when the rest of the family is in bed, you might find these features to be indispensable.

—*Vince Maskeeper-Tennant*

High Definition

Hacks 28–35

It's no exaggeration to say that high-definition programming has revolutionized home theater. Just a few short years ago, the quality of a good DVD was staggering; it was far superior to anything ever before seen on a television. However, high definition (HD) is head-and-shoulders above even the best DVD-quality pictures. Broadcasters are airing more and more programming in HD, and hit TV shows such as *Alias*, *CSI*, and *The Grid* are now more exciting than ever, with their true-to-life imagery.

Of course, with any technological leap comes complexity, and HD is no exception. In fact, HD is a big enough topic that it warranted its own chapter in this book; there was simply too much HD-specific content to squeeze into the video components chapter (Chapter 2). In this chapter, you'll learn about HD programming, how to get it, and how to ensure your TV will work with it. You'll also understand how important the old antenna has become again and how you can tweak your set to get maximum performance. So, buckle up, and let's go digital.

Ensure You Can Get HD Programming

Although high definition is certainly the future in video, it might not make sense for you to go to an HDTV unit. Learn what should push you over the edge—and what shouldn't.

There has literally been almost a run on HD-capable television sets in recent days. However, buying an HD set doesn't necessarily mean you'll get an HD signal. If you don't have HD channels available to you, the greatest set in the world is wasted on low-resolution standard-definition (SD) programming. But to determine what programming is available, you need to know a little more about HD programming.

The Telecommunications Act of 1996

The Telecommunications Act of 1996 was passed by Congress and signed by the president. One of its provisions requires all terrestrial TV stations in the country to convert to digital modulation; in other words, these stations will be required to broadcast a high-definition signal.

Contrary to a persistent rumor, the VHF channels won't be abandoned.

The deadline for this switch is a little fuzzy, but it currently is set for around the end of 2006. To stay competitive, all cable systems are rapidly converting to digital, but there is no deadline for that.

The preexisting TV technology in the United States is called *analog*. It also is called NTSC (National Television System Committee), after the people who defined the standard itself. The NTSC specification was created in 1946, updated for color in 1953, and updated for stereo in 1984. Both of these updates were backward compatible, which avoided rendering anyone's TV set obsolete. But the new digital standard is totally different; the only thing it has in common with NTSC is the 6-MHz channel width. To continue using an NTSC TV after 2006, you might have to buy a converter box, which probably will cost about $200. At the time of this writing, these boxes were not yet available. You won't need such a box if you can rely on a cable or satellite box that has an NTSC output.

The new digital standard is called ATSC (Advanced Television Systems Committee). Soon the government will require that all new TVs be able to receive ATSC channels. The ATSC standard includes multiple formats, from 640×480 pixels to 1920×1080 pixels. All TVs must receive all of these digital formats and display them suitably. Then the broadcaster can choose from any of these formats.

To make the transition manageable (and, theoretically, gradual), the FCC is temporarily giving all terrestrial TV stations a second channel, so they can broadcast a digital channel along with their analog channel until 2006. There are 1,500 terrestrial TV stations in the United States, and 1,000 of them have their digital channel on the air. Most of these transmit some high-definition programs. More than 90% of the U.S. population can receive some high-definition programming from these stations.

These numbers are accurate as of September 2003, which is the last time reliable information was collected.

So, given all of that, here's the bad news:

- The cost to consumers for the new hardware is still pretty high.
- Home TV systems can be especially complicated during the transition.
- The picture-in-picture (PIP) feature many people enjoy doesn't currently work with HD programming. Most TV set designers are currently concentrating on other features, but an HD version of PIP will eventually show up.
- NTSC images sometimes look worse on a big, high-definition set than on a small, standard TV.

On the other hand, the quality of TV reception will improve dramatically. Temporary inconvenience, at least in this case, results in permanent improvement.

Should I Buy an HDTV?

When it comes to actually deciding to purchase a TV, you must consider two main questions:

- Can I afford the step up to HDTV?
- How much HDTV programming is available to me?

Can I afford an HDTV? The top-of-the-line HDTVs go for $10,000 and up, but a minimal compromise in quality will put you in the $2,000 range (the first color TVs cost $500, which, adjusting for inflation, would be $3,200 today). This lower-range compromise means you lose screen size and horizontal resolution; full horizontal resolution for HDTV is 1,920 pixels. But many sets being sold today resolve to only 1,280 pixels, and it is often difficult to see the difference. As such, 1,280 is still considered high-definition.

Smaller HDTVs are now available in the $1,000 range. If this still is beyond your budget, you have two choices:

- Postpone the purchase. Set prices will continue to come down, although much of this decline will be from the introduction of sets with lesser features.
- Buy a cheap, standard TV and hope your finances improve with time.

How much programming is available? Table 4-1 lists the HDTV programming available at the time of this writing, aligned with the major satellite and cable providers.

Table 4-1. Current HDTV programming

Network	Hours/day	DirecTV	DISH network	VOOM	Cable providers
Discovery HD	24 (some repetitive content)	Yes	Yes	Yes	Time Warner Cable (TWC), Charter
HDNet	24 (some repetitive content)	Yes	Yes	No	TWC, Charter
HDNet Movies	24	Yes	Yes	No	TWC, Charter
INHD	24	No	No	No	TWC, Charter
HBO HD	18	Yes	Yes	Yes	TWC, Charter, Comcast
Cinemax HD	17	No	No	Yes	Comcast
Showtime HD	10 (plus some DVD-quality content)	Yes	Yes	Yes	TWC, Charter, Comcast
Starz HD	8 (plus some DVD-quality content)	No	No	Yes	Comcast
The Movie Channel	8 (plus some DVD-quality content)	No	No	Yes	None
Bravo	Undefined	No	No	Yes	None
Encore	Undefined	No	No	Yes	None
Playboy	Undefined	No	No	Yes	None
TNT	Undefined	No	Yes	No	TWC
CBS	4	Yes	Yes	No	Check local availability
NBC	3	No	No	No	Check local availability
ESPN	Undefined	Yes	Yes	No	Charter, Comcast
Pay-per-view	Undefined	Yes	Yes	No	None

For quality and quantity, the best networks tend to be HBO, HDNet, Discovery HD, and CBS.

Terrestrial Broadcasts

Long ago, many people switched from roof antennas to cable service because the picture quality was a little better. This argument no longer applies. ATSC channels are like satellite TV in that, if you get a channel, the picture will be perfect, snow-free, and ghost-free.

You might not want to give up cable because of the many other channels it offers. But in many locations, over-the-air (OTA) broadcasts will be the largest source of HDTV programming for the next three years. All the network stations in most large cities are broadcasting HDTV. The web site *http://www.AntennaWeb.org* can tell you what DTV stations are available in your area.

In smaller cities, some DTV stations might not be passing along network HD material. The easiest way to determine the HD programming available to you is to go into a store selling HDTVs and ask for the "HDTV expert."

You need an antenna to get these ATSC signals. Information at *http://www.AntennaWeb.org* can tell you the compass directions to the transmitters in your area and recommend an antenna. Their recommendations are close, but not perfect, so you might want to see what others in your neighborhood have done.

If you have been told you can't erect a small outdoor TV antenna, that is probably incorrect. The Telecommunications Act of 1996 has a provision that preempts (overrules) nearly all local restrictions such as deed restrictions, homeowners association rules, renters contracts, and so on.

For more details, see the FCC Fact Sheet at *http://www.fcc.gov/mb/facts/otard.html*.

If you are in an area where reception is difficult, you might see occasional distortions in the image, and dropouts lasting five seconds or so. If these occur often, a bigger antenna might help. Note that as antennas become bigger, though, they become more directional, making aiming more sensitive. Additionally, nearby trees affect UHF much more than VHF. If putting a UHF antenna on your roof doesn't raise the antenna above the trees, you must find a place to mount it where it can see the horizon in the direction of the station. A UHF antenna should be at least 8 feet above ground, but mounting it higher doesn't always get you a stronger signal. VHF antennas always should be mounted as high as possible. The best weak-signal UHF antennas are the multi-bowtie reflector antennas, such as the Channel Master 8-Bay.

Cable TV. The cable TV industry was slow to take up the transition to DTV, but is now charging ahead. Digital cable is now being introduced in many areas, and some of these are carrying from 6 to 10 HD channels. About 40% of households can receive some HDTV programming via cable, although a special HD cable box is usually necessary. Ask your cable company what HDTV channels it carries. At present, HDTVs don't have built-in digital cable receivers, but they will soon.

Some analog cable systems have added a few ATSC (8VSB) channels to their lineups as well. You can receive those channels by connecting an OTA DTV receiver to the cable system. This is temporary, as the whole cable TV industry is converting to digital cable.

DirecTV. *To use DirecTV, you must have the oval dish that receives three satellites: 101°, 110°, and 119° west longitude. DirecTV and DISH Network are moving very slowly toward HDTV because they are spectrum-limited. None of the local channels is HD. DirecTV has stated that it is working toward HD local channels, but has not estimated when this will happen. It could take years.

DISH Network. Presently, all the HD channels except CBS are carried on the DISH Network satellite at 110°. Thus, you need a dish with two low noise blockers (LNBs).

There was an announcement from DISH that the HD channels will be moved to a new satellite at 105° and that a new three-LNB dish will be required. This author doesn't know the present status of that move.

None of the local channels is available in HD.

VOOM. This is a new Direct Broadcast Satellite service, presently in start-up phase. Some Wall Street analysts have predicted this company will fail due to insufficient funding; currently the company is funded by Cablevision, a cash-rich company, and it certainly will succeed if Cablevision wants it to. VOOM service is sold through Sears, although most Sears salespeople seem surprised when asked about VOOM. Reports from the few people who have signed up for it tend to state that the service is either "fantastic" or "horrific," but not much in between. It seems most VOOM personnel are learning as they go.

VOOM provides about 100 channels, approximately 30 of which are high-definition. About 20 of those HD channels originate at VOOM and are exclusive to VOOM.

No independent reviews of those channels were available at the time of this writing.

VOOM bills customers yearly instead of monthly; it costs $750 per year, and the receiver and dish are free. Receiver and dish installation also is free. Plus, VOOM will provide an OTA antenna for free, and install it for free (although probably not in weak-signal areas). VOOM has been giving free months to many customers to compensate for a rough start-up. I'm still neutral on whether signing up for VOOM is a good idea. The company probably will succeed, in which case VOOM would be the leader in satellite HD.

CBS national stations. DirecTV and DISH Network carry WCBS-HD (New York) and KCBS-HD (Los Angeles). However, they are available only to viewers in cities where the network owns the local CBS station, and even then you might have to apply for a waiver from the station. If your local CBS station is privately owned, DirecTV and DISH are not permitted to offer you these channels. You should call DirecTV or DISH to find out if you qualify for CBS-HD.

The DISH Network carries these channels on a satellite at 61.5° for East Coast viewers and on a satellite at 148° for West Coast viewers. In addition to the regular DISH Network dish, you need a second dish for the HD satellite. But the need for that second dish possibly will be short-lived.

C-band 4DTV. *C-band* and *Ku-band* refer to the satellite systems that require an 8-foot dish. *4DTV* is a digital service available on these bands. The high-definition channels on C-band are HBO East and West, Showtime East and West, Starz East and West, Encore East and West, Discovery HD Theater, and Nebraska Educational TV (PBS). About 150 DVD-quality digital SD channels also are available.

—*Kenneth L. Nist*

Get the Right Type of HD Set

As with any other component, tons of options are available when it comes to buying HD televisions. Learn which is best for you.

Just as when it came to choosing a TV in the general sense [Hack #9], an HD set offers an almost dizzying combination of display technologies, set types, screen sizes, and options. You'll need to know which set fits your budget, the size of screen you want (and can afford), and how lighting affects your room to make the right choice.

Display Technologies

The first thing you need to understand is the technologies involved. Although the overall set might determine what technology you choose, understanding how each works is critical.

CRTs. CRTs, short for cathode-ray tubes, are used in direct view, rear projection, and front projection TVs. They have several advantages over other technologies:

- Best color fidelity.
- Excellent blacks, and fairly bright whites.
- Smooth blending of adjacent pixels.
- Some draw both 1080i and 720p, avoiding errors in format conversion.

Here are the disadvantages of CRT:

- Brightness drops as screen size increases.
- Few sets truly achieve 1920×1080 resolution, although some get very close. Projection TVs employing 9-inch CRTs get the closest.
- There is a constant need for convergence adjustments (making red, green, and blue coincide perfectly).
- Focus and size adjustments are required every couple of years.

Many experts think CRTs produce the best pictures, but other technologies are getting better every year. Relatively few CRT sets draw both 1080i and 720p, and most draw one and require the tuner to convert the other. For example, a 720p-only CRT set doesn't work with an external tuner that converts everything to 1080i.

Plasma. These flat panel displays are large and very bright. Their disadvantages are many, though:

- High price
- Poor blacks
- A particularly short life span (they dim with age)

Additionally, pixels can die, which often is not covered by warranty. Even worse, plasma isn't a good choice for video games. The popularity of plasmas, despite these disadvantages, is a testament to the space they save, and the decorative enhancement they provide.

DLP. DLP, short for digital light processor, is also called DMD for Digital Micromirror Device. This is a large chip with about one million tiny mirrors on its surface. The chip can tilt each mirror to vary the amount of light reflected off of it. DLPs are used in rear and front projection TVs. The advantages of DLP are several:

- Very bright pictures
- Good blacks
- No burn-in problems

The DLP disadvantage is the "rainbow effect" (explained later in this chapter). Texas Instruments is the only maker of DLP chips. At the time of this writing, chips with two million mirrors, capable of 1080i resolution, were scheduled to be available in late 2004.

LCD. LCD stands for liquid crystal display. Polarized light shines through a sheet of glass, onto which is deposited a liquid crystal array. Each pixel changes polarization to either pass or block the light. LCDs are used in front and rear projectors. Large LCDs are used in flat panel displays. Their most notable disadvantage is that their blacks are not as dark as on a CRT. Additionally, on cheaper models, pixel response can be a little slow, causing blurred motion.

LCOS and D-ILA. LCOS stands for Liquid Crystal on Silicon and it's made by Toshiba and RCA, and D-ILA stands for Direct-drive Image Light Amplifier and it's made by JVC. Both of these are reflective chips with a polarizing LCD layer. Each pixel changes polarization to either reflect or block a light beam. Their biggest problem is that their blacks aren't as dark as on a CRT. These chips are used in front and rear projector TVs.

Set Types

Once you understand the technologies involved, you can start to hone in on the type of set you want. Several options are available and often, size and budget will determine which are available to you.

Direct view. This category includes flat panel and CRT TVs. A CRT produces a bright, sharp image you can view from any angle in a fully lit room. The largest of these sets are about 38 inches, measured diagonally. You might find that, due to size constraints, no more than two people can sit in front of one of these sets. Another negative is that widescreen CRTs [Hack #13] come in cases that are exceptionally deep.

Plasma and LCD flat panel TVs also are available. Plasmas are large and expensive, while LCDs are small and cheap ("cheap" means $1,000, at least right now).

Rear projection. Rear projection sets come in sizes ranging from 40 inches to 73 inches, and are the most popular HDTV technology. Most are based on CRTs, but there are other competing technologies.

Typically, these sets have three CRTs (red, green, and blue) hidden inside a box. The CRTs point upward, into a flat mirror, and then reflect onto a diffuser screen. The CRTs have lenses that focus the image at the screen. Adequate intensity is a problem, so the diffuser screen is designed to not radiate light in directions where there are no viewers. A fully bright image is visible to about 45° to the left and right, but only to about 10° to 15° above and below any spot on the screen.

Typically, if you sit closer than three times the screen height, you can't see the top and bottom of the screen at full brightness. Although three times the screen height is a correct distance, if your head moves vertically even a few inches, you might lose brightness at the top or bottom of the screen. Sitting at 3.5 times the screen height reduces this problem (see Figure 4-1). If you don't mind the loss of brightness, you can watch standing up.

Room lighting must be controlled for RPTVs, but an absolutely black room isn't necessary. Indirect ceiling lighting works well, but usually you can see the reflection of lamps in the large, flat screen; ditto for windows. You probably will want dark shades for windows that reflect in the screen. If the salesperson tells you the screen is nonreflective, be sure to check it out for yourself.

Front projection. For pictures larger than 73 inches, front projection is the way to go. Usually the projector is ceiling-mounted or placed on a shelf

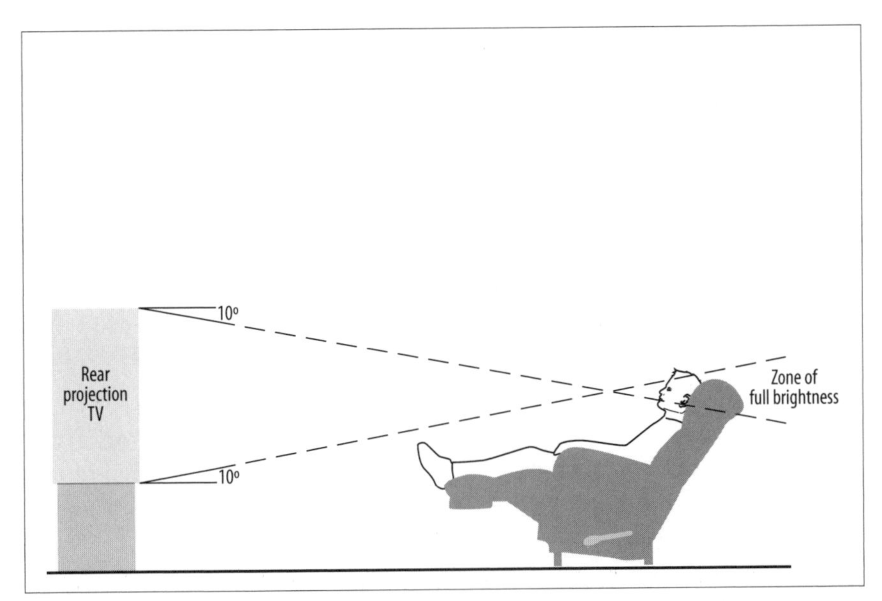

Figure 4-1. Sitting correctly in front of an RPTV

behind the viewers. Because the screen is white, a completely blackened room is necessary; otherwise, black appears as gray.

Some systems are bright enough to project onto a gray screen, reducing this problem somewhat.

Set Configurations

Another consideration is the configuration of the set you're looking at. There are three basic categories.

Monitor
: This is a set without a tuner, and usually without any audio. This is presently the hottest-selling HDTV configuration, preferred either because external OTA receivers are thought to be better or because the owner plans to rely on cable or satellite. Many of these sets presently are being used just for viewing DVDs.

Integrated TV
: This is a complete HDTV set with NTSC and ATSC tuners, and also some speakers. An external tuner always can be added.

HD-ready TV
: This is a TV with an NTSC tuner. The set will draw 1080i or 720p if an external tuner is added. The FCC has mandated an end to this category

[Hack #28], and will require an ATSC tuner with any device that has an NTSC tuner. This requirement is phasing in gradually, applying to large TVs now and to the smallest TVs in 2007.

Burn-in

Another issue in today's newest HD sets is *burn-in*. The phosphors used in CRTs and plasma displays become less bright with use. The phenomenon is a lot like tire wear; if you drive fast, the wear per mile increases, but there is still some wear at any speed. The speed of a car corresponds to white in a TV image.

CRT burn-in used to be rare, but the demand for brighter images has made manufacturers less conservative. Now, CRTs that have been showing a Windows desktop for a couple of years often will show a lightly burned-in task bar when the screen is painted entirely white. The CRTs in big-screen TVs are pushed even harder, especially in the largest sets.

All CRT and plasma sets dim with use. Making the screen age evenly is the user's responsibility. The user must ensure that a fixed, unmoving shape is not displayed for many hours, or that shape will slowly become burned into the screen. LCD, LCOS, and DLP sets don't suffer burn-in.

The Rainbow Effect

One-chip DLP sets employ a rotating color wheel. Thus, the three colors are delivered to the screen sequentially. Suppose the image comprises white text on a black background. If you shift your gaze rapidly across the image, the white lines will decompose into the primary colors (until your eyes stop moving). Most people don't notice this, and most of the people who do see it learn to ignore it. But a few people can't get past being distracted by this rainbow effect.

Set makers can reduce this problem by changing the colors faster (using color wheels with 6, 9, or 12 color segments). The rainbow effect is eliminated in three-chip DLP sets, which have no color wheel.

Some CRTs have a similar problem. My set employs a green phosphor that stays lit four times as long as the red and the blue. A rapid eye shift will reveal some flashes of green in an image that has only black and white.

—Kenneth L. Nist

Add a Set Top Box

Once you've got an HDTV set, you'll need something to pull in HD signals and pass them onto your television. Learn what to look for, and what to buy.

Set top box, or *STB*, is a term that can include any type of accessory that can connect to the HDTV. Common STBs are satellite receivers, cable TV receivers, OTA receivers, DVD players, VCRs, and so on. Generally, though, an STB is a device that pulls in a high-definition signal, instead of just pushing pictures through to your TV.

Choosing a set top box is like choosing any other component; you need to find one that has the right functionality, and more important, the right connections for your gear. For an STB, that means ensuring that both the video and audio it receives can be passed on to the rest of your gear.

STB Video Output Options

Unfortunately, a single universal standard for unit-to-unit video connections doesn't exist. Eventually, through competition, the best of the following will survive. Any STB you acquire probably will have more than one of these output connectors. When you buy an HDTV and an STB, try to select units that can connect to each other directly. Otherwise, you will have to pay for a transcoder or a video switch box. Here are your basic connection options.

CH3/CH4 output
: ATSC output via a preselected channel on the TV is one of the oldest connection methods. Obviously outdated, this almost never shows up in modern components.

Composite video
: This one-wire standard, in use for many years, conveys complete video images. It is designed for NTSC and can't transport HDTV images.

S-Video
: This two-wire standard is an improvement over composite video. But it was designed for NTSC and can't carry anything else.

Component video
: This three-wire standard, originally designed for DVD players, can carry HDTV via three wires with phono plugs. The three wires carry analog raster (image-scanning) signals, either red/green/blue or Y/Pr/Pb (Y=intensity, Pr=Y-red, and Pb=Y-blue). Some units can handle either color scheme. You must verify that both units can use the same scheme.

Neither the red/green/blue nor the Y/Pr/Pb scheme is better than the other.

VGA

This five-wire standard, originally devised for computer monitors, carries HDTV raster signals, usually red, green, blue, Hsync (horizontal sync), and Vsync (vertical sync). However, in some units, Y, Pr, and Pb can substitute for the color channels. Usually the five wires are bundled into a single cable.

Five separate cables are advisable for runs longer than 12 feet.

The connector can be a 15-pin VGA connector or five BNC connectors.

Some HDTVs have VGA inputs that accept only computer formats, such as 600×800 and 720×1024. Many makers use the term RGB in place of VGA despite the confusion that causes.

DVI (Digital Visual Interface)

This connector conveys HDTV raster-like signals in binary data form. The data rate is very high (1.65 Gb/s). Monitors other than CRTs, such as plasma, LCD, DLP, LCOS, and others, prefer binary data. DVI comes with a decryption option called HDCP (High-bandwidth Digital Content Protection), which will decode encrypted programs such as first-run movies. However, there is a serious problem here: the motion picture industry might try to require distributors (HBO, Cinemax, etc.) to use HDCP encryption on all high-definition movies. HDCP decryption hardware is proprietary, and any hardware manufacturer must sign a contract to include it in his product. That contract forbids high-definition analog output (VGA or component video) when encryption is enabled, and allows HDCP decryption to take place only in the monitor. This is an attempt by Hollywood to prevent unauthorized copying and distribution of high-definition material. However, it means that millions of HDTVs already sold that have only analog inputs could become useless, except for viewing whatever sitcoms or dramas the networks allow. The FCC hasn't yet ruled on this, and doesn't seem to be in any hurry to step into the issue.

HDMI

This new miniature connector is intended to replace DVI. It is backward compatible with DVI, and an adapter will connect it to a DVI unit. It has 19 pins and carries DVI, plus digital audio. It also has a control line that allows the STB to sense the monitor's state and native formats.

IEEE 1394

Also called FireWire or i.link, this is a high-speed bus common in computers. IEEE 1394 is fast enough to carry compressed MPEG-2 video data plus audio and controls. There is an encryption standard for IEEE 1394, called DTCP (Digital Transmission Content Protection, and sometimes called *5C copy protection*). But because IEEE 1394 is an open standard, Hollywood has less control over it. Because it is a two-way bus, it could allow units to control each other. This holds out the promise of eliminating the need for 5 or 10 handheld remotes to control the home theater. IEEE 1394 is just a connector definition plus a software shell. Additional software is required for the units to talk to each other.

Home Audio Video Interoperability (HAVi) is such software. HAVi allows plug and play recognition of devices, interoperability, and brand independence.

If the STB has a CH3/CH4, composite, or S-Video connector, it is for standard-definition images only. When a high-definition program is being received, these connectors are either disabled or carry an image that has been down-converted to NTSC.

Neither VGA nor component video is superior to the other. For a cable length of six feet, VGA is more convenient. For longer runs, component video is usually more convenient.

When using DVI, VGA, and component video, very few sets will draw both 1080i and 720p. If you feed the set a mode that it can't draw, you will get either a blank screen or garbage.

The law requires a set to receive all 18 modes. However, the law only regulates tuners, not these intermediate inputs.

An exception to this is fixed-pixel displays that will redigitize component video.

DVI and 1394 are presently competing for the hearts and minds of the manufacturers, but which will win is unclear. A third possibility is that both will be adopted, DVI for video and 1394 for audio and control. All recording devices likely will use 1394.

More on DVI. DVI was originally developed for computer monitors, but has been adopted by HDTV. DVI comes in different versions. All versions use the same 29-pin connector. Sometimes you can tell which version you have by seeing how many of the 29 pins are missing.

DVI-D
: This is the version most commonly used for HDTV. The five large pins usually are missing. There is a single-link version of this that uses only 12 of the 24 small pins. Single link will work properly with all HDTVs.

DVI-I
: This version uses all 29 pins. The 5 large pins pass analog VGA signals. Presently, the computer industry primarily uses DVI-I, but front projector HDTVs, from a number of makers, support DVI-I. DVI-to-VGA adapters and adapter cables are available for these units. Front projectors from a couple of makers accept component video signals through their DVI connectors. These companies provide DVI-to-component adapter cables. However, this is nonstandard.

  These adapter cables work only with DVI-I. In most cases, if you want to connect a DVI unit to a VGA or component unit, these adapters won't work. That would require a transcoder circuit that can convert between analog and digital signals.

HDMI
: HDMI is a single-link DVI plus digital audio and a control line in a miniature connector. It carries no analog signals.

Avoiding (most) risk. Unfortunately, you can't avoid risk completely. When you select an STB, you must decide among DVI, 1394, or analog (VGA and component video are considered analog). There is no way to tell which will become the long-term winner. Presently, Hollywood doesn't want any DBS or cable set top box to have a 1394 connector passing MPEG-2 data. They even consider analog to be a piracy threat. If the DVI interface catches on big, Hollywood could order all DBS and cable companies to disable all STB analog or 1394 video outputs whenever a high-definition movie is showing. Some people think the FCC would delay that order by 10 years to allow depreciation of the millions of HDTV sets that would become OTA- or SD-only as a result.

I believe Hollywood will not carry out its threat anytime soon. What Hollywood is most concerned about is movie piracy via the Internet. Currently that is not practical at high definition because it takes too long to download huge movies in this resolution. However, if it should become practical and piracy proliferates, Hollywood will try to shut down those STBs that contribute to it. The FCC certainly will side with Hollywood if movie piracy makes movie-making unprofitable. This is not all bad, though, because it guarantees home access to first-rate films.

STB Audio Output Options

An STB is likely to provide one or more of the following audio outputs:

- Six-channel audio (six wires with phono plugs)
- Coaxial digital audio (one wire with phono plug)
- Optical digital audio (one TOSlink fiber optic line)
- IEEE 1394 audio and video
- DVI audio and video

Again, it is wise to plan for audio connectivity before buying. You could be in a real bind if the TV and STB don't have compatible connectors for video, or if your receiver/processor and STB can't speak audio to each other.

—*Kenneth L. Nist*

Properly Size Your HD Image

Once you've got your HD signal coming in, you'll need to ensure that it appears properly on your television set. Otherwise, you'll miss important detail and lessen the HD experience.

Many are the stories of a consumer buying an expensive HD-ready widescreen TV [Hack #29], spending hundreds on killer programming [Hack #28], and still seeing only 70% or 80% of the overall image. The choice seems to be between black bars on the side [Hack #13] or an image that clearly extends below the bottom of the screen and above the top. Some careful tweaking can take care of this once and for all.

HD programming needs to form an exact 16×9 image, form-fitted to your TV's 16×9 screen. Unfortunately, this almost never is set up for you out of the box. RPTVs are *overscanned* quite heavily from the factory, meaning you are missing substantial parts of your picture. This process of reducing the overscan is not for the faint of heart, however, because doing so hoses your geometry and convergence, all of which have to be totally recalibrated [Hack #62] afterward.

You'll need to enter your television's service menu, and then manually set both the vertical and horizontal size.

As always, when dealing with service menus, write down everything before changing anything.

When you've found the menu options you want, load up a true HD image. You can usually get one of these by setting your set top box [Hack #30] to PBS or HDNet; both channels have true HD broadcasts most of the day. Then, as the broadcast is showing, squeeze in the edges of the screen vertically until you can just see the edges of the picture on the top and bottom of your TV. Then you can just nudge the size back up, and you're all set. Repeat the same process horizontally: squeeze in, and then set the overscan.

As you're setting the size, you might find that the picture is also off-center. You can see this is happening when one side of the picture is flush with the edge of the display on your set, but the other side is not. You'll have to move the image toward the side that shows the extra edge and then resize again. This is a trial-and-error process, and can take some time.

For those of you used to working with computer monitors, this is the same way you ensure your monitor is displaying all of the video image and not cutting some off (or leaving wasted display space along an edge).

If you're having trouble getting the edges aligned, focus on the center of the edge you're working on; in other words, don't worry about the corners. When you rework the convergence and geometry of your screen via calibration, these corners will take care of themselves.

Finally, with the picture just where you want it, you'll have to go back to calibration, working on the color levels, convergence, and the like. This can be a lengthy process; the more the screen is off when you get it, the larger the error your adjustments can cause. Still, it's well worth it to see all of the image you're working with rather than just part of it.

—*Robert Jones, Image Perfection*

Get the Right Antenna

An over-the-air antenna can provide you crystal-clear HD broadcasts. Learn how antennas work, where to mount them, and what commercial offerings are good buys.

Even if your cable or satellite provider offers great HD programming [Hack #28], you'll still probably want an OTA antenna. In some cases, the antenna can provide local channels your provider might not yet offer; more important, though, an OTA antenna is the key to HD broadcasts of local networks. Until cable and satellite providers offer you local channels in HD, an OTA antenna is an essential part of a high-end home theater.

Antenna Basics

If you're going to get into antennas, you're going to need some basic background. This section gives you that background, while still keeping things at a fairly mild technical level.

The TV channels. *Hertz* (Hz) means cycles per second.

Heinrich Hertz was the first to build a radio transmitter and receiver, at least while understanding what he was doing.

The radio frequency spectrum is divided into major bands, as shown in Table 4-2.

Table 4-2. Major radio frequency bands

Abbreviation	Meaning	Frequency	Wave length (in meters)
VLF	Very low frequency	3 kHz–30 kHz	100 Km–10 Km
LF	Low frequency	30 kHz–300 kHz	10 Km–1 Km
MF	Medium frequency	300 kHz–3 MHz	1 Km–100 m
HF	High frequency	3 MHz–30 MHz	100 m–10 m
VHF	Very high frequency	30 MHz–300 MHz	10 m–1 m
UHF	Ultra high frequency	300 MHz–3 GHz	1 m–100 mm
SHF	Super high frequency	3 GHz–30 GHz	100 mm–10 mm
EHF	Extremely high frequency	30 GHz–300 GHz	10 mm–1 mm

The terms kiloherz (kHz) means 1,000 hertz, megahertz (MHz) means 1 million hertz, and gigahertz (GHz) means 1 billion hertz.

A TV channel in the United States will always occupy 6 MHz of this spectrum (see Table 4-3).

Table 4-3. Spectrum occupied by channel groups

Channels	Spectrum occupied
2–6	54 MHz to 88 MHz (with one small gap)
7–13	174 MHz to 216 MHz
14–69	470 MHz to 806 MHz

Channels 2–13 are the VHF channels. These are split into two groups so that antennas will work better: in general, an antenna designed for frequency N also works well at 3N, but very poorly at 2N.

The wave length of a radio wave is defined as:

```
λ = 300/F
```

Here, F is the frequency in MHz, and λ is the wave length in meters. Antenna elements are usually about half a wave length long.

Decibels. Decibels (dB) are commonly used to describe gain or loss in circuits. The number of decibels is found from this formula:

```
Gain in dB = 10*log(gain factor)
```

Alternatively, you can just use the diagram in Figure 4-2.

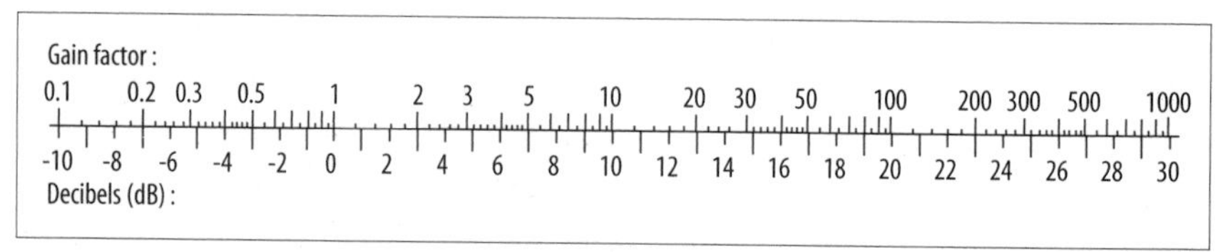

Figure 4-2. Determining decibels (dB) from gain factor

For example, suppose 10 feet of cable loses 1dB of signal. To figure the loss in a longer cable, just add 1dB for every 10 feet. In general, decibels let you add or subtract instead of multiply or divide. There are also some specific decibel markers, and the effect they have on gain, that you might want to memorize:

- 20dB = gain factor of 100
- 10dB = gain factor of 10
- 3dB = gain factor of 2 (actually 1.995)
- 0dB = no gain or loss
- –1dB = a 20% loss of signal
- –3dB = a 50% loss of signal
- –10dB = a 90% loss of signal

Noise. Whether a signal is receivable is determined by the *signal-to-noise* ratio. For TVs there are two main sources of noise.

Atmospheric noise
: This noise can come from many different types of sources. For example, a light switch creates a radio wave every time it's turned on or off. As another example, motors in some appliances produce nasty RF (radio frequency) noise.

Receiver noise
: Most of this noise comes from the first transistor the antenna is attached to. Some receivers are quieter than others.

Receiver noise dominates on the VHF and UHF bands, and atmospheric noise is usually insignificant. On an analog channel, noise looks like snow. If there is only a barely perceptible amount of snow, it corresponds to how noise-free a DTV signal must be for a DTV receiver to lock on to it.

Signal amplifiers and preamplifiers. Many people think that connecting an external amplifier to the antenna can improve the performance of the antenna. Unfortunately, this is usually wrong.

The signal-to-noise ratio is generally set by the receiver's first transistor. But if an external amplifier is added, the first transistor in the (added) amplifier determines the signal-to-noise ratio.

Because the signal and the external amp's noise are magnified greatly, the receiver's noise becomes insignificant.

Because there is no reason to think the external amp's first transistor is quieter than the receiver's first transistor, there is generally no benefit to the signal-to-noise ratio from an external amplifier.

However—and here's the good news—an external amplifier can compensate for signal loss in the cable if the amplifier is mounted at the antenna. Without this amplifier, a weak signal, just higher than the noise level at the antenna, could sink lower than the noise level due to loss in the cable, and be useless at the receiver.

RG-6 cable loses 1dB of the signal every 18 feet at channel 52. For a DTV channel, 1dB can be the difference between dropouts every 15 minutes (probably acceptable) and every 30 seconds (unwatchable). I recommend a mast-mounted amplifier whenever the cable length exceeds 20 feet.

If you are in a good-signal area or you have no high-numbered UHF channels, you can, to an extent, ignore this advice.

The amplifier should have a gain equal to the loss in the cable (for your highest channel) plus another 10dB (to keep the receiver's first transistor out of the picture). You generally can overshoot this target by 10dB without causing any trouble.

When figuring the cable loss, be sure to include the loss in any splitters and baluns. If a 2-to-1 splitter were 100% efficient, you would figure a 3dB loss because each TV gets half of the power. More realistically, splitters are usually 80% to 90% efficient.

The antenna and the amplifier both have gains measured in dB, and many people add these two numbers (and then maybe subtract the losses) to find the strength of the signal at the receiver. However, this calculation has little real value; you always should keep the net gain in front of the amplifier separate from the net gain that follows.

Receiver noise. Actually there is a reason to think the external amplifier is quieter than the receiver is. Long ago, designers made an effort to make the TV's first amplifier stage very quiet. But now 90% of homes use cable or satellite boxes (strong sources) and most of the rest are rural homes using antennas that have mast-mounted amplifiers. So, the TV's noise is rarely a factor. Many TV makers no longer put any effort into making their sets quiet.

Suppose you live in an apartment 15 miles from the transmitter of the signal you're trying to capture. Your indoor antenna mostly works, but you are troubled by dropouts. Will adding an amplifier right at the TV improve things? Yes, if it is quieter than the TV. Unfortunately, TV makers see no reason to publish the noise figures for their receivers. So, buying an amplifier for an indoor antenna is a total crapshoot. I recommend that you try a Channel Master Titan or Spartan amplifier, but make sure you can return it if it is of no help.

Transmission cable. Twinlead (ribbon cable) used to be common for connecting TVs to antennas, and it does have its advantages. However due to its unpredictability when positioned near metal or dielectric objects, it has fallen out of favor.

Such objects, even if not touching the cable, cause a portion of the signal to bounce, return to the antenna, and get retransmitted.

Coaxial cable is recommended instead. It's fully shielded and not affected by nearby objects. Coaxial cable has a feature called *characteristic impedance*, which for TVs always should be 75 ohms.

Fifty-ohm coaxial cable is also common. Avoid that cable!

Although rated in ohms, this measurement has nothing to do with resistance. A resistor converts electric energy into heat. The "75 ohms" of a coaxial cable don't cause heat. Where it comes from is mathematically complicated, and beyond our scope here.

But coax also has ordinary resistance (mostly in the center conductor: see Table 4-4) and thus loses some of the signal, converting it into heat. The amount of this dissipation (loss) depends on the frequency as well as the cable length.

Table 4-4. Cable diameter and center conductivity

Type of cable	Center conductor	Cable diameter
RG-59	20–23 gauge	0.242 inches
RG-6	18 gauge	0.265 inches
RG-11	14 gauge	0.405 inches

The cable loss is shown pictographically in Figure 4-3.

The table in Figure 4-3 is only approximate. There are many cable manufacturers for each type, and there is no enforcement of standards. If the mast-mounted amplifier gain exceeds the cable loss, it shouldn't matter what cable type you use. But there are two problems with this.

- Some cable has incomplete shielding. This is most common for RG-59 and is another reason to avoid it.
- When the cable run is longer than 200 feet, the low-numbered channels can become too strong relative to the high-numbered channels. In this case, RG-11 or an ultra-low-loss RG-6 is recommended (these alternatives are expensive). Alternatively, frequency-compensated amplifiers will work.

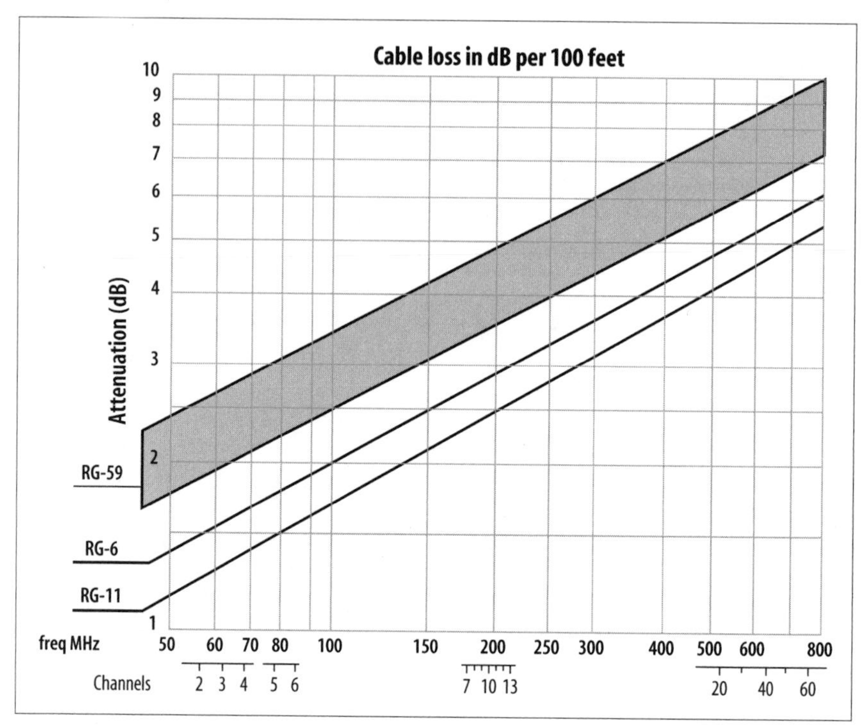

Figure 4-3. Cable loss per 100 feet

I usually recommend RG-6 for all TV antennas. It can be stapled in place using a staple gun with common 9/16-inch T25 staples. How long the cable lasts depends solely on how long you can keep water out of it. 3M Vinyl Electrical Tape is a good waterproofer. Even better is an asphalt putty called "Coax Seal" (RadioShack 278-1645), but it is so tenacious it should not be used for temporary connections. Cover the connectors completely.

Receiver overload. Signal amplifiers are supposed to be linear. That is, the output is a magnified but otherwise unaltered version of the input. But too much signal can make an amplifier nonlinear, usually clipping off the tops and bottoms of the sine waves. When this happens, all channels are affected, not just the one that is too strong. In fact, the too-strong signal usually is not a TV station. A close FM station or police station is more likely.

If you add a good amplifier to your antenna system and your results get worse instead of better, you have overload, and you need to reconsider more carefully what you are doing.

An attenuator is a resistor network that can reduce the gain of an amplifier; 6dB attenuators are available at RadioShack. Insert the attenuator between the TV and the power injector. If you are close to an FM station, there might be a narrow range between too much and too little gain. You can make that range larger by using an amplifier with an FM trap or by using a more directional antenna.

Types of Antennas

First, you'll need to have some more basic terms under your belt.

Gain
: A measure of how much signal the antenna will collect.

Beamwidth
: A measure of how directional the antenna is.

Bandwidth
: A measure of how the gain varies with frequency. A narrow-band antenna will receive some channels well, but others poorly.

The dipole antenna. This is the simplest TV antenna. Variations on the dipole are the bowtie (which has wider bandwidth), the folded dipole (which can solve an efficiency problem), and the loop (a variation on the folded dipole). All four have the same gain and the same radiation field: a toroid (doughnut shape: see Figure 4-4). The gain is generally 2.15dBi, which means "dB of improvement over an isotropic radiator." That's a lot of verbiage that simply means an antenna that radiates equally in all directions.

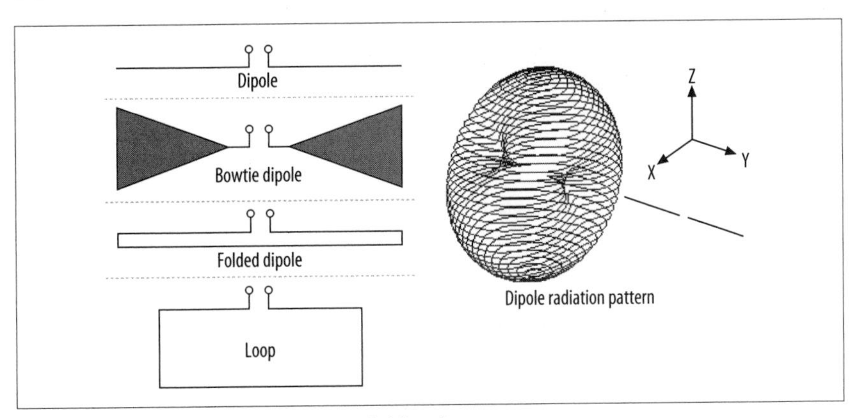

Figure 4-4. Dipole antenna radiation field and variations

The dipole has positive gain because it doesn't radiate equally in all directions. This is a universal truth. To get more gain, an antenna must radiate in fewer directions. Imagine a spherical balloon. Now press on it from opposite sides with a finger of each hand. Push in until your fingers meet. The result looks like the toroid, as in Figure 4-4. But more important, the balloon expanded in the other directions. Aha! Gain! That's the way antennas work.

Keep this balloon analogy in mind. More complicated antennas work by reducing radiation in most directions. They distort the balloon considerably, but the volume of the balloon remains constant.

Another rating system for antennas uses dBd, which means "dB of improvement over a dipole antenna." To convert dBd to dBi, just add 2.15. Antenna makers specify their gains in dB. They actually mean dBd, but given the way they exaggerate their claims, dBi is usually closer to the truth.

In the United States, TV antennas are always horizontal. If you rotate an antenna about the forward axis (a line from the transmitting antenna) the signal strength will vary as the cosine of the angle. In other words, when the antenna elements are vertical, no signal is received because TV signals have horizontal polarization.

Stacked dipoles. Two heads are better than one, and so it is with dipoles. N dipoles will take in N times as much RF power as one dipole, provided they are not too close to each other. Thus, a four-dipole antenna would have a gain of 8.15dBi. (That is 2.15dBi doubled once [plus 3dB] and doubled again [plus another 3dB].) This assumes their positions and cable lengths are adjusted so that their signals add in-phase. This explanation of gain might seem at odds with the balloon explanation, but ultimately they are equivalent.

Adding dipoles doesn't increase the volume of the balloon because phase cancellation occurs in some directions.

Dipoles are commonly stacked horizontally (collinearly), vertically (broadside), and in echelon (end-fire). Figure 4-5 shows examples of all three configurations.

When dipoles are stacked horizontally, the horizontal beamwidth becomes very narrow. This is because they don't add in-phase for directions that are not straight ahead. Similarly, when stacked vertically, the vertical beamwidth becomes narrower.

Let's say you are 20 miles from a city, and TV transmitters are scattered all over the city. A medium-gain antenna might be too weak, but a high-gain antenna would be so directional you'd need a rotor. Solution: a bunch of dipoles stacked vertically can give you the gain you need (as illustrated in

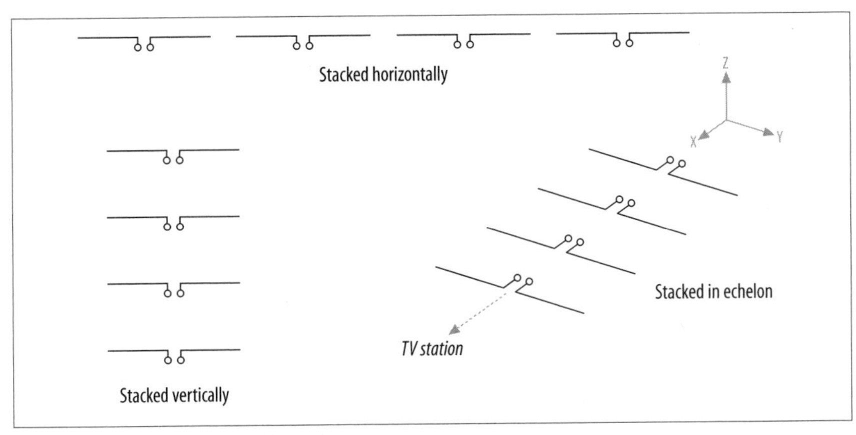

Figure 4-5. Various stacked dipoles

Figure 4-6). The vertical narrowness of the resulting beam is of little importance, but the horizontal broadness of the beam means no rotor is needed.

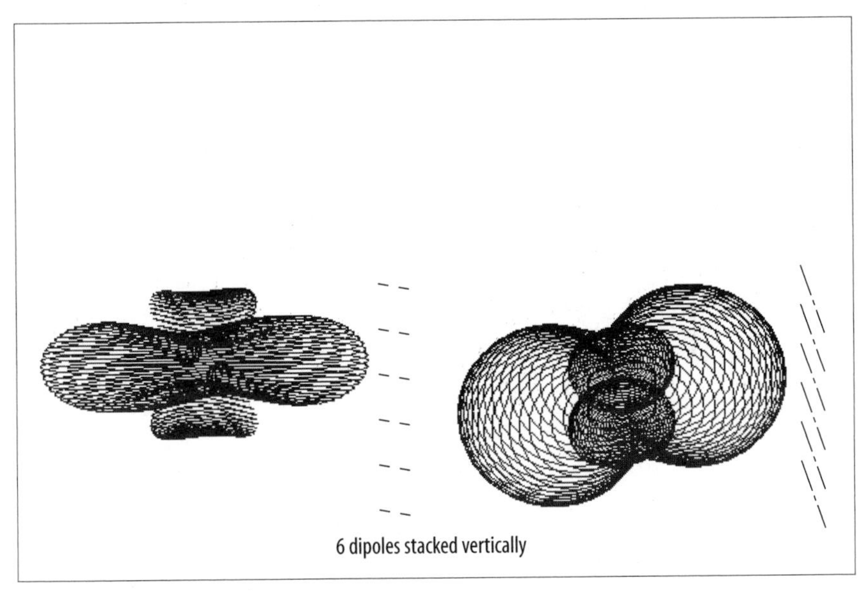

Figure 4-6. Stacked dipoles

Reflector antennas. Radio waves reflect off a large conducting plane as if it were a mirror. A coarse screen can serve as well. Reflector antennas are very common. The double bowtie in Figure 4-7 has gain of 5–7dBi. With a bigger screen, it would have more.

The parabolic reflector shown in Figure 4-8 focuses the signal onto a single dipole, but its bandwidth is a little disappointing. The corner reflector (in

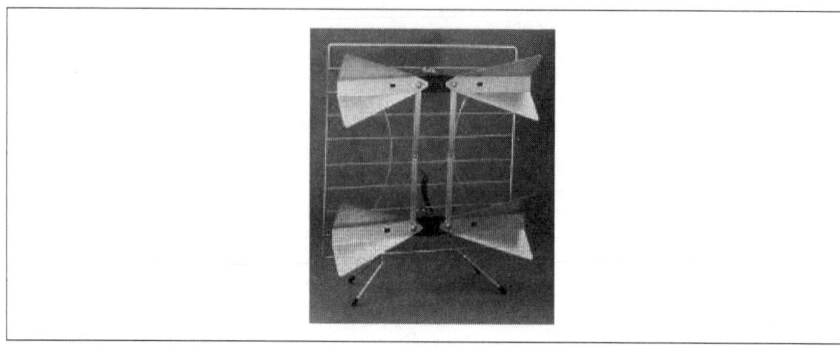

Figure 4-7. Double bowtie reflector

the same figure) has a little less gain but much greater bandwidth. The corner reflector has roughly the gain of three dipoles. It is a good medium-gain antenna, widely used for UHF.

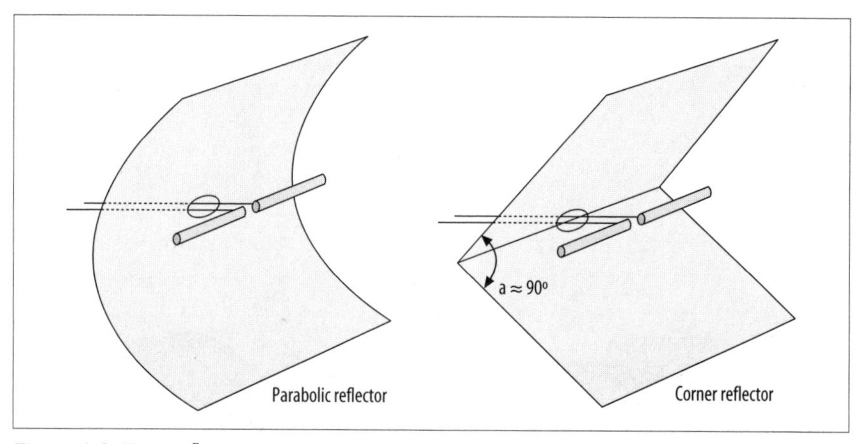

Figure 4-8. Two reflector types

If you need more than 25dBi, the paraboloid dish shown in Figure 4-9 is the only practical choice.

Log-periodic dipole arrays (LPDAs). The LPDA, shown in Figure 4-10, has several dipoles arranged in echelon and crisscross-fed from the front. The name comes from the geometric growth, which is logarithmic.

This is a very wide-band antenna with a gain of up to about 7dBi. For any frequency, only about three of the elements are carrying much current. The other elements are inactive. As frequency increases, the active elements "move" toward the front of the array. Most VHF antennas are LPDAs.

Figure 4-9. Parabolic reflector

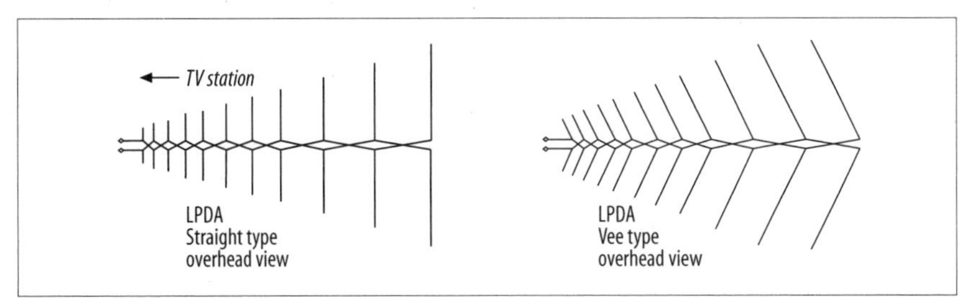

Figure 4-10. LPDA antenna types

TV LPDAs come in two types: straight and Vee. The Vee type has just a slightly higher gain for channels 7–13. But I usually favor the straight type because it has nulls 90° to each side that can be used to cancel out ghosts.

Yagi antennas. A Yagi antenna has several elements arranged in echelon. They are connected together by a long element, called the *boom*. The boom carries no current. If the boom is an insulator, the antenna works the same. Figure 4-11 shows the overhead view of a Yagi antenna.

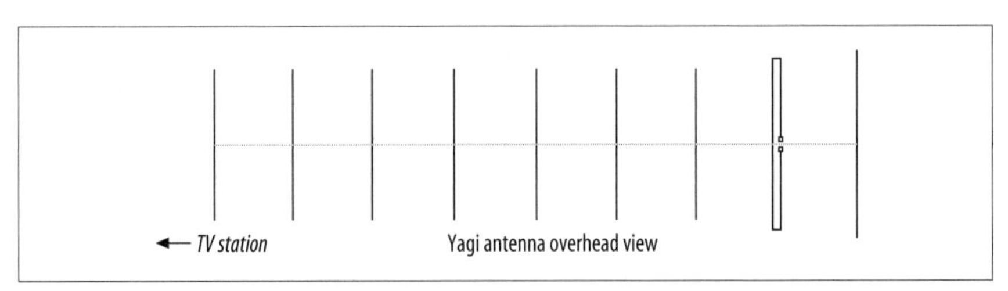

Figure 4-11. Yagi antenna, overhead view

The rearmost element is called the *reflector*. The next element is called the *driven element*. All the remaining elements are called *directors*. The directors are about 5% shorter than the driven element. The reflector is about 5% longer than the driven element. The driven element is usually a folded dipole or a loop. It is the only element connected to the cable. Yet the other elements carry almost as much current.

The Yagi is the most magical of all antennas. I won't attempt to explain why it works; the math simply gets way too complex. Suffice it to say that the more directors you add, the higher the gain becomes, and gains higher than 20dBi are possible. But the Yagi is a narrow-band antenna, often intended for a single frequency. As frequency increases above the design frequency, the gain declines abruptly. Below the design frequency, the gain falls off more gradually. When a Yagi is to cover a band of frequencies, it must be designed for the highest frequency of the band.

A UHF Yagi today is designed for channel 69. If you see an old Yagi, it might be intended for channel 82. In the future they will be cut for channel 51. It's not possible to tell by looking at a Yagi which era it belongs to, so be careful. Often the reflector element is replaced by a corner reflector, as shown in Figure 4-12.

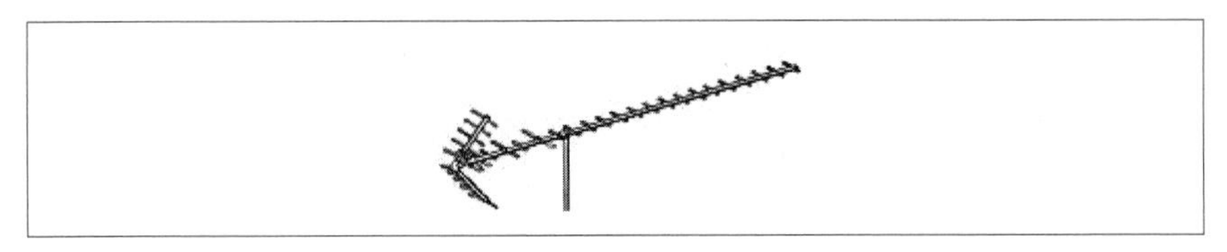

Figure 4-12. Corner reflector on a Yagi antenna

This corner reflector makes up somewhat for the poor performance on the lower-numbered channels. Although the Yagi/Corner-Reflector is not the best antenna, it is the most common UHF TV antenna, mainly because it can be mounted on the front of a VHF antenna without degrading the VHF antenna.

An antenna also has an *aperture area*, from which it captures all incoming radiation. The aperture of a Yagi is round (see Figure 4-13) and its area is proportional to the gain. As the leading elements absorb power, diffraction bends the adjacent rays in toward the antenna.

Comparing antenna types. The graph in Figure 4-14 shows the gain functions for four TV antennas:

- Plot A is the Channel Master 4228 8-Bay, a stacked dipole reflector antenna.
- Plot B is the Channel Master 4248, a Yagi/Corner-Reflector.

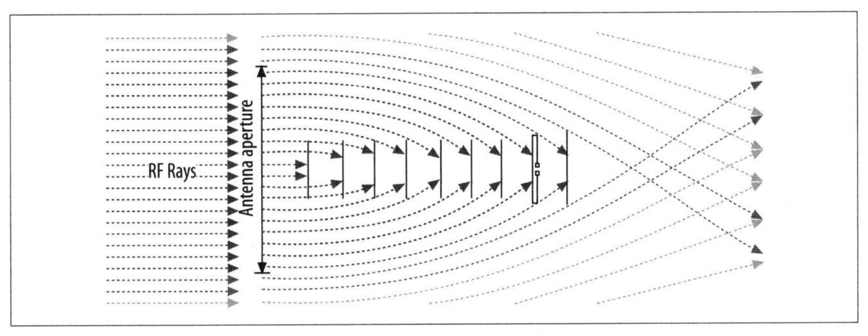

Figure 4-13. Aperture area

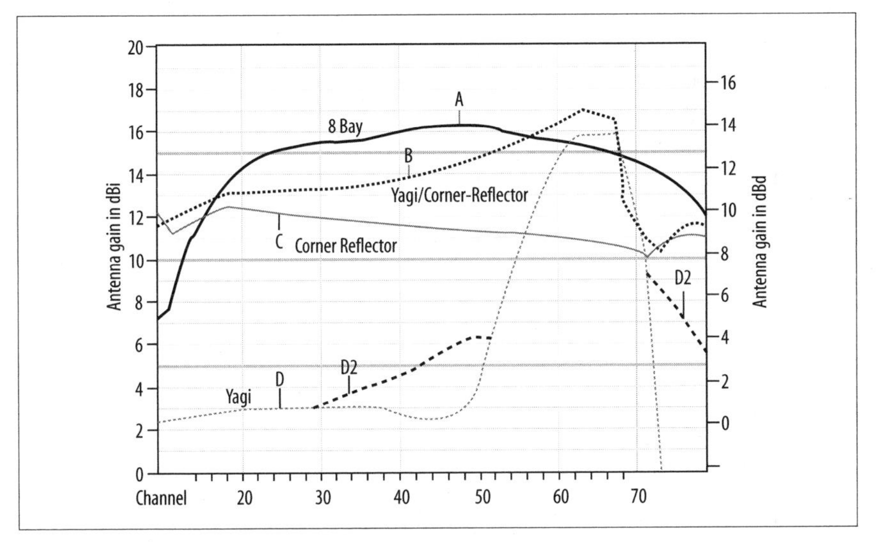

Figure 4-14. Comparing antenna types

- Plot C is the 4248 with all of its directors removed, making it a pure corner reflector antenna.
- Plot D is the 4248 with its corner reflector removed and replaced by a single reflector element, making it a standard Yagi. The D2 plot shows the backward gain where this exceeds the forward gain.

The point of this graph is that a Yagi/Corner-Reflector performs like a Yagi for the high-numbered channels, and like a corner reflector for the low-numbered channels. For the middle channels it outperforms the sum of the two types. Clearly, the 8-Bay and Yagi/Corner-Reflector are the favorites here.

Radiation patterns. The last thing to be aware of is the radiation patterns of these antennas. First, Figure 4-15 is the overhead view of an 8-Bay antenna.

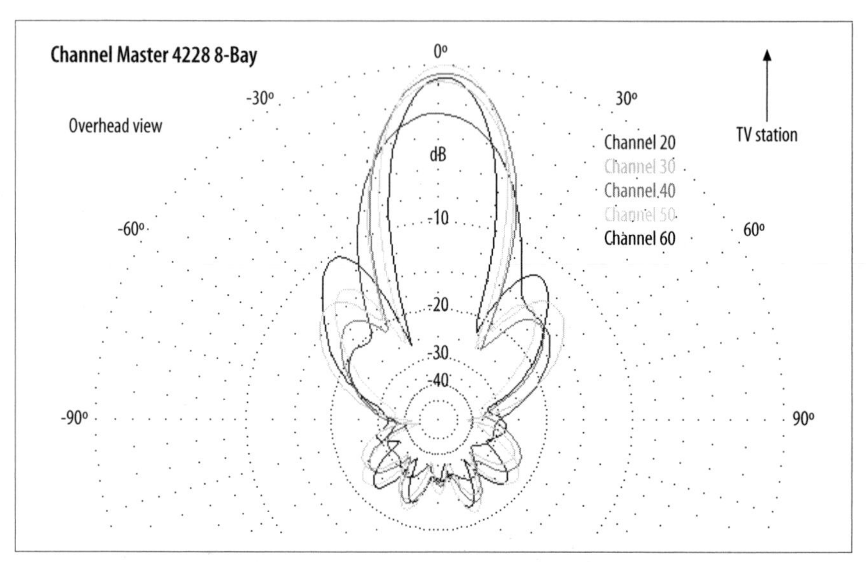

Figure 4-15. Overhead view of 8-Bay antenna

Figure 4-16 is the elevation view of the same antenna.

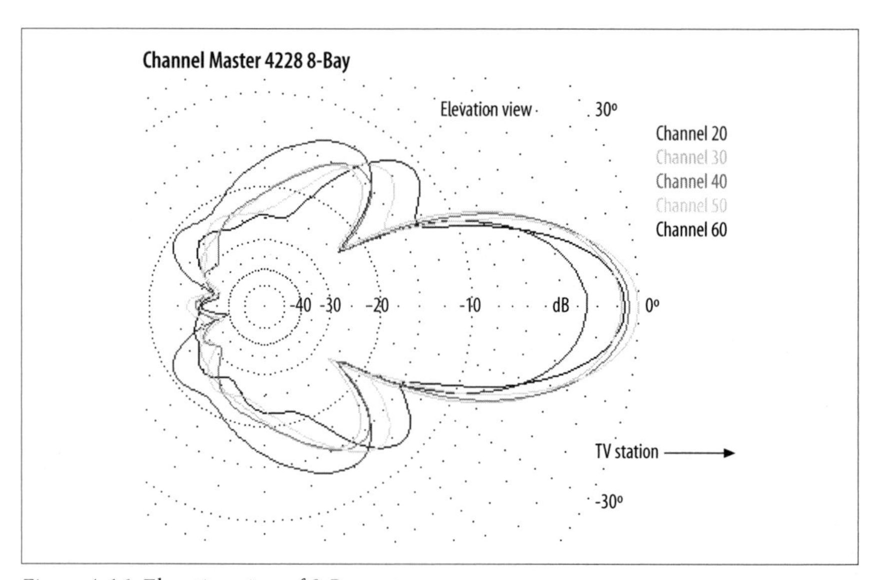

Figure 4-16. Elevation view of 8-Bay antenna

As you can see, the 8-Bay is a very directional antenna. If the aim is off by 5°, you can lose 1dB of signal. If the horizon is more than 5° above horizontal, you should tilt the antenna up to point at the horizon.

The overhead view shows nulls at 30° and 90° to both sides. These can be used to eliminate multipath (ghosts) or interference. You simply rotate the antenna until the offending signal is in one of the nulls.

Compare that to Figure 4-17, the overhead view of a Yagi/Corner-Reflector antenna.

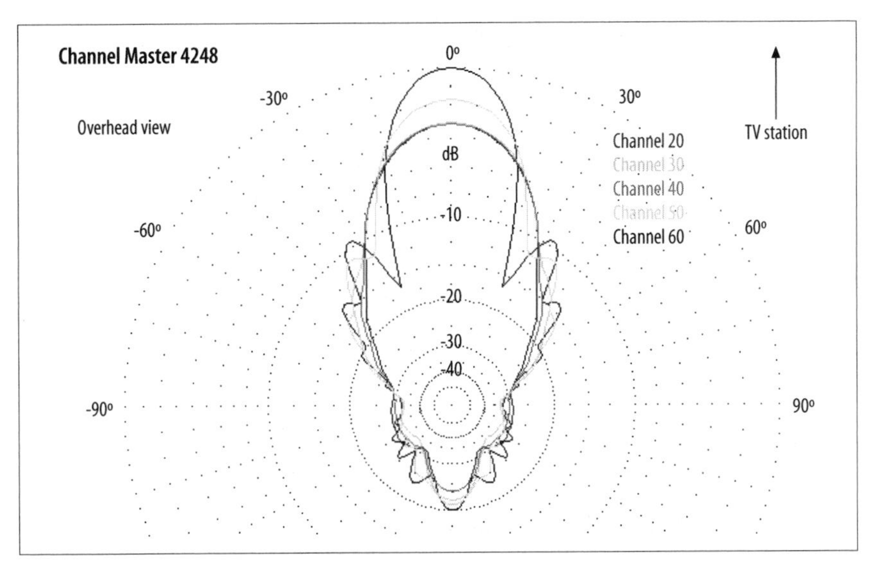

Figure 4-17. Overhead view of Yagi/Corner-Reflector

Figure 4-18 is the aerial view of the Yagi/Corner-Reflector.

A Yagi also has some forward nulls that can be used as ghost killers. But a Yagi/Corner-Reflector acts more like a corner reflector for most channels, and has no nulls. At channel 60 you can finally see the Yagi pattern start to emerge.

I prefer the 8-Bay to the Yagi/Corner-Reflector for the following reasons:

- It has high gain.
- Its gain is evenly distributed over the channels.
- It has nulls that can eliminate multipath.
- It has a rectangular aperture that permits efficient stacking when more than eight bays are necessary **[Hack #81]**.

However, the high gain means it is hard to aim. In good-signal areas, avoid high-gain antennas.

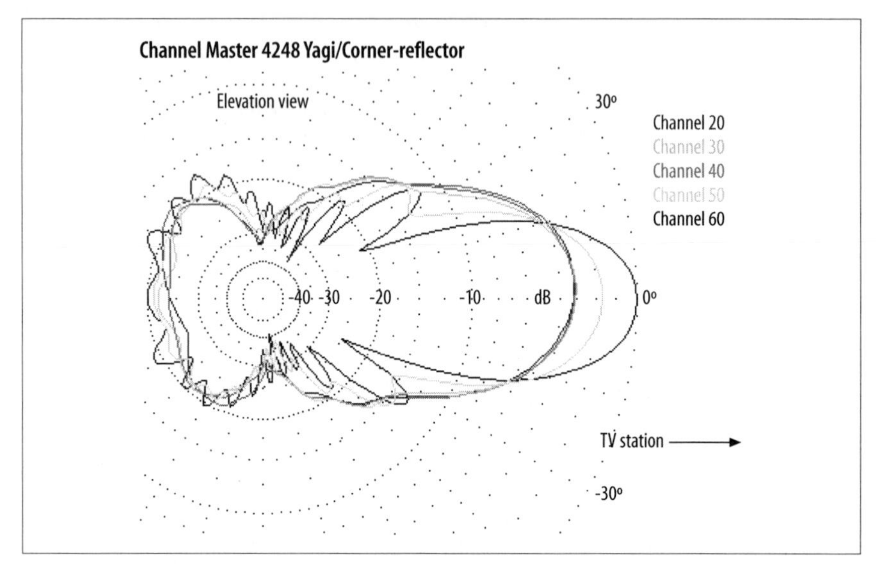

Figure 4-18. Aerial view of Yagi/Corner-Reflector

Commercial Antenna Types

Antenna marketing is a racket in that the less honest you are, the more antennas you sell. Gain figures published by antenna makers are mostly useless, except maybe for comparing antennas by the same maker. The data for all the charts in this section came from computer simulations of the antennas that I performed.

UHF antennas. *Raw gain* is the true gain of an antenna—using the "pure" definition of gain (as seen in Figure 4-19).

However, a fraction of the power is going to be rejected by the transmission line because of an impedance mismatch. This rejected power gets retransmitted. What is left is the *net gain*. Therefore, the graph in Figure 4-20 is the one you should pay the most attention to.

Here's the legend for both graphs:

- A: Channel Master 4228 8-Bay
- B: Channel Master 4221 4-Bay
- C: Channel Master 4248 Yagi/Corner-Reflector
- D: Televes DAT-75 Yagi/Corner-Reflector
- E: Winegard PR-8800 8-Bay
- F: Winegard PR-4400 4-Bay
- G: Channel Master 4242 VHF/UHF Combo

Why Computer Simulations?

A few years ago, *QST*, which is the principal publication of the HAM radio community, announced it would no longer accept advertising for antennas if the ads contained gain figures that were measured experimentally. Henceforth, any such gain figures would have to be the result of computer simulations. There were two big problems with the experimental data:

- The experimental antenna is affected by its surroundings. Computers can do true "free-space" modeling.
- The process of choosing the surroundings encouraged overly favorable choices. Most of us would call it cheating, but they justified it to themselves with the belief that their competitors were doing it.

The program used for these simulations was NEC-2/EZNEC-M by Lewallen (W7EL), which is one of the programs approved by *QST*. Data points were produced for all the even-numbered channels. Some problems were encountered that necessitated approximations, so the data isn't highly accurate. But I stand behind the principal features of the graphs and resulting conclusions. I will make the antenna model files available to anyone who requests them.

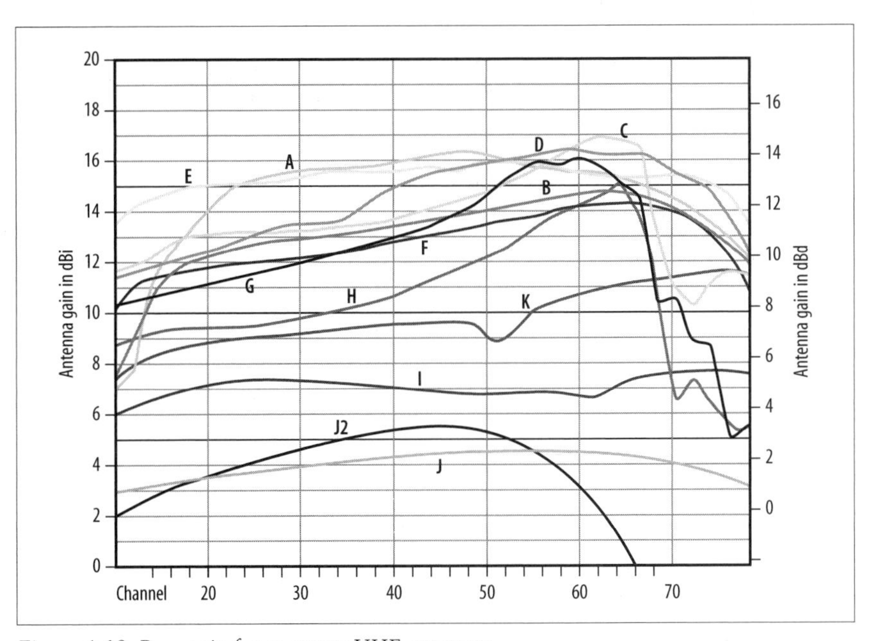

Figure 4-19. Raw gain for common UHF antennas

Raw Gain Still Matters

Why show raw gain when it is net gain that is important?

- If atmospheric noise exceeds receiver noise, the raw gain is what counts. This is rare for VHF or UHF but does occur in some neighborhoods.
- There are cable-matching methods that make the net gain as good as the raw gain for any channel. But although these methods make some channels better, they make other channels worse, and there is seldom an overall improvement. At the time of this writing, no hardware is available that lets consumers improve the match, except for some indoor antennas.
- The program that predicts net gain isn't very accurate. Net gain is affected by minor details in the way the cable attaches to the antenna. The raw gains are very accurate.

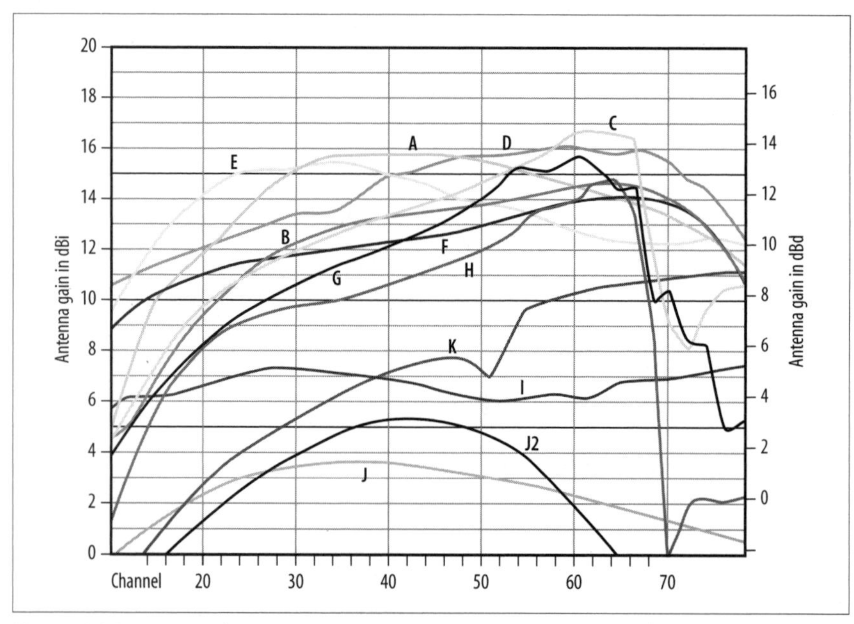

Figure 4-20. Net gain for common UHF antennas

- H: Channel Master 3018 VHF/UHF Combo
- I: Zenith Silver Sensor indoor LPDA
- J: Small indoor loops
- K: Double-Bow

Some of these antennas are UHF-only. Although you might not need VHF presently, you probably will after 2006.

The 4242 and 3018 represent typical Yagi/Corner-Reflector UHF antennas that are part of a VHF/UHF combo. You can estimate any other unknown such antenna from these two. Just find the length of the UHF part of the boom of the unknown antenna (measured from the intersection of the corner planes to the frontmost director). Compare this length to the 4242 and 3018 to estimate where the plot for the unknown lies in the preceding graph. The 4242 has an 87-inch boom; the 3018 has a 57-inch boom.

VHF antennas. Here are the same graphs, but for common VHF antennas. Figure 4-21 shows the raw gain for common VHF antennas.

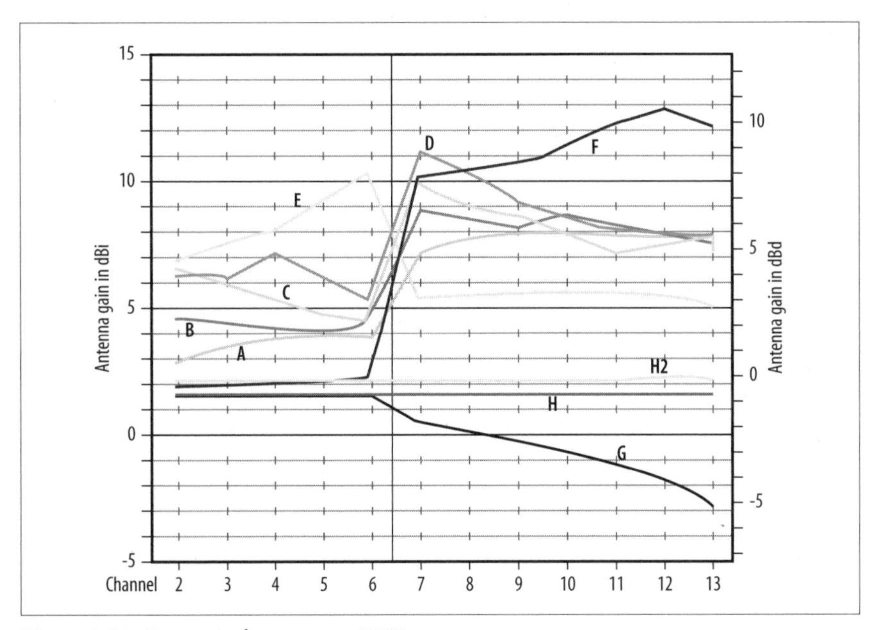

Figure 4-21. Raw gain for common VHF antennas

Figure 4-22 is the net gain for these VHF antennas.

Here's the legend:

- A: RadioShack VU-75XR VHF/UHF combo
- B: RadioShack VU-90XR VHF/UHF combo
- C: RadioShack VU-120XR VHF/UHF combo
- D: RadioShack VU-190XR VHF/UHF combo
- G: Rabbit ears – 40° 45°
- H,H2: Getting the most out of rabbit ears

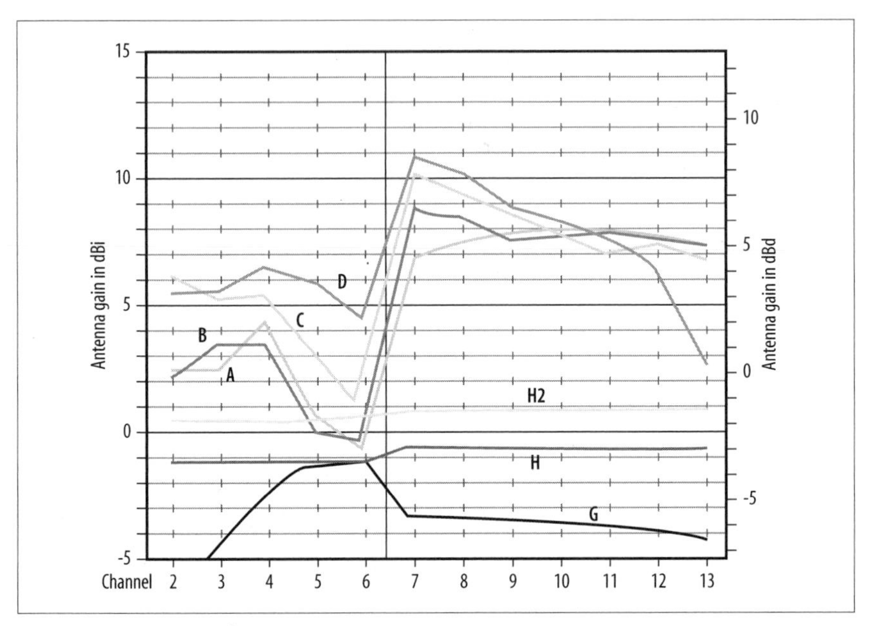

Figure 4-22. Net gain for common VHF antennas

If all the elements are parallel (as in a straight-type LPDA), there will be nulls at +90° and –90° that might be useful for eliminating ghosts and interfering signals.

I realize that's a lot of information, but I'd rather you make an informed decision than just take my word for it (or, even worse, some salesperson's opinion). With these graphs and the other information in this hack, you should easily be able to determine which antenna is best for your house, location, and broadcasting needs.

—*Kenneth L. Nist*

HACK #33 Erect an OTA Antenna

Once you've got an antenna, there's as much knowledge involved in mounting it correctly as there was in selecting i. If you don't hire an installer, you're going to need some basic knowledge to get things right.

Once you've got the right antenna **[Hack #32]**, you've got to put it up. I've separated selection of an antenna from installation, as many folks will allow someone else to take care of installation once the initial purchase has been made. Still, installation and mounting are common tasks, and assuming you understand the possible problems, you're ready to get an antenna set up by yourself.

Proceed at Your Own Risk

There is a chance that the first antenna you install won't meet your expectations. Once an outdoor antenna has been installed, the seller won't take it back; even RadioShack won't take back an installed outdoor antenna. The cost of a second antenna might wipe out any savings you hoped for by doing the job yourself. Further, an installer won't charge you for two antennas if he is wrong on the first try.

If you aren't a do-it-yourself type, you can find an installer in the Yellow Pages under "Antennas" (or possibly "Televisions—Dealers and Services"). The cost ranges from $100 for an easy install to $800 for a difficult install, with $300 being the most typical bill. If you do it yourself, you will pay almost $200 just for the hardware:

- Antenna: $70
- Amplifier: $70
- 50 feet of RG-6: $30

You might be able to get some free advice or a free rough estimate over the phone or by visiting the installer's shop. If he comes to your home, the estimate won't be free, but it will be accurate.

And, it would be irresponsible to not mention the following:

- Every year, people get killed while erecting antennas.
- There are places within the station's broadcast radius where reception isn't possible, no matter how good your installation is.
- There are places where reception is so difficult that the challenge can outwit the installer.
- Although the dollar cost of an antenna system is modest, a lot of your time might be required.

With all that said, if you're willing to be careful, patient, and diligent, it's fun and inexpensive to install an antenna. So, take these warnings seriously, but don't let them scare you off.

Choosing a Mounting Site

Once you've got these terms and concepts down, your first step in erecting an antenna is to choose a site to mount the antenna. There are several considerations; here's the rundown.

Diffraction. *Diffraction* is the ability of a wave to bend around into the shadow formed by an obstruction. It doesn't matter if it is an absorbing or reflecting obstruction. Most OTA viewers depend on diffraction for their reception. The only exceptions are:

- Where the transmitting tower can be seen.
- Sometimes in cities with tall buildings, reflection is more effective than diffraction.

Low frequencies diffract efficiently, but VHF diffracts poorly (see Figure 4-23).

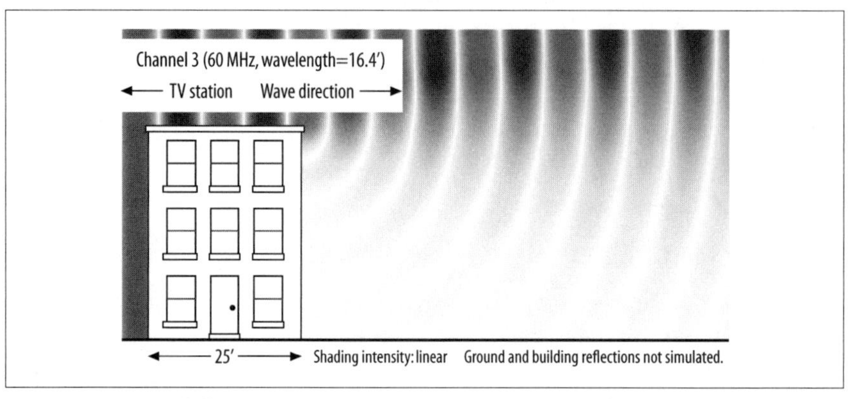

Figure 4-23. VHF diffraction

UHF is another 10 times worse (see Figure 4-24).

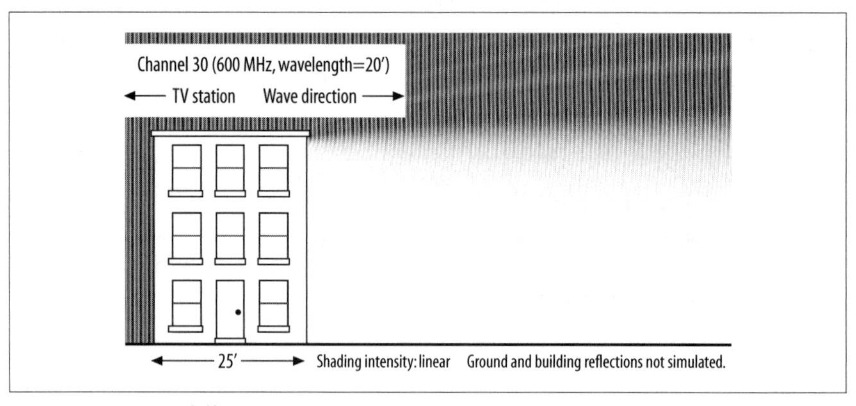

Figure 4-24. UHF diffraction

These diagrams use linear shading and thus are perhaps overly pessimistic. Reception might be possible where these diagrams show no signal. Logarithmic shading would convey more optimism.

To make up for the poor performance of UHF, the FCC allows UHF stations to broadcast a much stronger signal, as Table 4-5 shows.

Table 4-5. TV transmitting power allowed by the FCC

Channels	Flat region	Hilly region
2–6	50 kilowatts	150 kilowatts
7–13	150 kilowatts	500 kilowatts
14–69	500 kilowatts	1.5 megawatts

The numbers in Table 4-5 are approximate. Stations often argue for and get a higher limit. But the goal in most cases is a 60-mile reception radius

Trees. If the antenna is behind a tree, it is in overlapping fields: a weak field that passes through the tree plus a weak field that is diffracted around the tree. Overlapping fields are complicated, with strong spots and weak spots. If you get a UHF antenna to work behind a tree, you likely will see dropouts when the wind blows because the strong and weak spots will move around.

Many people install antennas in the winter and think they were successful. Then in the summer they wonder why the antenna totally quit working. It's the trees!

Is a higher antenna always better? For VHF, higher is always better. An antenna should be four wave lengths above the ground to be unaffected by the ground. For channel 3, this would be 70 feet (see Table 4-2 to check the wave length for each channel). Of course, for most houses, 70 feet is unreachable.

The rules for UHF are a little more complicated than for VHF, though. UHF is affected more by obstructions and less by height. For UHF, four wave lengths is only about seven feet. However, a UHF antenna should be higher than this in the following cases:

- If at all possible, get the antenna above any obstructions.
- If your horizon is less than 200 yards away, raising the antenna makes a significant difference. (You would be a candidate for a tower.)
- As Figure 4-25 showed, even if you can see the transmitting tower, if the signal skims over the top of some obstructions, there is a stronger field just a few feet higher.

You probably will want to attach a VHF antenna to your chimney. That is also likely the best place for a UHF antenna. But if your chimney mount is still obstructed (by trees, etc.), an unobstructed site closer to the ground will work better for UHF. The essential goal is to find a spot where your UHF antenna can see the horizon in the direction of the station (see Figure 4-25).

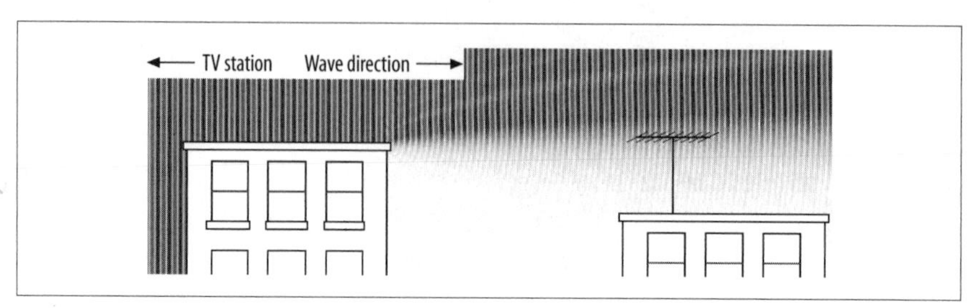

Figure 4-25. Obstructions and their effect on antennas

Keeping the antenna above four wave lengths minimizes the effect of the ground reflection. Actually, the ground-reflected signal could either add to or subtract from the direct signal, so being close to the ground is not always a disadvantage. I've seen a situation where the strongest UHF signal was found 3 feet off the ground.

Adding to the confusion. Diffraction over the horizon ridge often results in overlapping fields. Overlapping fields will result in weak signal spots (cold spots) and strong spots (hot spots), arranged in a regular pattern. For UHF, the hot and cold spots are often 5 to 20 feet apart.

If you are in a neighborhood with overlapping fields, moving your antenna a few feet can make a huge difference in signal strength. The chimney might seem like the perfect site, but if the chimney is in a cold spot it's a mistake.

To make matters worse, the pattern of hot and cold spots are different for different frequencies. You will want to find a spot that is hot for all the channels you want, but such a spot might not exist above your roof. In this case you need an antenna with higher gain than is otherwise recommended.

To make matters even worse, you likely won't discover that you are in such a neighborhood until after you have purchased and installed the antenna. To prove that you have hot and cold spots, you move the antenna (forward and backward, left and right, higher and lower) while keeping it perfectly pointed at the signal and watching the DTV signal strength indicator. It is hard to keep a large antenna pointed correctly while devoting half of your attention to not falling off the roof, but a smaller antenna might not achieve a digital lock.

At this point a professional installer starts to look like the smart choice. But will he stick with it, or will he, too, quickly declare further improvements impossible and walk away? He will hesitate to raise his estimate, but he will not work at a loss.

These problems are all UHF problems. VHF does the same thing, but with hot and cold spots 50 to 200 feet apart they aren't evident and there usually isn't much you can do about them.

Attic antennas. If an indoor antenna isn't as reliable as you want, an attic antenna is the next step up. If you are in a neighborhood with moderately strong signals, an attic antenna might work. But you are wasting your time installing an attic antenna in a poor-signal neighborhood. Most successful attic antennas are within 20 miles of the transmitter.

Thirty miles often works if you are on the crest of a hill.

The problems with attic antennas are:

- The antenna might not be high enough above obstacles outside the house, such as trees.
- It is hard to estimate the signal loss caused by wood and other construction materials.
- Metal objects in the attic can block the signal.

Estimating the signal loss in ordinary construction materials requires knowledge of their water content. Exceptions are aluminum siding, stucco (which has an embedded metal screen), and foil-backed insulation, all of which totally block all signals. Concrete and most bricks have moderate water content, but their thickness is enough to block all signals. In a desert, plywood becomes so dry that it causes no signal loss at all, even for UHF. In any other place, there will be some moisture. Exterior wood is generally always wet inside, especially in north-facing surfaces. The amount of water varies with the weather. Asphalt shingles are mostly transparent to UHF, but the way they overlap encourages water to persist between them. The vapor barrier is often wet on one side or the other. The bottom line is that there is no way to quantify the signal loss in these materials.

Metals reflect signals. A metal object eight inches long is big enough to reflect UHF. Smaller objects, such as nails, are of no concern. Wires and metal pipes effectively reflect VHF, as do plastic pipes containing water. If these reflecting objects are positioned to the side, to the rear, above, or below the antenna, they will have little effect on it, provided they aren't too close. These objects should be further away than two feet for UHF, four feet for VHF-high, or six feet for VHF-low, and an even larger separation can help a little.

You might wonder why these numbers aren't proportional to the wave length. It is because the lower-frequency antennas are lower in gain. An antenna's aperture depends on the gain as well as the wave length.

There should be no horizontal or diagonal wires or pipes in front of the antenna. A perfectly vertical metal vent pipe is invisible to TV signals, but its flashing at the roofline might not be.

—Kenneth L. Nist

HACK #34 Don't Use Portable Signal Strength Meters

When working with an OTA antenna, you'll waste your time by trying to use a portable signal strength meter. Understand why, and avoid this common mistake.

Sometimes readers ask where they can get a portable signal strength meter, thinking this will allow them to make objective studies of the signal from their OTA antenna [Hack #33]. However, a signal strength meter is not as useful as you might think. It will work fairly well for normal standard-definition broadcasts, but the fact that you're here probably means you're looking for HD broadcasts.

For HDTV, a strong meter reading is no guarantee that you have found a good reception spot. The ultimate arbiter of where the signal is good for your receiver is the HD receiver or set top box [Hack #30] itself. Therefore, I recommend that the signal strength readout provided by the receiver—not a meter—be used to search for the best antenna spots. The quality of the result justifies finding ways around the small problem you will discover when you try to do this: your antenna is rarely located where your receiver is! This means you (obviously) can't see the signal readout while mounting the antenna. This is still fairly simple to solve, though. A spouse and a mobile phone, or a buddy with a walkie-talkie is more than sufficient to get around this issue.

This same principle applies for pulling in signal from a satellite dish, such as when installing DirecTV or DISH Network. There are some great gadgets that will help you get things pointed in the right direction, but there's no substitute for checking the receiver to ensure you're getting signal.

Keep in mind that "receiver" in this hack always refers to the cable or satellite receiver, and not an audio receiver. All cable/satellite receivers will have some sort of system menu where you can check the signal strength.

—Kenneth L. Nist

HACK #35 Resolve Problems After Buying an HDTV

There are a handful of tips you can follow and tweaks you can make to your setup after adding HD-capable devices. These are fairly random but all together they can make a huge difference in quality and operability.

No matter how much research you do before buying a TV, you're still almost guaranteed to have a few surprises and problems when you break the set out of its box. This hodgepodge of tips and technical explanations will help you solve the most frequent issues you'll run into.

Bridging Component Video and VGA Connectors

You've just brought home a new unit, and you've discovered that your HDTV requires a VGA cable, while your receiver offers only component video connections (or the other way around). Some manufacturers have anticipated this situation, and have designed their sets to break the rules, in a manner of speaking. That is, some VGA ports might allow syncs on the green wire.

For VGA, Hsync and Vsync are the fourth and fifth wires (remember that VGA connectors have five cables). In component video, these syncs are multiplexed onto the green wire (labeled Y in a Y-Pr-Pb system).

For example, my TV set will accept component video force-fed into its VGA port. There was nothing in the set's instructions about this, but I found a menu item that allowed the set to accept green-wire syncs.

If the unit literature says nothing about this, ask the "expert" at the store where you bought the TV. If the sales staff doesn't know anything about such an option, you can experiment. Adapters that might work are available via the Internet. An alternative to the adapter is to get a VGA cable that has five BNC connectors on one end, and then get three BNC-to-RCA adapters from RadioShack. These cables probably will cost around 50 bucks, but you can often return the cables if things don't work out.

Understanding Subchannels

Let's say the FCC has given the NTSC station on channel 3 permission to use channel 41 as a digital (HD) channel. So, you tune to channel 41, and your new receiver says you are now on channel 3-1. To add to the confusion, you also have discovered there is a channel 3-2. What are these channels and how did they get there?

You've discovered *virtual channels*. A virtual channel is a physical channel with a different name or number. The *physical channel* refers to the actual RF spectrum being used. The virtual channel could be called almost anything. In this example, 3-1 and 3-2 are virtual channels, and also are referred to as *subchannels* of virtual channel 3, which is physical channel 41. And just to confuse matters a little more, these virtual channel 3s have absolutely nothing to do with an analog station that is on the physical channel 3.

The data stream of DTV channel 41 has data blocks called *PSIP data*. The PSIP data tells the receiver that channel 41 has two subchannels: 3-1 and 3-2. The channel 3 people choose these subchannel names to remind you whom you are watching. However, not every station follows the example of this hypothetical channel 3; a different management might choose 41-1 and 41-2 for physical channel 41's subchannels.

Your remote control will let you key in either the physical or the virtual channel number, but there are some differences between manufacturers. For example, some cable/satellite receivers will assume 3-0 means analog channel 3, while others will try to locate a virtual channel 3 for the same key sequence.

Why Can't I Get My Local DTV Station?

Your first days with your new HDTV can be a very confusing and frustrating time, particularly where OTA stations are concerned. You can't tell your receiver that a channel is digital; the receiver has to figure out for itself whether a physical channel is analog or digital. If the antenna is getting a marginal signal or is mis-aimed, the receiver often guesses incorrectly. You then can't aim the antenna because the receiver thinks the channel is analog, and you can't convince the receiver to switch because the antenna is mis-aimed. In strong signal areas, the receiver might eventually right itself. Otherwise, you might have to figure out how to make the receiver unlearn a channel (consult your menus and documentation). Even if you get this fixed, though, you're still going to have a mis-aimed antenna.

Nearly all DTV receivers have a signal strength meter of some type. Most of these meters read zero until the signal is good enough (or almost good enough) for reception. In weak signal areas, these meters won't tell you much about whether you need your antenna to aim more to the right or to the left. When you get no reception, you are left not knowing whether your antenna is just mis-aimed, or if the signal strength is inadequate. As you might guess, this leaves a lot of room for guesswork and can lead to plenty of frustration.

The good news is that once the receiver has learned all the channels correctly, these problems are gone forever. In fact, people in areas with strong signals will never see most of these problems to begin with. All receivers have a *channel learn* sequence, in which the receiver will search for and learn all available OTA channels at once.

When you initiate this learning sequence, some receivers will forget everything they learned previously, which creates problems for users who use a rotor or who switch between two antennas. These users will need to learn how to add channels manually.

Picture Quality

The image quality of an HD picture isn't affected by a low to moderate level of noise in the signal. This is true for both satellite and OTA DTV. Yet some people can't resist wondering whether some antenna tweaking would improve the signal strength and result in a better picture. The answer, without any hedging, is that this sort of tweaking will have no effect on picture quality.

When the signal for a channel becomes too weak to display, you will see *macro-block errors*—parts of the screen will be shifted or out of place—and you will experience sound dropouts lasting a few seconds, as well as image freezes lasting a few seconds. All of these errors are crude, unsubtle errors. If these are not present, your image is perfect. You either get the picture, or you don't.

Once you're at this stage, there is still one reason you might want to try to improve the signal: you might be able to decrease the chance of dropouts in bad conditions, such as heavy rain. Rain can affect DBS and UHF reception, but not VHF. In some places, wind can affect UHF.

If you get sound dropouts but not image dropouts, or vice versa, the fault is not a reception problem. Usually the station is at fault, but occasionally it is the set top box [Hack #30].

Determining Display Resolution

It's often hard to tell if you've got a picture in full-blown 1080i, or if you're seeing 720p [Hack #1], or just plain old 480 (progressive or interlaced). When a TV station decides to provide an HD subchannel, that subchannel is normally 1080i (or 720p) all the time at the transmitter, even if some of the programming originates from NTSC cameras that can't capture HD images. There is no technical requirement for this, but it seems to be nearly universal practice. Thus, your receiver's HD detector is not a reliable indicator of whether the program is actually HD. It will pick up the transmitter's information rather than the format of the source material.

NTSC 4:3 images often have black bars on the side that you might not be able to eliminate because they are part of the 16:9 transmitted image.

So, what is the most reliable way to tell if you are seeing true HD? If the image is 16:9, the image is not stretched, and there are no black bars on the sides, you're almost certainly looking at an HD image. ABC and ESPN typically are sending in 720p, and most other carriers are sending 1080i images.

Waiting on Local Networks to Broadcast in HD

All DTV transmitting equipment can handle HD at no extra cost. The only extra cost associated with passing on HD information is that associated with staffing two sets of transmission procedures. However, the cost of the DTV transition has hit the local stations hard, and some are resisting even this small expense.

Also, many of the newest DTV stations are rural stations. Rural stations often don't have HD taping equipment, let alone a staff that can operate this higher-end gear. So, if you're not in a larger city or television market, you might be stuck pulling in HD from a satellite provider for a while.

Grabbing HD Local Channels Through Satellite Providers

As you've no doubt figured out, most satellite providers give you local channels through their feeds, usually for a nominal cost. It would stand to reason, then, that HD versions of these local channels would soon follow. That turns out not to be the case, though, and probably won't be for several years. An HD channel requires about five times the bandwidth of an SD channel. To convert all their local channels to HD, DirecTV and DISH Network would each have to launch several more satellites.

Both companies are very secretive about their future plans.

Another seemingly good idea would be for satellites to simply pass through local channel HD feeds, supplying consumers with a single source for all programming (the satellite dish, as opposed to requiring both a dish and an OTA antenna [Hack #33]). However, the National Association of Broadcasters lobbies effectively for local stations. As a result, Congress has legislated that these stations continue to enjoy their monopolies, and disallows satellite providers from passing on their feeds. In most cases the satellite operators are forbidden to offer viewers feeds or stations that would compete with the local channels.

The only exception in this area, at least right now, is CBS. CBS-HD is available on DirecTV and DISH Network. But to qualify for this, you probably will need a waiver from your local CBS station.

Getting Rid of Artifacts

In image processing, an *artifact* refers to any predictable flaw in the image that results from shortcuts or shortcomings in the processing technology. In HDTV, most artifacts result from compromises that have to be made when the picture changes too rapidly and requires more than the allowed bandwidth. Sometimes the solution to these bandwidth limitations is to delete frames, while other times the choice is to randomly delete 16x16 macroblocks. There are also a number of common artifacts that result from converting 24 frames/sec films to 30 frames/sec TV broadcasts; this category of artifacts generally is not something related to bandwidth, and probably will be solved as conversion processes improve.

Snow and interference generally aren't called artifacts.

It's also not uncommon for a particular HDTV set to introduce some artifacts of its own. If you see these consistently, on multiple channels, contact your set manufacturer and see if you can get some resolution. You might also be trying to view broadcasts at a higher resolution than your set supports (although this is rarely the case with any but the first-generation HDTV sets).

The Problem with SD Programming

In cases where SD programming is pushed out via an HD broadcast, you can get an almost unheard of phenomenon, at least in HD programming: snow. Remember I mentioned earlier that either you get an HD picture, or you don't. That's true, but SD programming over an HD transmission is a bit of an exception. This is a consequence of the change in the bandwidth between recording and playback. Although the bandwidth of SD is said to be 5 MHz, it is not a sharp cutoff. The roll-off starts before 5 MHz, but some image information above 5 MHz survives the recording process, albeit mixed with electrical noise (which shows up as snow). The broadcaster has to decide if she wants to filter out this snow or leave it in. If the broadcaster decides to filter out the snow, some image information is lost, resulting in increased blurriness. Because this is a trade-off, different broadcasters will make different choices.

There has been a significant improvement in the average quality of NTSC broadcasts over the last three years. There are two reasons:

- True SD resolution (640×480) used to be a target TV production crews believed they had to aim for but didn't really have to meet. Their product often was well below 640×480, for a variety of reasons. But now, with large, high-definition sets becoming common, these production people are seeing how bad their product can look, and are paying more attention to the details (camera focus, circuit noise, cable reflections, filter circuit selection, etc.).
- Many people are watching NTSC broadcasts of shows that were shot with high-definition equipment. These situations always reveal any deficiencies in the SD source material.

—Kenneth L. Nist

Speakers and Wiring

Hacks 36–47

It might surprise you that it took five chapters to actually get to the subject of speakers, and even with that, this chapter begins with organization of your racks and gear before jumping into speaker selection. If you're an old home-theater guru, this might not be that odd. But newbies rarely realize that killer speakers without the right gear are just expensive paperweights. Don't get me wrong: the reverse is also true; high-end gear can't put out great sound on its own. However, if you get your components set up and working well, selecting, installing, and testing speakers becomes far, far easier.

In this chapter, you'll get a handle on how to keep your wiring orderly. That rather simple-sounding task turns out to be daunting, and a constant maintenance issue. (If you think it's not a big deal, wait until the first time something goes wrong and you've got to trace a cable through 20 twisted feet of components.) Then I'll dive into speaker selection and really focus on the surround speakers that are so critical to home theater. Finally, you'll learn more than you ever really wanted to know about wiring, from choosing the right kind to using banana connectors to bi-wiring and bi-amping. Get ready for some sound: this is the chapter that makes it happen.

Organize Your A/V Racks

With hundreds of cables involved in connecting your components, you'll need to be careful and logical in placing each audio or video unit on shelves or racks. You also should take care to label and group cables and components.

When it comes to setting up your home theater, you'll find that you'll spend as many hours connecting cables and wires as you spend tweaking the picture and adjusting the sound. And, as you'll always find some new piece of killer gear later on, this isn't a one-time operation; you'll be back in that morass of cables again and again. Taking some time to figure out where best to place your components will save you frustration, hassle, and wasted dollars on additional, longer cables down the line. From there, you can use labels, a bit of Velcro, and banana plugs to take the mess that is your theater setup and turn it into an organized, easily maintained audio/video system.

Component Placement

As a general rule of thumb, put your amplifier on your very lowest shelf. It's an incredibly heavy unit, and you don't need it breaking through and crashing on top of another unit. Place your audio processor or receiver on a shelf just above the amplifier, for the same reason. Additionally, you'll have a lot of connections between your processor and amplifier, and this will allow for shorter cables to be used. If you have a receiver instead of an amplifier/processor combination, just place it on the lowest shelf you've got.

Moving upward, add other devices that you rarely touch. If you have a video scaler or equalization unit, for example, these should fit nicely just above your processor. Because you probably won't fiddle with them much, they can sit lower on the shelf and not be a bother. Continuing to move upward, begin to place devices that you do touch a lot—your VCR, your DVD player, and any game systems that you have.

Because it's easy to get caught up in the heat of a game of *Halo*, you might want to add some sticky tape to the bottom of your game consoles to avoid them flying out of the shelf when you yank on the controller to 180 your Warthog.

Although this might seem a bit pedantic, you get some significant advantages from this layout.

Speaker wires flow outward
: Your speaker wires aren't going to be hanging over components; they're coming out of the amplifier or receiver on the lowest shelf. This means you'll never accidentally yank that left-rear surround wire and miss out on a cool effect.

Less chance of crosstalk
: Your speaker wires are now moving toward the floor, while your interconnects are flowing upward from the processor. This avoids any possible crosstalk between the two. Although crosstalk is less common in today's nicer cables, it's still something to watch out for.

Less connection strain
: If you've ever cut your speaker wires too short, you know what can happen when strain is put on connections. Either the signal begins to degrade, or the wire pops completely out of the terminal post and

you've got to dig in and reconnect it. With your speaker wires dropping straight to the floor, presumably less than a foot away from the binding posts, this problem is nicely avoided.

Rack stabilization
: Heavy units such as your amplifier and processor anchor your rack, and ensure it doesn't move or shift.

Increased ventilation
: Your receiver/processor/amplifier now has more vent space above it. Although some will tell you that ventilation is improved with height, this is a misconception. Heat will dissipate better if it can "see" empty space above it.

You might even want to consider raising the shelf above the hottest unit (your amplifier or receiver, generally) so that there's even more room to ventilate.

Better ergonomics
: If you've got devices you use at eye level, you won't have to bend down just to put a DVD in the player. And when you're rewiring, your home theater brain (the processor or receiver) is low on the ground, where you'll be lying down with pliers and zip ties in hand anyway!

Label Your Components

Labeling is the only thing that will save you when you have to rearrange components, swap out a receiver, or reset a satellite receiver. Although you can buy little label makers for 40 or 50 bucks at most office stores, you're just as well off getting some white paper and Scotch tape. Write the name of the label on a scrap of paper, wrap the paper around the cable you're labeling, and then wrap the paper and cable with Scotch tape. This will ensure you can read the label, and that it stays where you put it. If you do purchase a label maker, you still might want to wrap the labels in Scotch tape, as the label adhesive typically is not cut out for sticking to curved surfaces (such as cables).

Begin labeling with your power cords. Create two labels with an "A" on each, and then attach a label on each end of a power cord. Apply these labels about 3 inches from the end of the cable; this allows room for the cable to plug in, and the label to still be visible behind your component racks or shelves. Create additional labels for each power cable, labeled "B," "C," "D," and so forth. Finally, hook up all your power cables, and wrap them together with a zip tie, or even better, Velcro or split-loom tubing. These make it easy to add or remove cords from the bundle.

Repeat this process for each type of cable. In other words, gather all your S-Video cables, and start again with "A." Do the same for coaxial cables, component cables, and so forth. Connect all of these to your components as well. Also consider bundling cables by type rather than device. A good organization is to bundle your power cords in one group, your speaker wires in another, and interconnects (RCA cables, S-Video cables, component cables, etc.) in a final group. This makes it easier to locate a specific type of cable quickly.

At each stage, ensure that your labels are visible from behind your system (or from wherever you access your components). This might involve relabeling cables, and moving the labels further away from the cable endpoints.

Many folks will advise you to label the cables based on their usage: "DVD Coax" and "VCR 1 In." This is usually a waste of time, and often is more trouble than it's worth. When you upgrade your DVD player to a unit that supports coaxial output instead of optical output, you're going to have to throw away that "DVD Optical" label. In many cases, this creates a trickle-down effect, as you then use that optical on another component (and have to relabel it again), remove a cable from that component (there's another label wasted), and then go on down the line. With simple letter labels, this is never a problem.

Use Banana Plugs for Connectors

It's a nightmare to thread thick speaker wires into the tight cluster of binding posts on most receivers. Not only is it awkward, but also, you cannot leave any tiny strands of copper sticking out. Stray threads can cause short-circuits, which will overheat your receiver and cause damage, or reduce the lifetime of your amplifier. Wiring is one of those places where neatness counts.

I recommend the dual banana plugs from RadioShack, part #278-308, available online at:

`http://www.radioshack.com/product.asp?catalog%5Fname=CTLG&product%5Fid=278-308`

These units have a solid black spacer between connectors, which ensures the wires don't touch. They also have a nice oversize hold, allowing for a 12-gauge wire to fit through easily. Finally, there is a knob for tightening, meaning you can get a solid, tight fit for a great mechanical connection.

However, the spacing of the binding posts on these is nonstandard. Before buying 10 or 15, try out one pair with your amplifier or receiver to ensure it fits properly. These units also add 3 inches of depth, which might prevent your components from fitting into your cabinets. Consider all these factors before committing to the dual banana plugs.

If these don't serve your needs, consider single plugs from RadioShack, part #278-306, available online at:

http://www.radioshack.com/product.asp?catalog%5Fname=CTLG&product%5Fid=278-306

These also are easy to use: unscrew the back, insert wire into the red section, and fold the cooper strands over the lip of the plug. Screw the banana back down, and you're all set.

Finally, don't be swayed by the naysayers of banana plugs. There are some "golden-eared" reviewers who swear they can hear the difference between speakers wired with banana plugs and those wired without. These same folks also will insist that your speaker wires all be the same length [Hack #45], which is also largely worthless advice. Banana plugs provide ease of use with no perceptible sonic change; Figure 5-1 shows some cables with banana connectors, all soldered together and ready for use.

Figure 5-1. Banana connectors on speaker cables

Use Velcro for Cable Management

There are lots of good solutions for cable bundling. The most popular are probably zip ties, which you can buy in bulk for about $5 at good hardware stores. Zip ties provide an easy way to bundle cables, however, they can't be undone and redone. This means you're going to spend a lot of time cutting ties and rebundling when you change, add, or remove cables. Even worse, zip ties (and bread-wrapper twist ties) sometimes can cut into the rubber casing of your cables.

A better solution is to visit your local fabric store and buy two-inch-wide Velcro. You can buy Velcro in a prepackaged form, but you'd be better off just buying it by the foot. Buy plenty: it's too inexpensive to count pennies over an extra foot or two. Then you can then organize your cables into three groups: power cables, speaker cables, and interconnects. Cut the Velcro into two- or three-inch strips, and you should be all set. It's also a good idea to wrap cables loosely, again you don't want to cut into your cables' casings.

—*Brett McLaughlin and Robert McElfresh*

Get the Right Speakers for the Job

Speakers are the most obvious features of any home theater, and they can turn even a lousy movie into an in-depth, surround sound experience. Learn how to choose the best speakers for your room.

Speakers will stand out in any theater, even if you've got the coolest HD set with all the toys. Ultimately, home theater means surround sound, and it's your speakers that will bear the task of pushing out this sound. So, take some time to understand the technical issues, and then go forth and buy!

Understand Speaker Crossovers

A crossover is probably the most important part of a speaker. A bad crossover design can cause a speaker that uses the best drivers in the world to sound like junk. An extremely well executed crossover network, however, can enable some very inexpensive drivers to sound excellent. The intricacies of crossover design are way over my head (and certainly not all that helpful in a newbie document), but a basic understanding of what a crossover is and what it does can help immensely in understanding home theater and, in particular, subwoofers.

Crossover basics. In a loudspeaker (such as your home stereo speakers) you almost always have more than one driver. In a "two-way" speaker system, you have a tweeter (usually 0.75 inch or 1 inch in diameter) and a woofer (usually 4 to 8 inches in diameter). The tweeter covers the higher-frequency sounds while the mid/woofer covers the lower frequencies.

The crossover is what makes the tweeter get only high frequencies, and the mid/woofer get only low frequencies.

A *crossover* consists of two filters: a high-pass (HP) filter and a low-pass (LP) filter. When you combine the two, you have a crossover. An HP filter allows higher frequencies to pass through it, while attenuating lower frequencies, and is therefore connected to the tweeter.

This should make sense: a high-pass filter allows only high frequencies to pass through.

An LP filter allows lower frequencies to pass through it, while filtering away higher frequencies, and is therefore connected to the mid/woofer.

Digging into the technical details. To understand the finer points of a crossover, you need to know what an *octave* is. An octave, at its simplest, is a doubling of audio frequency. So, when you see someone complain about subwoofers not being able to play the first octave, they usually mean 16 Hz to 32 Hz. The next octave is 32 Hz to 64 Hz, the next is 64 Hz to 128 Hz, etc.

The order of the crossover is how steep the slope is; in other words, it is a measure of how quickly the crossover filters away audio. A *first-order crossover* filters the signal gradually as you move away from the crossover point, while a *fourth-order crossover* filters much more drastically.

For anyone interested in the specifics, a first-order filter attenuates the input signal 6dB/octave. A second-order filter attenuates at 12dB/octave, a third-order filter at 18dB/octave, a fourth-order filter at 24dB/octave, and so on.

It's very important to understand that the chosen crossover point frequency mentioned earlier isn't a brick-wall divider. Let's say the chosen crossover point is 2,000 Hz. This doesn't mean that the tweeter plays the frequencies from 2,001 Hz and up and the mid/woofer plays the frequencies below 2,000 Hz. What it does mean is that below 2,000 Hz, the HP filter starts to filter off the lower frequencies at a specific rate. The farther below 2,000 Hz you go, the more the signal is filtered away. The LP filter is the reverse. Above 2,000 Hz, the LP filter starts to attenuate the high-frequency signal at a specific rate. The farther above 2,000 Hz you go, the more the signal is attenuated.

The neat part is that, given how the dB scale works, when you have a tweeter with an HP filter and a mid/woofer with a comparable LP filter, the frequencies where the filters overlap (frequencies that both drivers are playing) will be filtered in such a way that you get an even level across the entire frequency range the two drivers are capable of!

So, that's a two-way speaker. A three-way speaker just has two crossovers. It has an HP attached to a tweeter. An HP and an LP are attached to the midrange driver and an LP is attached to the woofer. There are also four-way speakers, and I'm sure some fool somewhere has designed higher-way speakers.

You might also have seen speakers listed as 2 1/2-way. This means you have a two-way speaker with an HP attached to a tweeter and an LP attached to a mid/woofer. But you also have a second mid/woofer—or just a woofer—with another LP filter. This second LP filter starts to attenuate higher frequencies at much lower frequency than the first LP filter.

Your receiver also has a crossover in it over which you have some configuration options. The purpose of this crossover is to do the exact same thing a normal two-way speaker's crossover does between the tweeter and mid/woofer. The receiver's crossover just does it between your subwoofer and the other speakers.

Choose the Speakers with the Best Music Playback

Many of you might be balking at this hack already; isn't this a book about home theater? Yes, but this advice is still warranted. When choosing speakers, it's common to be able to quickly narrow down your speaker choices to just a few brands, and sometimes even to just a few specific models within the same brand. However, decisions at this level become harder to make, and too often, price becomes the only factor. Although price is important, add the musical listening experience to your thought process.

A DVD movie soundtrack is very empty and is highly compressed [Hack #26], at least compared to a typical music CD. Music is a much harder job for a speaker to reproduce; a few minutes with music will show you things about a speaker that a movie will not. This is one of the reasons there are many budget home theater systems that sound really good: speakers that have to serve in just a home theater environment don't have to be highly accurate audiophile-grade units.

Along these lines, be sure to choose a good *two-channel* music CD. SACD (Super Audio CD) and DVD-Audio both provide multichannel music, and are great secondary choices for musical auditions of speakers. However, they still are going to focus sound in all speakers, and you want to really test those front two speakers (the front left and front right), as they will bear the brunt of the load in all your music and movie applications.

Five Mini-Speakers Trump Two Towers

We all love big speakers. Our fathers had them if we were lucky, and we grew up with the idea that a good music system had to include large speakers that resemble the monoliths from *2001: A Space Odyssey*. But these music systems tried to fill the corners of several adjoining rooms with sound. A home theater speaker system has a very different mission.

A good home theater tries to surround a few chairs with a circle of speakers. You don't care what it sounds like outside the circle, and you really don't want the sound to go into the next room (waking the kids and ruining your evening). Big tower speakers can be a part of the speaker array, but they sometimes are overkill for even medium-size rooms. They take up the most room, are the most expensive, and if not properly matched with the rest of your speakers, can actually worsen the home theater experience.

Try to obtain five identical speakers in your home theater so that as special effects jump from speaker to speaker, those sounds don't change tone and break the illusion of movement. It is a lot easier to set up monitor-style speakers; you are guaranteed a tone match, they have a higher SAF (Spousal Acceptance Factor), and they are usually about half the price of their taller siblings with identical inner workings.

Monitor speakers have another advantage: they don't have woofers. The woofers in large speakers take up a lot of power. When you position the circle of speakers around your room, the locations are almost guaranteed to be bad for the low-frequency sounds that the woofer produces. With five monitor-style speakers, you are forced to add a self-powered subwoofer to your system. This external sub can now be put in a better location in the room for low-frequency sounds without disturbing the other speakers. Although towers are great if you have a big room and lots of bucks, consider getting smaller, matched speakers for most cases.

The Importance of Brand Matching

If at all possible, you should buy all of your speakers (front left, front right, center, surrounds, and rears) from the same manufacturer. This will ensure that they are *tone-matched*, allowing sound to move evenly and seamlessly from one to the other. Additionally, many manufacturers provide specific lines that go together; these speakers will work even better than mixing and matching speakers within different lines from the same manufacturer. Better yet, you often can get deals on buying a complete matched set at the same time.

In cases where you don't have the budget to buy all at once, consider getting the front three speakers (front left, front right, and center) at the same time, in a matched set. Even if the surrounds and rears for that set change later, you've got the front of your theater, which drives most of the sound, perfectly matched.

However, subwoofers are an exception to this rule. A subwoofer produces a lot of *indirect sound;* this means the sound bounces off your walls, floors, and ceiling before it hits your ears. Human hearing is very poor at subwoofer frequencies. In fact, many of the better subwoofer manufacturers don't even build speakers, and instead focus on just subwoofers. Don't get the idea that you need to buy matching speakers and subwoofers. Sometimes they are sold in sets, but that's just to make it easier on some people. Concentrate on matching your speakers, and then pick the subwoofer you like the best, regardless of brand or manufacturer.

—Brett McLaughlin and Robert McElfresh

Select the Perfect Rear and Side Speakers

The primary decision in choosing a rear or side speaker is selecting a monopole, bipole, or dipole unit. Learn the difference, and when each is appropriate.

Although the side and rear speakers generally are used only for effects, and then only a small portion of the time, they make a tremendous difference in the overall feel of a theater's soundscape. There's nothing as impressive as hearing a shuttle fly overhead, as well as seeing it move across the screen. The type of speaker—monopole, dipole, or bipole—determines much of the quality of these effects.

Monopole Speakers

A speaker that fires sound in a single direction is called a *monopole* speaker. The most conservative (and arguably the best choice) for rear and side speakers are monopoles. Rear- and side-channel effects are usually simple, direct, and best served by being fired straight out.

Dipole Speakers

Another speaker that fires sound in a single direction is called a *dipole* speaker. In the days of stereo videotape, the engineers at Dolby Labs figured out how to take sounds and direct some to the front speakers, while sending others to rear speakers. This eventually evolved into Dolby ProLogic decoding. Although this was a tremendous leap forward, there were some serious drawbacks:

- Sound came out of both rear speakers, instead of being directed to just one rear or the other.
- Sound couldn't play below 150 Hz.
- Rear sounds were limited to ambient noise, as engineers could not count on people having rear speakers in the typical listening setup.

Dolby eventually decided that rear sounds, especially as most were ambient noise, should be vague and nonlocatable. The engineers recommended turning the rear speakers away from the listener, allowing sound to be bounced off of walls for an indirect listening effect. Alternatively, one part of the speaker should fire at the listener and another part should fire the same sound in a different direction, to "hide" the location of the sound (although this would not technically be a dipole).

These effects all became the domains of dipole speakers, which can effectively hide sound location. However, as Dolby engineers (and listening environments) advanced, the ability to direct specific sounds to only one rear speaker became possible. Additionally, listeners were assumed to have rear speakers, and ambient noise was no longer the only sound sent to rear channels. As a result, dipole speakers went the way of ProLogic—to the big pile of technology labeled obsolete.

There is another category of dipole speakers represented by high-end acoustic speaker companies. These speakers are ideal for music reproduction and are anything but obsolete. Before making fun of your buddy for blowing four grand on a set of dipoles, make sure what category of speaker he's talking about!

Bipole Speakers

A bipole speaker fires out sound from both the front and back of the speaker. If you can create enough delay time for the back-firing sound to hit your ears with the front-firing sound, your ears will be fooled into believing the sound is much further away than it really is.

In the psycho-acoustic world, this is known as the *Hass Effect*.

Several good companies are producing bipole front speakers. Some also are now producing bipoles for rear and side speakers. However, these are ideal only if you can locate your rear channels three or four feet away from any walls. As this is impractical in most home theaters, bipoles are best left for the front channels.

—*Robert McElfresh*

HACK #39 Little Speakers Can Create Big Problems

Speaker manufacturers are enticing more and more consumers to go with small, mini-, or micro-speakers for even the front and center channels. Although these speakers might look great, they just can't do what the large speakers can do.

These days, many home theater speaker systems feature very small speakers for the front, center, and surround channels. This has the advantage of allowing you to fit a home theater into a closet, if you should happen to use one as your viewing room. Such small speakers also are very unobtrusive and can be more acceptable to other family members. However, if you check carefully into the specifications for what is actually tucked inside those small speaker enclosures, you might find only one small (like two inches!) speaker.

Although that might not sound like a big deal, remember that one small speaker is being asked to cover all the frequencies from where the speaker system's subwoofer stops, on up to the very top frequencies delivered by your system. Due to their small physical size, these small speakers will not do a good a job of delivering real power at midrange frequencies, let alone the extreme edges of its range (the lowest lows the speaker is asked to handle, and the highest highs).

Some manufacturers design these little speakers with "long throw" capability, which will help them move a little more air for their size. Still, the best analogy I can think of is comparing a big, powerful, eight-cylinder car doing 90 mph to a little, four-cylinder economy car trying to go the same speed. The automobile with the big engine will be cruising along, while the little car will be thrashing as hard as it can just to keep up! Sure, the Volvo looks good, but it's just not going to get you where you're going as fast as the car with the big engine. The same is certainly a good (albeit imperfect) analogy for many of the micro-speaker sets that are in vogue these days.

In addition, bigger speakers usually have at least one tweeter while smaller speakers don't. Tweeters are specifically designed to perform well at the high frequencies, and easily outperform small speakers trying to be all things to all people. So, now you've lost some high-end quality and some midrange quality. Guess what's next on the "not as good" list. Yup: the lows don't sound as strong, either.

The home theater speaker manufacturer might try to make up for the lack of low end in the little speakers by designing a "matched" subwoofer to work at frequencies up to 120 Hz, 150 Hz, or even higher! This subwoofer essentially becomes required buying if you go with the smaller speaker set. If you check the frequency range of such subwoofers, though, you might find that although they can play some higher frequencies, they end up bottoming out around 40 or 50 Hz, which doesn't make for much deep bass. Besides, who wants the main sounds of a movie to come out of the subwoofer in the first place?

Further, another difficulty arises from allowing the subwoofer to handle frequencies above 80 Hz. As you move upward in frequency from 80 Hz, it becomes easier to tell where a sound is coming from. Below 80 Hz, it is much harder to tell where a sound comes from. This allows a subwoofer that doesn't produce frequencies above 80 Hz to be located where it is convenient [Hack #50], or where it makes for the best bass [Hack #52]. This is why Lucasfilm's THX standard uses 80 Hz as the crossover point between the subwoofer and the rest of the speakers in the system. It allows for varied placement of the sub, and it lets the sub focus on what it should do best: produce bass.

If your system uses a crossover frequency much higher than 80 Hz, make every effort to put the subwoofer right near the television, so the illusion that the sound is coming from the screen is maintained even at the higher subwoofer frequencies. If the home theater speaker set manufacturer is careless, or cares more about delivering deep bass, the subwoofer supplied might not reach "up to" the frequency where the little speakers start to be able to produce a useful signal. The result is a hole between the top end of the subwoofer frequency range and the bottom end of the little speaker's frequency range. Low male voices will sound thin, and music with low to midrange frequencies also will be weak.

Most likely, this hole will lie between 100 and 150 Hz.

Because our hearing is so adaptable, many folks can listen to home theater speaker systems with small speakers and not know what they're missing. Before you purchase a home theater system using small speakers, listen to the same movie or music on a good home theater speaker set where each center and front speaker has at least a four-inch diameter midrange speaker—in addition to at least one tweeter. Then compare that sound to what you hear on the small speakers. This way you will be aware of all the trade-offs in using small speakers.

—David Gibbons

HACK #40 Add Bass Shakers to Feel the Lows

Everyone's home theater has bass, but with bass shakers, you can ensure that your listeners not only hear the bass, but also feel it.

Bass shakers could very well be one of the best "bang for the buck" additions to any home theater. Yes, a 5.1 home theater audio system gives you the full range of aural dimension but adding these bass shakers, more formally called *tactile transducers*, can bring a new and exciting dimension of feeling to the sights and sounds of your home theater.

Not only will you see and hear action in whatever you are watching, but you will also feel it! Bass shakers allow you to experience every thump, thrust, and shake...the way bass sound was originally intended.

What Is a Bass Shaker?

As already mentioned, a bass shaker is a tactile transducer. Tactile refers to touch, and a *transducer* is simply a device that converts energy of one form (sound) into another (motion). So, a tactile transducer converts audio into something that you not only hear, but also feel.

In the case of a bass shaker, I'm talking about an electromechanical device that shakes (yes, it literally shakes). It's similar to a loudspeaker woofer driver, but without the cone. The bass shaker is connected to an audio amplifier and mounted to a solid object such as your sofa, loveseat, wall, or even floor (see Figure 5-2). When the low-frequency signals from your home theater are fed to it, the vibration is transmitted to the object it's mounted to, hence the tactile sensation.

Buying a Shaker

Several manufacturers make bass shakers. The following list details the most popular brands:

- Aura Systems (*http://www.sensaphonics.com/aura.html*)
- RBH Sound (*http://www.rbhsound.com*)
- Rolen Star (*http://www.invisiblestereo.com*)
- Clark Synthesis (*http://www.clarksynthesis.com*)
- Buttkicker (*http://www.thebuttkicker.com/*)

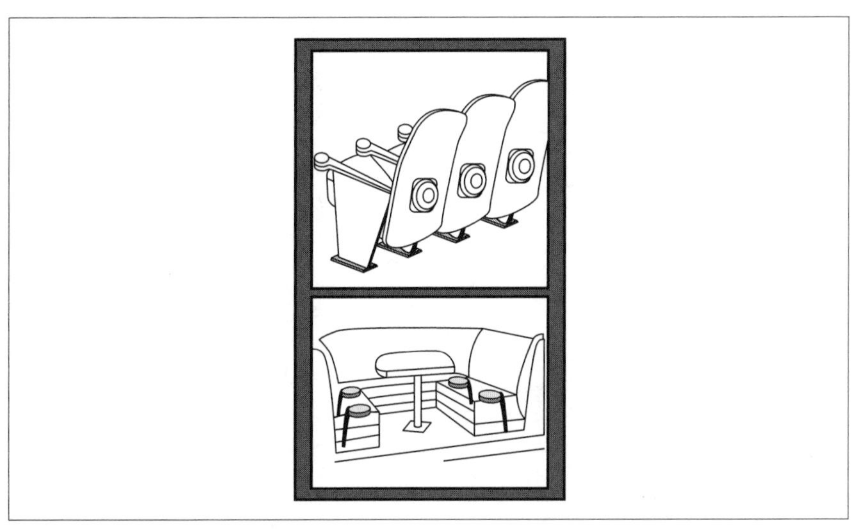

Figure 5-2. Bass shakers on seat backs or in seat cushions

The Auras are the least expensive (and a great value); the Buttkickers usually are rated as the best, but also are the most expensive.

> If you love this idea but just can't afford the Aura model, you might want to consider using a sub to produce tactile bass sensations **[Hack #42]**.

Sample Installation: Aura Systems

The Aura Systems bass shakers look like Figure 5-3.

Figure 5-3. Aura Systems bass shakers

When you get your kit, you'll have the parts shown in Figure 5-4.

Installation is pretty straightforward:

1. Turn your sofa or loveseat upside down. Carefully remove the fabric cover on the back or bottom. Find two places on the sofa frame that allow for some resonance and mounting of the shakers (D).

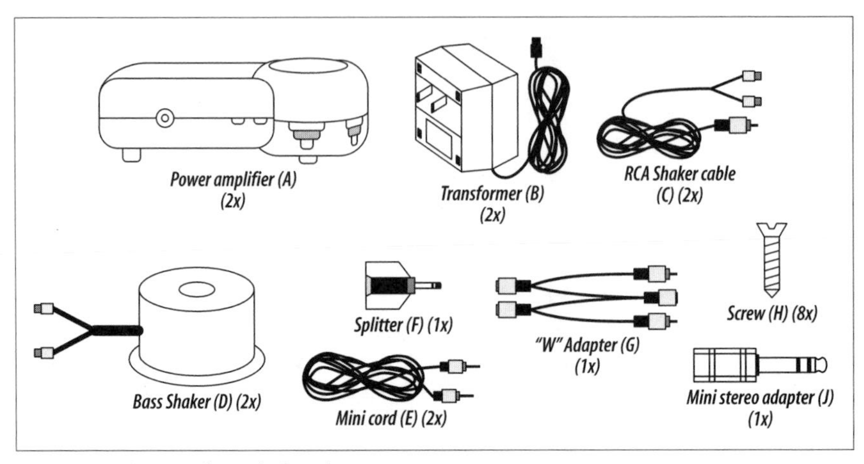

Figure 5-4. Parts in bass shakers kit

2. Mount the shakers (D) from the inside of the frame, as shown in Figure 5-5. Use the provided Phillips screws (H).

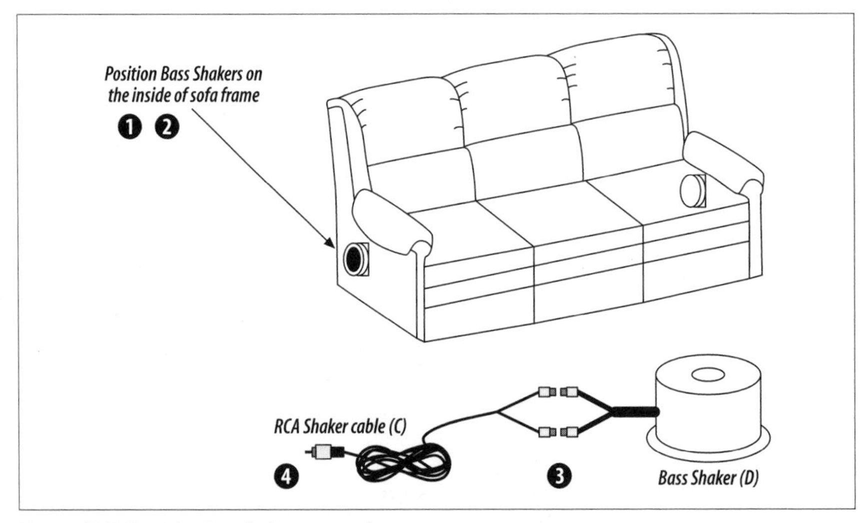

Figure 5-5. Positioning shakers on sofa

3. Attach the RCA cable lugs (C) to the shakers (D) and lead the other end of the cables out of the back of the couch before reclosing the cover fabric with new staples (see Figure 5-6).
4. Plug the RCA connector (C) into the OUT jack of the amplifier (A) and the transformer (B) connector into the POWER PACK jack of the amplifier (A), as shown in Figure 5-6.

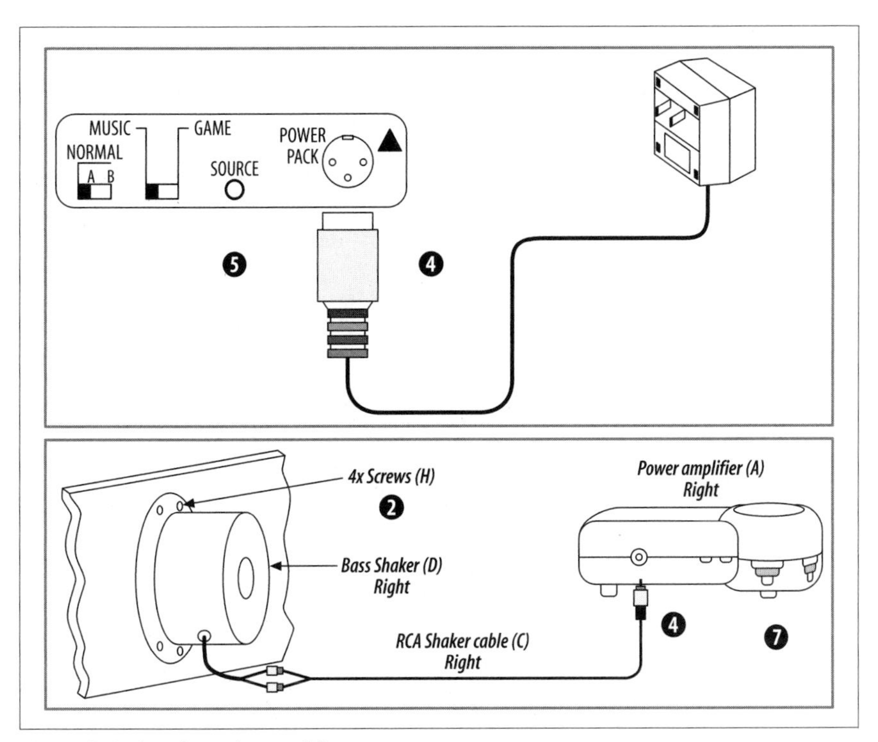

Figure 5-6. Attaching the amplifier

5. Plug one end of the first 3.5mm mini cord (E) into the SOURCE jack of the amplifier (A) and the opposite end into the splitter jack (F), which you then must into the second amplifier (A). Attach the second mini cord (E) to the music source and to the splitter (F).
6. Plug the amplifier transformer (B) into the nearest AC power outlet.
7. Turn on the amplifier (A) by setting the VOLUME and FILTER levels to your desired intensity of bass vibrations. Level 5 is suggested. For bass only, set the filter level to 10.

For optimum performance and volume, the green LED should light (signal input indicator), with occasional flashes of the red LED (clipping indicator).

Figure 5-7 shows a single-sofa installation.

Figure 5-8 shows how to set up an installation with a sofa and two loveseats; this is a bit more complicated, but really gives a nice group effect.

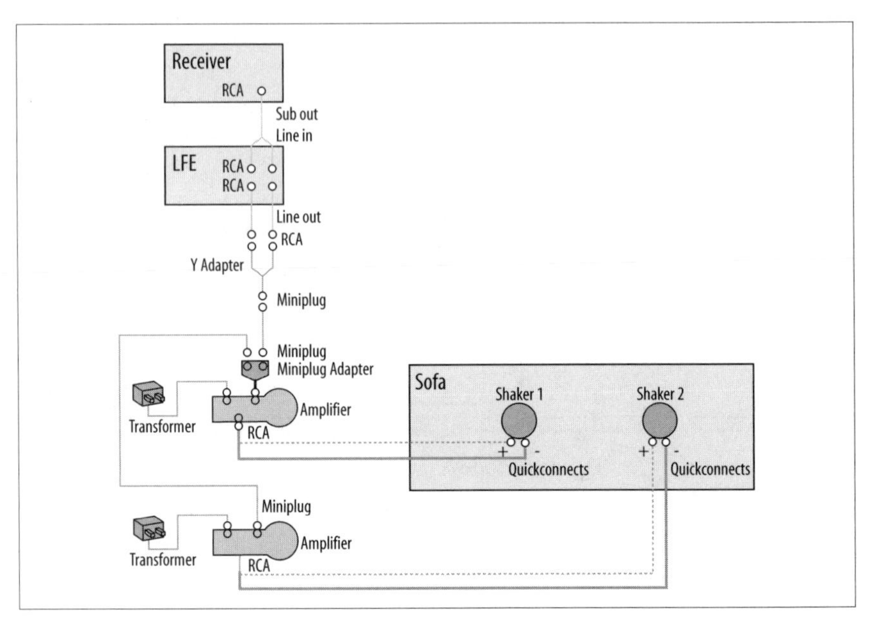

Figure 5-7. Single-sofa setup

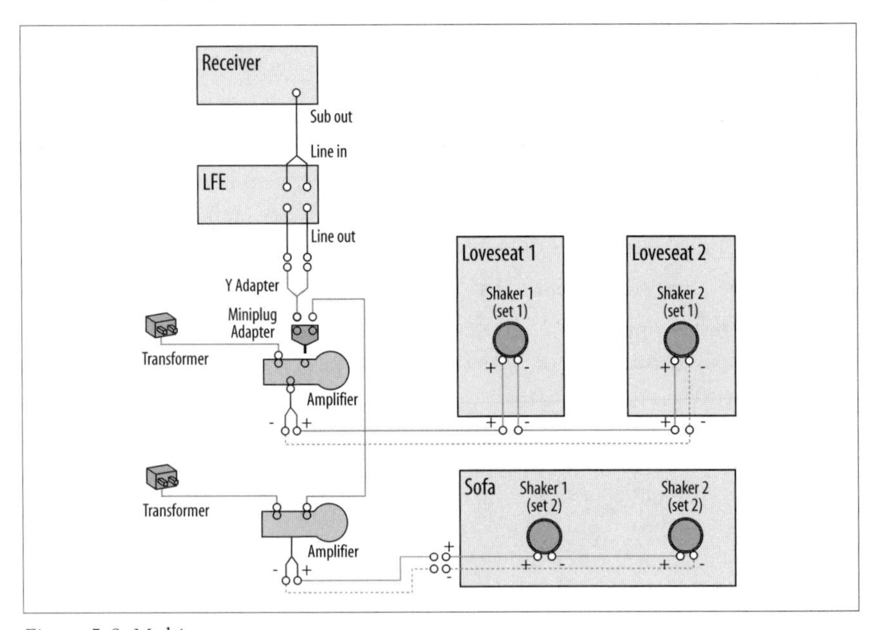

Figure 5-8. Multiseat setup

Filtering

Because most transducers cut off around 50 Hz, they can produce a buzzing effect at higher frequencies. This can be annoying: you're watching *The Lord of the Rings*, and it sounds like there are flies behind your seat. There are a couple of ways to eliminate, or at least reduce, this buzz. The first is to source the signal for the shakers from a subwoofer output via the receiver's tape or headphone output jack. Because you're pulling from a sub signal, you're not going to get any high frequencies. The drawback here is that many receivers won't provide an output of this type; the only headphone out is the general output, which still transmits high frequencies along with the lows.

The other option is to make use of an inline low-pass filter, which blocks the higher frequencies. You can get a 50-Hz low-pass filter at Parts Express (*http://www.partsexpress.com*) (part #266-250) for about $25 or at Accessories4less (*http://www.Accessories4less.com*) for only $14.99. Both are great buys and are perfect for this purpose.

Lower the Resonant Frequency of Aura Bass Shakers

Although bass shakers in their default configuration add some punch to your home theater, you can tweak the resonant frequency of the Aura models to create an even deeper shake and better performance.

I have a pair of Aura Pro Bass Shakers **[Hack #40]** attached to my couch. The shaking they produce is always a hit with visitors and is vital to my listening experience because my wife hates loud noises, and I love LF (low-frequency) effects. I can't crank up my subwoofer, but the shakers keep us both happy. Anyway, they normally have a peak effect in the low 60s—measured in hertz—but that's just a tad too high for best effect. Some people use bass equalization to tame this peak, but I decided to lower the resonant frequency of the shakers themselves.

It's easy to open the Aura Pro Shakers using a #20 Torx driver. Once you open them, you'll see a three-legged plastic spiral spider that supports the moving magnet/mass. The voice coil is fixed and on the periphery of the unit. To change the resonant frequency, you can either increase the mass, or make the spider more compliant.

If you're not used to seeing the word *compliant* in an audio context, it simply means to make something more yielding to pressure.

The legs of the spider are quite stiff; they are more than able to support the mass. Because there was little room to add more mass, I used a Dremel tool and cutter bit to trim the spiders on both speakers so that they were thinner, narrower, and slightly longer. In my case, I took off nearly half the thickness of the spiders, and also cut into the periphery to lengthen them slightly. Obviously this can shorten the shaker life if you take off too much plastic, but they are cheap enough to replace if you goof up. By the way, don't cut too far into the peripheral or you might damage the voice coil!

I probably took off just more than half of the spiders' plastic, mostly in thickness, and was careful to smoothly contour all of my trimming. Maybe they'll break a bit sooner, but I suspect they'll do fine. In the meantime, they are more fun.

Don't forget to double-check the phasing of your connections when you wire the shakers back together.

When I was done, the spiders still firmly supported the magnet. Checking with the LFE low-frequency sweep in AVIA **[Hack #62]** the shakers now peak at just under 45 Hz instead of about 60 Hz. They also now produce output down to the high 20s instead of giving up in the 40s. I needed to turn down my shaker amplifier slightly, as the shakers are now more efficient. The effect is definitely deeper and fuller-bodied. After going through the opening scene in *Lost in Space* and the helicopter scene in *The Matrix*, I am convinced my hour of work paid off. Of course, all the usual caveats apply; this definitely voids any warranty on the shakers, but they're pretty inexpensive.

A side effect of lowering the resonant peak to the 40s is that less high-frequency audio leaks through. The shakers get a low-pass signal, but their former peak—just over 60 Hz—made them respond to things that shouldn't be felt in a "bass" shaker. Now they intensify low bass effects, but the annoying "voice in your butt" effect is greatly lessened.

Going Further

I decided to further deepen the response of one (and only one) of my Aura Pros by taking off even more of the spider, and adding a bit of mass. After extra trimming of the spider and careful epoxying of heavy gauge solder to the magnet, the resonant frequency dropped to about 33 Hz. With the two shakers at differing resonances, the shaking effect begins to intensify at 55 Hz and stays strong down to 28 Hz. There is now palpable effect down to 22 Hz.

Trimming the spider even more does decrease the maximum intensity of shaking, which avoids chatter because the mass can move further. It still maxes out at much more effect than I'd ever want, but this is an issue which you should consider if you trim the spiders down to 1/4 their normal size.

Well, the modified shakers are still going strong. I think they are much better now at avoiding the upper end of low bass from leaking into the seating. The newfound depth of "bass" extension blends well with my subwoofer's falloff. Before the modification they didn't extend much below the sub. Now they definitely do.

Feeling the water rushing around in *Titanic,* or the *Apollo 13* launch sequence, is really impressive, and you won't go deaf going after the effect.

—*Keohi HDTV*

Use Subwoofers as a Poor Man's Bass Shaker

If you love the idea of shaking to low-bass frequencies but can't afford a "real" bass shaker, you can make do with some low-end subwoofers and clever placement.

The Aura bass shakers **[Hack #40]** can easily run more than $200, and some folks won't want to mess with the extra cabling and connections **[Hack #41]**. For those of you who want bass shaker theaters without going the extra mile, you can use a second sub to accomplish the task.

I discovered this trick as a result of my house being built on slab. Unlike wood subfloors, concrete slab doesn't conduct bass very well. My seating (a loveseat) was not shaking as much as I wanted, but I didn't have the dough for professional bass shakers. As an experiment, I moved my 15-inch front-firing Velodyne sub from the front corner of my listening room to where I could point it directly at the loveseat we use to watch movies. I played parts of *Star Trek: Insurrection* that contained low-end bass and I was nearly thrown onto the floor with all the shaking!

Then I turned the sub down to one-quarter volume and turned the adjustable crossover down to 80 Hz. This produced incredibly deep shaking, but the bass was lacking elsewhere in the room. At this point, I realized I needed my trusty Velodyne to produce bass and a smaller sub to handle bass shaking.

It wouldn't take much of a powered sub to accomplish the task. I went to Best Buy and found two that I hoped would do the trick—a $149 Infinity 8-inch 75W and a $189 KLH 10-inch 100W, both of which were front-firing. The choice became simple: the KLH was in stock, and the Infinity was not.

For less than $200, I got a sub with a (built-in) 100-watt amp, an adjustable crossover, auto power on, a phase reversal switch, and a level control! Place this sub in your room, pointed directly at your seating—preferably as close as possible (see Figure 5-9).

Figure 5-9. Bass shaker pointed at loveseat

After some experimentation, I finalized on these settings for the sub:

- 60-Hz crossover setting

> You really don't want to hear this sub—you just want to feel it—so setting this lower than your main sub is a good idea.

- Phase-inverted
- Level at 65%

The crossover also is set so low because I didn't want the second sub to compete with my 15-inch Velodyne. I wanted it to bolster the "feel" frequencies only. The phase reversal sounded a little better to my ears; your mileage may vary.

Also, the sub needed to practically touch the loveseat for best performance. This allowed for maximum conduction from the sub to the couch. I'm tempted to make a foam rubber collar (and somehow paint or encase it in something black) to seal where the two meet, but it is working so well now, I'm going to let well enough alone.

In retrospect, I wish I had found a used sub for sale instead of buying a new one. I'm sure there are many used, smaller-powered subs for sale by those who upgraded to larger ones; that would practically give them away.

—Tim Procuniar

Convert In-Wall Speakers to In-Ceiling Speakers

Sometimes that great set of in-wall speakers from your last house just won't work in your new house. However, it's easy to convert in-wall speakers to in-ceiling speakers, saving hundreds of bucks on a new set.

Have you ever set up the perfect home theater and then had to move? I've had this happen twice now, and both times I lugged my expensive in-wall speakers with me. However, at one home I owned, there was no room on the walls for these speakers, at least not where I wanted the sound to come from (see Figure 5-10 for what these looked like before modification). In these cases, many consumers would bail on the in-walls and buy a new set of speakers, or have expensive remodeling done to support the in-walls. That's awfully wasteful, though; with a little caution you can convert a good in-wall speaker to a good in-ceiling speaker. You usually can position these anywhere on a ceiling (short of where a fan is mounted), and angle them just as you want.

Figure 5-10. In-wall speakers from Dayton (8-inch, three-way speakers)

That said, you can't just pull the speakers out of the wall and mount them on the ceiling. There are three problems when using in-wall speakers for ceiling installation:

- The weight of the speakers can cause the ceiling drywall/Sheetrock to bow.
- Ceiling insulation, especially if it's the blown-in type, can actually touch the speaker cones and cause nasty problems.
- Cold and hot temperature extremes in an attic can cause a speaker to fail.

I was able to solve all three problems by building plywood enclosures to surround the backs of the speakers. I found that 1/2-inch plywood fit perfectly into the groove of the in-wall frames on my speakers; you might find this size works great for you, or you might need to alter that measurement slightly (see Figure 5-11).

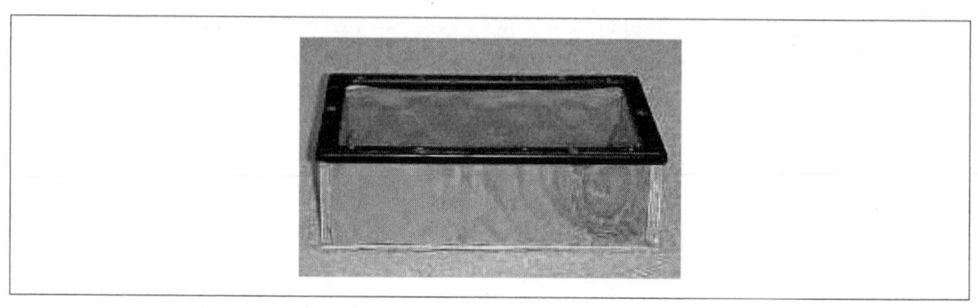

Figure 5-11. Speakers fitted into plywood enclosures

I then mounted the boxes in the ceiling, butting them against a ceiling joist. I drove screws from the inside of the box into the joist, which helped support the weight of the speaker and wood enclosure. To be honest, you probably don't need to offset the entire weight of the speakers/enclosures, but stability is better than nothing. The ceiling can handle some added weight, but I played it safe, especially because I built the enclosures with plywood that was as thick as the frame channel would allow. The drawing in Figure 5-11 shows my completed setup.

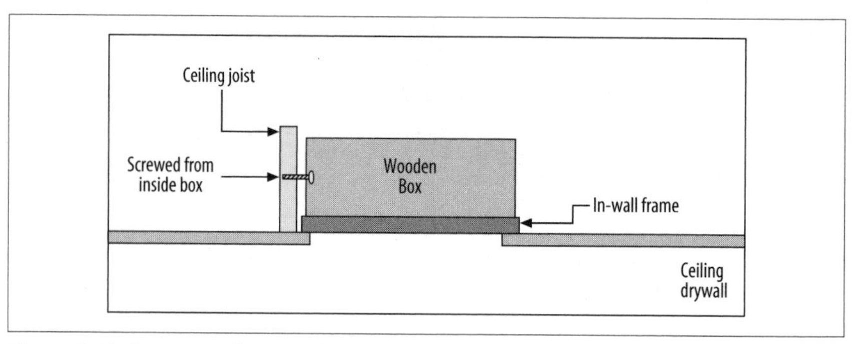

Figure 5-12. Diagram of mounted in-ceiling speaker

This quick and painless procedure saved me a bundle of dough on new speakers, and I still get my unobtrusive wall speakers.

—*Tim Procuniar*

Banana Plugs Trump Bare Wires

If you want the absolute best connection between your receiver/amplifier and your speakers, use one piece of cable and connect the bare wires at the ends to the terminals. However, there are times when a whole lot of convenience trumps a tiny bit of signal.

Any audiophile will tell you that the best way to get a clean signal from one place to another is with an unbroken signal path. The fewer breaks in a connection, the less likely it is that the wiring will oxidize, and the less opportunity for signal to be lost in transfer. In the home theater realm, this means that running a single wire from your amplifier or receiver into your speakers will result in the best signal transmission. I've seen too many homes with multiple strands of speaker wire spliced together, either because the wiring through walls [Hack #46] was done poorly, or because the person making the connections was just careless.

Further, you'll get the best connection between your binding posts on your speakers with bare wire; there is no signal jumping from wire to wire, wire to banana plug, or anything else. You've created the cleanest, simplest signal path possible. That said, you'll almost never find a high-end home theater that doesn't use banana plugs! That might seem strange, but it turns out to be a simple case of trade-off.

The binding posts on receivers, and most amplifiers, are usually in a small, tight area. It is a real challenge to get 12-gauge wire [Hack #46] threaded into the small holes on these posts without leaving strands of copper sticking out. If you do have wire sticking out, you can end up with short-circuiting, overheating, and all sorts of nasty problems. Furthermore, have you tried connecting 10, 12, or even 14 speaker wires to binding posts, while lying on your back, often behind a cabinet already stuffed with cabling? It's not so easy (although if you have your receiver on the bottom shelf [Hack #36], it's at least workable).

Given all of these negatives, it's a simple decision to give up a fraction of a fraction of your signal for the sake of banana plugs. They're easy to connect to speaker wire (even when you're lying behind that equipment rack), they ensure you don't have any stray wire sticking out, and they are simple to connect to your binding posts (see Figure 5-1 earlier in the chapter for a picture). Additionally, friction-fit bananas have a good chance of pulling out without taking your receiver with them should your dog or young child trip over (or rip out, in the case of the child!) your speaker wire.

As mentioned in an earlier hack [Hack #36], I recommend the dual banana plugs from RadioShack, part # 278-308. These are paired units, which keep a pair of speaker wires together, making organization easier than using individual plugs. It's easy to fit 12-gauge wire through them, and the knob on the side makes for easy tightening. However, the spacing of the binding posts on these is nonstandard. Be sure to try out one pair with your amplifier or receiver to ensure it fits properly. If these don't serve your needs, consider single plugs from RadioShack, part #278-306.

—Robert McElfresh

HACK #45 Use the Same Speaker Wire Lengths (Not!)

As long as the length of wire connecting your various speakers is fairly similar, you'll get great results and sound, without interference or adverse delays.

One of the more popular home theater myths going around today is that the length of speaker wire connecting your receiver or amplifier to your speakers must be identical, especially for rear and surround speakers. The prevailing thought is that a longer wire introduces additional delay in transmitting sound, resulting in an uneven production of audio between two speakers (left and right). Although this sounds good (and probably will make your friends think you're really smart when you dispense this as advice), it's absolute nonsense, at least outside of a laboratory.

We did a calculation to try and determine how much speaker wire it would take to get a 1% phase shift (time delay) between the left and right speakers. These calculations demonstrated you needed a difference of about 80 feet of speaker wire before this would happen; keep in mind that most home theaters use between 80 and 100 feet of speaker wire for the entire room!

It's also worth keeping in mind that this test assumes you could actually hear a 1% phase shift in sound. That's a highly suspect assumption, even in audiophile situations.

At the same time, this isn't an admonition to cut your speaker wire to the absolute shortest length, either. If you are installing a new speaker system, cut your wires to length, but add 3 to 6 feet of extra wire to each run. This will give you room to play with speaker positions. Once you have decided on the final locations, cut the wires off to the required lengths, leaving about 2 feet of slack. You want this additional slack so that you can move a connected speaker around a bit, and so that you have some extra to trim back as the exposed copper oxidizes over the years.

—Robert McElfresh

Use Thicker Wiring for Longer Runs

If you're running cable for any length greater than 20 feet, use at least 12-gauge speaker wire.

Many home theater newbies get tied up into knots when it comes to determining the difference between 12-, 14-, and 16-gauge speaker wire, and that assumes they're even noticing the gauge of wiring to begin with. To make matters more confusing, smaller gauge wiring is actually thicker than higher gauges. Using gauge as the unit of measurement comes from when a rod of metal was run through rollers to squish it to a smaller size. A rod that ran through the rollers 12 times was thicker than a rod that ran through 16 times. This is why the higher numbers mean thinner wires.

To put this into perspective, here's a typical recommendation for speaker wire gauge versus the length of the run.

1 to 10 feet
: Use 16-gauge wire (the thinnest acceptable wire for any home theater usage).

11 to 20 feet
: Use 14-gauge wire. You'll commonly find 14-gauge wiring in large electronics stores **[Hack #3]**.

20 or more feet
: Use 12-gauge wire. You might have to shop at an electronics boutique **[Hack #4]** or go online **[Hack #6]** to find quality 12-gauge wiring.

Using this and the definition of wire gauge, you can see that you'll need thicker wire for longer runs; however, many neophytes actually read this as just the opposite and use their thickest wire for the front and center channels, where it's needed the least.

Rather than spending your free time with a tape measure and these measurement guidelines, buy a large amount of 12-gauge wire and use it everywhere. It will drive your rear and surround channels with no problem, and of course it will work great with your front channels. As an added bonus, you won't have to worry about the length of your runs; 12 gauge works fine all the time.

It is worth pointing out why some will claim you need different gauges of wiring, especially for the odd occasion when someone turns up his nose at your setup and criticizes your thinking. People who claim to hear the sound change caused by long, thin wires are not talking about the rear speakers of most home theater systems. They are referring to a setup using, at a minimum:

- A high-end CD transport
- A preprocessor/amplifier system with around 200 watts per channel
- Very accurate electrostatic or panel speakers with a huge radiating surface, compared to traditional cone speakers
- CD music that has been listened to so many times that even miniscule changes in sound are noticed

This is a very different situation from the rear speakers of a home theater system. Rear speakers are active only about 20% to 40% of the time, largely carry special effects, rarely transmit dialog, and usually echo what the front speakers are doing. I doubt even the audio gurus could notice the effect of 16- versus 12-gauge wire going to the rear speakers in this setup.

—Robert McElfresh

Bi-Wiring and Bi-Amping Speakers

You've probably been told by your home theater buddies about how they've bi-wired their speakers—or maybe they've bi-amped them—or both. Learn what all these "bi" terms mean, and which you want to put into play in your system.

The best way to figure out if you want to bi-wire or bi-amp your system is to understand what each is.

Active Bi-Amping

Active bi-amping occurs when you have an active *crossover unit* between your preamplifier and amplifier that splits the full-range signal coming out of the preamp into a high- and low-frequency signal. So, right off the bat, we're talking about a pretty high-end system.

Then these split signals are fed from the active crossover to two separate amplifiers (or two different channels on a single amplifier) and are amplified completely separately.

This separation of amplification is why even the best receivers can't perform active bi-amping; they don't have two channels for any one speaker.

By separating amplification in this manner, your high- and low-frequency signals stay completely independent from the very beginning of the signal chain. They get amplified separately and passed down a dedicated speaker cable—one cable for the highs and one for the lows.

When these two signals are fed into a speaker, they must go into a speaker specially wired to accept two discrete signals and route them to the proper drivers in the speaker. These speakers will have no internal crossover circuitry—all signals sent to the speaker must be prefiltered—and the speakers offer direct electrical connections to the components inside. So, in addition to a high-dollar preamplifier, this involves high-dollar speakers as well. Your manufacturer's specifications will be very clear on whether your speakers support this feature.

This is true bi-amping, and it is very rare in home theater applications. It is ideal, as it keeps the frequencies discrete before amplification, and gives you complete control over the signals. However, it's very, very expensive to actually implement. You'll find this type of setup in some movie theaters, some studio monitoring situations, professional PA systems, and the home theaters of the very well off!

Passive Bi-Amping

Passive bi-amping is possible if you don't have an active crossover unit to split your high- and low-frequency signals. Instead, you feed identical full-range signals into two amplifiers (or two channels of a single amplifier). This is more feasible, as it doesn't involve the expensive, high-end preamplifiers that active bi-amping requires.

The two full-range signals are amplified separately and then fed into a speaker, with each signal on its own wire. Again, the speaker must be designed to accept these two discrete full-range signals. So, you're still going to need somewhat specialized speakers for this application. A passive crossover filter inside the speaker filters out the high-frequency signal from the material destined for the woofer, and filters out the low-frequency signal from the material destined for the tweeter.

The crossover circuit inside the speaker is not connected between the high and low signals; each band pass is essentially discrete. It's only in the speaker that filtering occurs, whereas in active bi-amping the signal was split before being amplified.

Passive bi-amping is also uncommon in home theater, although more common than true bi-amplification. Even though passive bi-amping is less costly than active bi-amping, it still requires that you have two amplifiers (or channels) to drive each speaker and that the bi-amped speakers be equipped to be wired this way.

Bi-Wiring

Bi-wiring is when you run two speaker cables from a single amplifier channel, and hook the cables to two separate inputs on the speaker.

Because bi-wiring doesn't require separate amplifier channels, it works with receivers as well as a preamplifier/amplifier combination.

If you have special speakers designed to accept bi-wiring (such as the ones explained in the passive bi-amping section), the passive crossover filter inside the speaker filters out the high signal from the material destined for the woofer, and filters out the low signal from the material destined for the tweeter. As in passive bi-amping, this circuit is not connected inside the speaker, and each band pass is essentially discrete from the amplifier. However, because only a single amplifier is used, the signal is shared from the amplifier into the preamplification (whether that's internal, as in a receiver, or external, as in a separate preamplifier).

I'm not a fan of bi-wiring; I believe it is of no real advantage. However, some certainly will argue otherwise. With bi-wiring, the advantage of having a discrete signal chain is lost when the speaker wires are connected at the amplifier channel. You no longer get more power to the drivers and no longer have discrete signals going into the speakers. Electrically speaking, there is very little difference between bi-wiring and just feeding a single full-range signal to the speaker and letting the speaker split the signal itself. Usually, any improvement in sound from bi-wiring seems to be the result of using two speaker wires, which doubles the amount of signal wire available for information.

—Vince Maskeeper-Tennant

Subwoofers

Hacks 48–54

Ah, the rumble of a '65 Mustang's engine; the sound of a jet fighter flying overhead; nuclear explosions that rattle your teeth. These are the effects that most of us live for in a movie theater and desperately try to reproduce in a home theater. The key to all of these effects is, of course, the bass in your system—specifically, the subwoofer (or subwoofers) that you're using. Even more exciting, room size (and a healthy dose of physics) allows you to go beyond sheer reproduction. Commercial theaters rarely can produce clean bass at under 30 Hz due to their size, but smaller home theaters actually can produce bass at 30 Hz, and even below 20 Hz, with authority. Of course, all of this depends on the right sub (woofer) and the right setup.

In the last chapter, you got the lowdown on speakers—floor-standing speakers, surround channels, placement, wiring, and more. However, you might have noticed that the big square box that all the bass comes out of was left out of those discussions. That's because subwoofers are really a topic completely unto themselves. You'll recall that when choosing speakers, you'll want to buy the same brand [Hack #37], except in the case of a sub. That allows you more flexibility in choosing the best sub for your room, but also means you're going to have to know even more about subs than you already do about speakers. We'll start with some technical terms, and then get into the hacks.

Learn Sub Talk

Before diving into subwoofer placement and optimization, you need to understand the basic terminology. This is more than just background; it will help you isolate problems with your subs, and even spur you on toward considering building your own sub one day.

More often than not, home theater consumers think of a subwoofer as just another speaker. Although that might be true in some senses, you're better off considering subs as a completely different category. This perspective dovetails with the very nomenclature of the system; it's not a six-channel system, it's a 5.1 system (with the subwoofer being the ".1"). You need to consider different factors when choosing a sub, but first, let's get some basic terms under your belt.

Subwoofer Parts

The first thing to get into your head is the actual part list, so to speak, of a typical subwoofer.

Subwoofer
: A *subwoofer* is a speaker specialized for producing bass (usually frequencies below about 120 Hz). They can be active (where a plate amp is built into the enclosure) or passive (requiring a separate amp). A subwoofer is made up of an enclosure, a driver, and, if it is active, an amplifier.

Plate amp
: A *plate amp* is an amplifier built on a plate that can easily be mounted to a subwoofer enclosure. It can have speaker-level and/or line-level inputs, and possibly outputs. It can have one crossover or no crossover, as well as rumble filters, phase controls, and an EQ (equalizer).

Driver
: A *driver* is the part of a subwoofer (or any speaker, for that matter) that actively produces the sound.

Baffle
: The *baffle* is the surface to which the driver is mounted.

Enclosure
: An *enclosure* is what "encloses" the driver. An enclosure can be a box, a cylinder, just a plain baffle, or even an adjacent room. Enclosure is a very broad term that can be used for any type of subwoofer.

Port
: A *port* is usually a pipe with one end open to the inside of the enclosure and one end open to the outside of the enclosure. However, a port doesn't have to be round. A port can be pretty much any shape; it's just that round ones have certain airflow advantages over other shapes.

Passive radiator
: A *passive radiator* serves the same purpose in a subwoofer as a port does. They do the same thing in slightly different ways. A passive radiator looks like a driver, but it doesn't have a motor attached to it (the motor is the magnet structure you find on the back of a driver).

Subwoofer Connections

Subwoofers often have different connection types than normal speakers. Most accept line-level connections, but there is the possibility that your subwoofer will accept a speaker-level connection.

Line-level connection
: Line-level is a nonamplified signal. You use coaxial cable terminated with standard RCA plugs for these connections. Line-level outputs on plate amps aren't for daisy-chaining subwoofers together because they will filter out the low frequencies (unless specifically labeled as a daisy-chaining output).

Speaker-level connection
: A speaker-level connection is an input or output on a sub to which you hook up speaker wire. This connector is set up to accept an amplified signal. If the subwoofer is active (in other words, it has a plate amp), the plate amp speaker-level output will pass the signal on to a speaker (possibly with the low frequencies the sub will play filtered out). The speaker-level input will convert the signal to a line-level signal that the plate amp will then reamplify for the subwoofer's driver. If the subwoofer is passive, there will be two speaker wire terminals that you hook up in the same way you'd hook up any other speaker.

Miscellany

Once you've got your head around the parts involved, there are several terms that just defy categorization. Still, you'll want to understand these to the same degree that you have your head wrapped around plate amps and line-level connections.

Gain control
: This is usually labeled the "Volume" control on a plate amp. It determines how much the input signal to the amp is boosted before it is amplified. It is not an absolute volume control, where one limit is no sound and the other limit is the amp's maximum output.

Excursion
: A driver moves its cone in and out. *Excursion* is how far the cone can move in and out. Two main excursion limits are defined for a driver: the *mechanical excursion limit* and the *linear excursion limit*. The mechanical excursion limit is how far the cone can physically move before another piece of the driver prevents it from moving any farther. The linear excursion limit relates to distortion; a driver's motor can move the cone only so far before the motion is no longer linear. When the motion

stops being linear, the driver will produce distortion. The linear travel limit of a driver always will be less than the mechanical travel limit.

Volume displacement (Vd)
: This is the amount of air that would be contained by a cylinder mapped out by the linear travel of the driver's cone.

Crossover
: A *crossover* is an electronic filter that splits a signal into two parts. This is done with two filters: one that cuts out the low frequencies (called a *high pass*) and one that cuts out the high frequencies (called a *low pass*).

Last octave
: When someone refers to the *last octave*, he usually means frequencies from 16 Hz to 32 Hz.

Front wave
: The air pressure wave that emanates from the front surface of a driver's cone.

Back wave
: The air pressure wave that emanates from the back surface of the driver's cone.

Choose the Right Subwoofer

Buying a subwoofer involves a lot of knowledge about enclosures, wattage, and drivers: get the scoop on all of these in this hack.

So, now that you've got the basics of subwoofers [Hack #48] in your head, there is still an important question to answer: "What should I buy?" This hack will give you some insight on what to look for; sometimes it's more than just raw power or how cool the enclosure looks.

Displacement

When it comes to high-output subs, there simply is no replacement for *displacement*. What is meant by displacement? Displacement is the volume of air the driver can displace. Take the effective radiating surface of the driver (usually abbreviated in driver specs as Sd) and multiply it by the distance the driver can linearly move its cone (usually listed as the driver's Xmax). This determines the volume of a cylinder that will be mapped out by the cone's full range of motion. This is the Vd, or *volume displacement*, of a driver. The higher the volume of air that the driver can displace, the better.

Table 6-1 supplies the approximate Sd values of the most common subwoofer driver sizes.

Table 6-1. Sd based on driver size

Driver Size (inches)	Sd (square centimeters)
8	225
10	325
12	475
15	775
18	1,150

As you can see, a 10-inch driver would have to move more than three times as far as an 18-inch driver to displace the same amount of air. Generally, bigger drivers can move more air. However, if the smaller driver can move more than three times as far as the 18-inch driver, the smaller driver works out just fine. In the end, don't make assumptions about a particular sub based on the size of the driver or any other one factor. Listen to it and trust your ears.

Power

How many times have you heard your buddy brag about the wattage on his newest speaker system? Probably a lot more times than he's explained how that wattage actually affects the sound produced! High wattage doesn't always translate directly to good sound, and you need to be clear on that before you spend thousands of dollars on a sub just because of its power rating.

Power with subs is something that is often very much misunderstood. Unlike Vd, more power isn't necessarily better. There are large, efficient subs that can do significantly more with 100 watts of power than a little 14-inch sealed cube with far more than 2,000 watts. How many watts a sub needs depends on a number of factors; the most important of these are the sensitivity of the driver and what enclosure it is in.

Another point not commonly known about the relationship between a driver and power is that there are two types of power-handling limits. There is a *thermal limit* and a *mechanical limit*.

Thermal limit
: The thermal limit is how much power the driver's motor can take before parts of the unit actually start to melt. With thermal power handling you also have to consider the difference between peak and continuous capabilities; for short periods of time, most drivers can take lots of power. Over lengthy use, though, thermal limits can become a real problem. Still, the amount of power a typical enclosure/driver

combination will need continually almost always will be far lower than the driver's thermal limit.

Mechanical limit
: This limit will vary depending on the enclosure, even when the same driver is being used. Mechanical power handling is the amount of power in a particular enclosure that it takes to make the driver travel to its mechanical excursion limit. In big enclosures, this always will be far lower than the thermal power-handling limit. In small enclosures you'll usually hit the thermal limit before the mechanical one.

The bottom line is that you shouldn't pay much attention to the wattage rating on a sub. The driver and enclosure combination might or might not be able to use that power in the first place. All that really matters are the frequency response of the sub, whether the power it has allows it to play as loudly as you need, and whether the distortion levels the subwoofer produces at those levels are acceptable.

One other item related to a subwoofer's power that often is very much misunderstood is the volume dial on the plate amp. The position of this dial is not an indication of what portion of the subwoofer's capabilities you are utilizing. It also is not an absolute volume control. For example, turning the dial to one-third of its maximum setting does not mean you have two-thirds of the sub's potential left. This dial is a gain control, and it controls how much the input signal is boosted before it is amplified. This means it is entirely possible that if the input signal is hot enough, with the dial in the one-third position, the plate amp could be putting out all it can. You should use the volume dial on a sub plate amp for rough level matching to your other speakers, and nothing more.

Subwoofer Alignments

A number of different types of enclosure designs are in common use for subwoofers. Each type of design and the variations within each design are called an *alignment*. They all have their pluses and minuses, and if you're serious about getting the right subwoofer, you'll need to be at least generally familiar with each.

Sealed. A *sealed* alignment is one of the two most common designs in subwoofers. It also is referred to as *acoustic suspension*. In this design, the driver is placed in an enclosure (the shape of which doesn't matter) that is airtight. The air trapped in the enclosure acts as a spring that resists the driver's movements and keeps the back wave isolated from the front wave (so that they don't cancel each other out). If the enclosure is too small for the driver being used, you'll get muddy, sloppy, boomy bass. There are some drawbacks to the enclosure being very large (you do have to fit the thing in your room), but being too big won't make a sealed sub sound bad. Sealed subs have a shallow and consistent roll-off that is approximately second order (12dB/octave). When this roll-off will start depends on a combination of the driver and the size of the enclosure.

Ported. A *ported* alignment also is known as *bass reflex*. Subs with ports and subs with passive radiators belong to this category. Bass reflex is the second of the two most common designs. With ported subs, the driver is installed in an airtight enclosure, except for a pipe. This pipe is the only means by which air can move in and out of the enclosure. With passive radiator subs the enclosure is completely airtight, but instead of using a pipe, a driver with no motor is installed. The advantage of a bass reflex design is that the port or passive radiator can push the point at which the sub's output starts to roll off down to a lower frequency. The trade-off, though, is that when the roll-off does start, it occurs at a much steeper rate: fourth order (24dB/octave) for ported and fifth order (30dB/octave) for passive radiator designs. Done incorrectly, bass reflex designs can have huge peaks in their response that make them sound very boomy and unnatural. Done correctly, bass reflex designs can sound just as good as sealed ones.

Bandpass alignments. There are many different bandpass alignments. These alignments involve multiple sealed or bass reflex enclosures, connected either by drivers or by ports. However, these alignments are very difficult to get right. Their biggest advantage is that they are quite efficient over their bandwidth. The problem is getting that bandwidth to be wide enough and the sound to not be boomy over it. You're better off staying away from bandpass designs unless you like a boominess to your bass.

Basshorn. When done right, basshorns can be spectacular. They are extremely efficient and produce the dynamics of music and movie soundtracks effortlessly. However, to get good extension out of a basshorn it has to be huge. How low a basshorn can go is determined by a combination of the horn's length and the width of its mouth. Typically basshorns don't get big enough to produce frequencies below about 30 Hz. However, if you place a basshorn properly against a boundary, you can get decent output down to below 20 Hz with a horn that would go down to only 30 Hz in a less finely tuned environment. Basshorns are big and expensive, but if you have the money and desire, they are definitely worth researching further.

Infinite baffle. Infinite baffle subs are currently the domain of the DIY (do-it-yourself) sub builders. An *infinite baffle* is essentially a really, really big sealed sub. One or more drivers are mounted to a wall or a manifold (yes, that's actually the wall of the room!). One side of the driver fires into the listening room while the other side of the driver fires into an adjacent space, such as an attic or basement.

The front and back waves of the driver need to be isolated so that they won't cancel each other out, and the adjacent space needs to be big enough.

If you choose the right driver, infinite baffles produce what is considered by many to be the best bass you can get. For output, you are limited only by how many drivers you have space and money for; use enough and you can even damage your house! Another big advantage is you can install them extremely discreetly (known as a *stealth install*). The only part that needs to be visible in the room is a hole in the wall, and it can be disguised in whatever manner suits your decorating tastes.

If you have some basic woodworking experience, some decent tools, and an appropriate adjacent space to your theater, I strongly recommend considering this option. The wiring is no more difficult than hooking up a normal speaker. The Internet is the best resource for learning more about implementing this type of sub. Pull up Google, and shortly you could be on your way to building a sub that easily will beat commercial subs costing many times what an infinite baffle costs.

Dipole. A *dipole* sub is like an infinite baffle, but it doesn't isolate the front and back waves. One or more drivers are mounted in a baffle. The baffle can be a big, flat board, or it can be a number of odd shapes that maximize the distance the front and back waves of the driver have to travel before they can meet. When the front and back waves do meet, they cancel each other out with a rather steep slope (the size of the baffle determines how low a frequency at which this starts). Dipoles usually counteract this low-frequency roll-off by applying EQ to boost the low end. However, this canceling also is on a plane that eliminates two dimensions of the room from causing problems with your sub's response. Dipoles require drivers that can move lots and lots of air to get loud. With the right drivers and EQ, they can sound great and provide decent output. Many swear they produce the best-quality bass of any sub design, period. If you don't need tremendously loud bass and you want the best in quality, dipole might be the way to go.

Hoffman's Iron Law

When talking about subs, there is one law that everyone should know. This law has many variations in many other fields, and all center on the fact that to get one thing, almost invariably you have to trade something else. This most definitely is true of subwoofers. Hoffman's Iron Law simply states that out of enclosure size (how big the box is), extension (how low a frequency the sub is capable of playing), and efficiency (how much power the sub needs to get to a volume you want) you can pick any two to control, but the third will be dictated to you. This means that if you want a small enclosure and good extension, you are going to need gobs of power. If you want great extension needing very little power to get loud, you are going to need a massive enclosure. Always keep this in mind when looking at subs: don't expect the tiny, one-cubic-foot subs to produce mind-blowing, loud, low, clean bass. No sub manufacturer has figured out a way to beat physics. Loud, low, clean bass requires big boxes with big drivers (or lots of medium-size ones). Small sub-power requirements are simply too high to achieve the same results. It's just not possible to build a driver that can take the kind of power a small box needs to produce loud, low, clean, effortless bass. There's just no way around it.

Match the Sub to Your Room

When you choose speakers, you spend as much time analyzing your room size and dimensions as you do choosing the actual speakers. When you're selecting a subwoofer, you need to do the same.

Your room will greatly affect how a subwoofer performs. However, before I get into how your room affects your subwoofer's playback, I'll address a couple of ways in which your room won't affect your subwoofer's playback.

Room Size Doesn't Limit Subwoofer Extension

In the continuing evolution of subwoofer speakers, the ability to play low (and lower-than-low) frequencies has become a blessing and a curse. It is a blessing because lower frequencies provide that room-shattering, earth-shaking bass that makes you believe a plane really is flying overhead, or that 10,000 horses really are galloping by. On the other hand, with increased frequency range, there is a tendency for the brainiacs to over-analyze the use of subwoofers and confuse the typical consumer.

As a prime example of this confusion, a number of fallacies seem to persist on the Internet about subwoofers. One of the silliest of these is the assumption that if a room's dimensions aren't large enough to contain a full wavelength of a frequency, the frequency can't be played in the room. In other words, a subwoofer in a small room is not capable of playing those super-low, long-wavelength frequencies. Of course, this is false, and really rather absurd. Any frequency can be produced in any size room.

To prove this, a little math is in order. The speed of sound is roughly 1,130 feet per second. Based on this, Table 6-2 shows some typical low frequencies and their resultant wavelengths.

Table 6-2. Frequencies and wavelengths

Frequency	Wavelength
16 Hz	70.6 feet
20 Hz	56.5 feet
30 Hz	37.7 feet
50 Hz	22.6 feet
70 Hz	16.2 feet

As you can see if frequencies could play only within rooms as large as their wavelength, home theater would be a terrible disaster! How many homes have a theater room (or any room!) more than 70 feet long or wide? If this myth were a reality, very few rooms would be able to produce bass to the last octave, car stereos would seem anemic, and headphones simply couldn't work.

Sound is just varying air pressure over time. As a sound wave passes by you, the pressure changes and you hear the sound. The driver of your speaker will start the wave propagating, and once it passes your ear, you will perceive the wave as sound. Not having enough space for the entire wave to exist in a straight line doesn't matter; these pressure waves will bounce off walls, no matter what their frequency. So, in the case of higher frequencies that have very short wavelengths, many full wavelengths will pass by your ear before you start to hear the first reflections. But with waves longer than room dimensions (such as bass frequencies) it's possible that the beginning of the wave will pass your ears twice before the end of the wave reaches it once. This doesn't mean you won't be able to hear the sound; however, it does mean you won't be able to localize the source of the sound (there is more to localization than this, but this is one of the factors that makes it difficult to localize low frequencies).

Subwoofer Orientation Doesn't Matter

Where you place a subwoofer in your room is just as important to the sound of your system as where you place your speakers. Oddly enough, though, one factor that you won't have to worry about is the direction that your subwoofer is facing.

Subwoofers in modern home theaters are playing extremely low frequencies. Because these wavelengths are so long, the orientation of the driver or the port won't have any effect on them. Compare this to the tweeter on a typical bookshelf speaker; it will produce wavelengths that are only a fraction of an inch long! If you change the orientation of the tweeter even a small amount on the enclosure, it will have an effect on the sound. The shortest wavelength a subwoofer will produce, though, will be many times the dimensions of its enclosure. This means that, relative to the wavelengths being produced, any effect a change in orientation will have on the wavelength's perception is negligible. The driver can fire up, down, to the front, to the back, or at an angle, and it won't matter. The same is true of the subwoofer's port.

In fact, you don't even need to worry about the surface the subwoofer rests on. Many people with carpet in their theater rooms have taken great lengths to get their sub off that carpet, with the idea that sound is somehow being muffled. Carpet would need to be ridiculously thick (on the order of two to three feet) to have any real effect on bass frequencies.

Downward-firing subs also aren't more likely to disturb downstairs neighbors than front-firing subs. Always remember the extreme length of a subwoofer's wavelength. A front-firing sub is just as effective at propagating these pressure waves to the floor as to the back wall, and downward-firing subs are just as effective at propagating these pressure waves to the back wall as to the floor. Without modifying the construction of the surface, there is only one way to reduce the level of sound that penetrates: increase the distance of the source to the surface. The change in distance between a front-firing and downward-firing sub is insignificant in this situation.

The one caveat to this is that the driver and port do need sufficient clearance to the nearest surface to prevent compression problems. This means that optimally, you want to have your subwoofer's sound-producing faces out a bit from any surface. As long as you stay four inches or more from any obstruction, you should have nothing to worry about.

Analyzing Room Effect

Although a room won't limit how low a subwoofer can play, it will have a rather large impact on a subwoofer's low-end response. There is a phenomenon known as *room gain* (in the car audio world it is called *cabin gain*, and the size difference is why car stereos can get so ridiculously loud). When wavelengths are longer than the room's dimensions, the room boosts those wavelengths. The smaller the room and the lower the frequency, the higher this boost. This means it is possible for a sub to sound simply amazing in a small room but absolutely pathetic in a big room. This results from the sub not having the headroom left to counteract the drop in room gain from the larger room.

The other major effect a room will have on a subwoofer is the standing wave patterns it will produce. A typical room will have three *axial room modes* (a room mode is the particular audio frequency a standing wave will form at in a room): one for the height, one for the length, and one for the width. At the exact frequency a perfect standing wave will form, one of two things will happen: the frequency will be reinforced, doubling its output, or the frequency will be completely cancelled out. As you move off the exact frequency, the amount of reinforcement or canceling will decrease. Reinforcement is called a *peak* in the subwoofer's response, while cancellation is called a *null*.

When you place a sub in the corner of a room, you will maximally excite the three main axial room modes (there also are tangential and oblique room modes and multiples of all three, but we won't worry about those in this discussion). You can roughly calculate what those frequencies will be by dividing 565 by each room dimension.

The measurement 565 feet/sec is roughly half the speed of sound; you are calculating the frequency of half a wavelength of your room's dimension by doing this.

Say your room is 24×18×8 feet. With your sub placed in the corner, you'll have room modes at roughly 24, 32, and 70 Hz (you'll also have a full-wavelength room mode of the 8-foot dimension at 35 Hz). If you had some extensive schooling in physics and acoustic theory (or access to the right computer software), you could get a pretty good picture of how all the different frequencies will interact with your room when produced from different locations. The problem is that the majority of us don't have access to the software, and fewer of us have the knowledge to work it all out!

The bottom line is that you have to experiment to find the best location in a room for a subwoofer. This location will be the one that causes the most even distribution of standing waves in your room. The more even the distribution, the more even your subwoofer's frequency response will be, and the better it will sound. See "Optimize Subwoofer Placement" **[Hack #52]** later in this chapter for one of the best and easiest methods to correctly place a subwoofer in a room.

Going Shopping

I hope now you feel you have a decent understanding of what a subwoofer is all about. So, now I'll look at actually buying a sub. First, think about a few things before you actually go demoing **[Hack #2]**:

- How much money can/should I spend?
- How close do I want to get to reference levels with movie soundtracks?
- How big is my room?

There are two ways to look at question one. Many enthusiasts ask, "How much money can I spend?" when a better question might be "How much money should I spend?" I recommend allocating 50% to 100% of what you spent on the other five speakers combined, if possible. If you think this is excessive, keep in mind that a subwoofer should be handling the low bass of every speaker in your system. You definitely don't want the subwoofer to be the bottleneck in your system, especially for home theater viewing.

You need to consider the second and third questions in conjunction. The second question almost always is answered with "I want to do full Dolby reference level." What most people don't realize is how few subwoofers can actually do full Dolby reference level, and what it will cost to get one of those few subwoofer(s) that can.

More often than not, you'll need multiple subwoofers to achieve Dolby reference level.

Full Dolby reference level in a typical home theater setup will require the subwoofer to be capable of putting out 121dB to the listening position all the way down to 20 Hz. Considering that most subwoofers in the $750 to $1,000 price range have difficulty passing 105dB at 20 Hz when measured 2m from the speaker in a medium- to large-size room, and that with this output capability it would take eight of those subs to hit 121dB at 4m, the financial reality of trying to achieve Dolby reference scares most folks away. The third question also has a big effect on how hard it is to meet the output goal you have for your system. If your room is 1,500 cubic feet, you'll be able to get away with a lot less subwoofer than if your room is 8,000 cubic feet. Keep this in mind when demoing subwoofers. If the demo room is bigger than your room, expect the sub to be more capable at home; if the demo room is smaller, the subwoofer won't be as loud at home. Most good audio stores **[Hack #4]** will let you have an in-home demo. Take advantage of this.

There is also some good news related to questions 2 and 3: most people don't realize just how loud Dolby reference level is. Most people will be more than happy at 10–15dB below Dolby's reference level; I know I am. Use an SPL meter **[Hack #63]** to determine what level you like to watch at and make sure the sub you decide on is capable of reaching the levels you want.

With this information in hand, start demoing some subwoofers. Check out the subwoofer(s) that fit within your budget and find the ones that will meet your output goals in your room. Then concentrate on the final and most important factor: do you like how it sounds? Preferably, make this decision in your own home theater, with your main speakers and components.

—Dustin Bartlett

Hook Up Your Subwoofer Correctly

Getting the best from your sub involves more than plugging in a cable or two and popping in a movie. Learn how to optimally set up your subwoofer, connect it to your audio system, and ensure it meshes with your existing speakers.

Now that you have a sub **[Hack #50]**, you need to know how to hook it up. Depending on the equipment you have, there are numerous ways to hook up a sub. I'll go into detail on the most common scenario. You have a relatively new surround sound receiver (or preamp/processor) and a 5.1 or 7.1 speaker setup. In this situation, the ideal hookup will involve making proper use of the receiver bass management features, a line-level connection from the receiver to the subwoofer's amp, and defeating the subwoofer's crossover.

Set Speakers to Small

The first hurdle you need to get over is a rather unfortunate naming convention used in receivers. The receiver will have speaker settings called "small" and "large." The knee-jerk reaction to these names is to set your tower speakers (or your big, capable bookshelf speakers) to "large."

> There's a lot more detail on this setting and how it affects your overall system **[Hack #67]**. The content in this hack focuses specifically on your sub.

You need to get over this reaction; even the most capable tower speakers out there can't produce low bass as well as a good sub. Setting all your speakers to "small" has two advantages, and both are worth far more than seeing "large" scroll across your receiver's display. The first advantage is that the "small" setting relieves your speakers and amp from the difficult task of producing the lowest few octaves of a soundtrack. An easier load is placed on your amp, and you'll find the midrange of your speakers improves when the last few octaves don't need to be produced.

The second major advantage goes back to how bass frequencies interact with a room. Finding the correct placement for one speaker to produce an even bass sound in your room is much easier than finding the correct combination of five, six, or seven bass sources; the speakers that now are producing bass in addition to your sub. To further complicate matters, the best location for those sources in relation to bass production rarely will match up with the best place to produce the midrange and high frequencies.

Setting the Crossover Frequency

The last thing to consider on this front is that a crossover isn't a brick wall (remember the section on crossovers in the primer [Hack #48]?). The rule of thumb is that your speakers should be capable of a flat frequency response to one octave below the crossover frequency. So, if you set your receiver to an 80-Hz crossover, your speaker should be capable of solid output down to 40 Hz. Because most good tower speakers are flat to just below 40 Hz (manufacturers' claims on frequency response almost always are exaggerated), this ends up working well. So, dig out your receiver's manual and figure out how to do the following:

- Set all speakers to "small."
- Enable the subwoofer (don't select an option that copies the LFE channel to the main channels, or any of the other bizarre setup options some receivers offer).
- Set the crossover to 80 Hz (some receivers don't allow you to vary this setting).

With some receivers you have to repeat these crossover settings for each channel.

Never touch these settings again!

Connecting the Subwoofer to the Receiver

Now you'll need to make a line-level connection between your receiver's subwoofer pre-out and your subwoofer's line-level input. The cable just needs to be a standard audio interconnect with RCA connectors on the end; you don't need a special subwoofer cable. Any decently shielded audio interconnect will do.

Some plate amps have a left and right input instead of just one line-level input. When using your receiver's bass management features with a sub such as this, you have two options. The first is to plug the cable into one of the left or right inputs (you don't have to use both); the second is to purchase a Y-adapter cable and plug your receiver's subwoofer pre-out into both. Either method works equally well on 99% of plate amps. On a select few, the input signal to the sub isn't strong enough unless you use the Y-cable. For most setups start without a Y-cable and if you get the output you're after from the sub, leave it alone. If you aren't getting the output you expect from the sub, you should try the Y-adapter. These are the only connections (besides plugging in the power) that you'll need to make to the subwoofer.

Defeating the Plate Amplifier's Crossover

With everything hooked up and the receiver configured, you have one thing left to do. The plate amp on your sub likely will have its own crossover, but because you are using your receiver's bass management, you'll want to defeat this "feature." Some subs have a switch to do this, but others don't. If your subwoofer has a switch, simply set it to the "off" position. If not, turn the crossover dial up as high as it will go, as this is the next best thing. Finally, make sure your sub is optimally placed [Hack #52], and you're ready to go.

All That Other Junk

What about all the other connections on the subwoofer's amp? If your receiver has a subwoofer pre-out and bass management (and almost all do these days), you should ignore all of these additional connections. If your receiver doesn't have these features, you should consult your subwoofer's manual for alternative connection options. The only time you should consider not using your receiver's bass management is if you are setting up a two-channel system, and the subwoofer's plate amp has two dials for the crossover—one that controls a high-pass filter and one that controls a low-pass filter. If this is the case, you've gone far enough into the world of two-channel reproduction that you'll know what to do with the two filter controls, or your dealer will have no problems giving you some advice (these are home theater hacks, remember?).

One final item I'd like to comment on with this topic relates to receivers that allow you to choose different crossover points for different speakers. If your receiver has this feature, don't use it; set all speakers to the same crossover point. To understand why, consider a simple example: assume you set your surrounds to an 80-Hz crossover, your center channel to a 60-Hz crossover, and your mains to a 40-Hz crossover. Each speaker will receive the appropriate high-pass information—surrounds 80 Hz and up, center channel 60 Hz and up, and mains 40 Hz and up. The subwoofer, though, will receive a signal that will be a summation of the surrounds', center's, and mains' signals, which then will be low-passed at 40 Hz. Do you see the problem? The main speakers will be the only properly produced channels! The center channel will have a hole between 40 Hz and 60 Hz in its response, and the surrounds will have a hole between 40 Hz and 80 Hz in their response.

—*Dustin Bartlett, Brett McLaughlin*

HACK #52 Optimize Subwoofer Placement

Once you've selected a subwoofer, you can figure out the best location for it with a little patience, a lot of trial and error, and a six-pack.

Although you don't need to worry about room size [Hack #50] when choosing a subwoofer [Hack #49], once you have your subwoofer, you should spend some time finding the best location for it. This often is the difference between loud bass and clear bass—an important distinction.

Basic Principles

A subwoofer for home theater tends to have the biggest impact if placed in a front corner with the longest unbroken walls on either side. This gives the sound the full length of the walls to reflect off of and reinforces the lowest possible frequencies.

However, many music lovers sometimes find the corner placement to have too much enhancement. What sounds great for movie special effects might overemphasize sounds for music.

Remember that the goal of a music system is to exactly recreate the original artist's sound.

To find a happy medium between powerful bass and maintain a good musical balance, you can put the subwoofer at one-third (or even three-fifths) down the length of the longest wall. This causes the subwoofer to "see" two smaller walls on either side. Although the sub no longer reinforces the lowest frequency, it still has the smaller wall sections to reflect from. This can create a more level response. However, even this is not a perfect solution in every case. The room walls and ceiling will reflect some sounds more than others. The interaction of all these reflections in the room can cause huge swings in sound depending upon where you sit. Moving your head six inches one way, or the sub six inches another way, can have a dramatic effect. So, instead of just accepting a rule of thumb, you should test out different locations.

The "Six-Pack" Method

One of the more popular techniques for sub placement is called the "Six-Pack Method." Here is what you need:

- A long piece of CATV coaxial cable with F-connectors on each end
- Two F-to-RCA-Male adapters from RadioShack
- A six pack

Attach the F-to-RCA adapters to the ends of the coaxial cable to make a long subwoofer cable.

You can make a 25-foot subwoofer cable for less than 12 bucks with this trick.

Take your subwoofer and put it right in the middle of your primary seating location (yes, move the couch out of the way). Use your homemade subwoofer cable to connect the sub to your receiver. Disconnect all your other speakers, and find a bass-heavy chapter on a favorite DVD. Use the A-B Repeat feature of your DVD player to play the chapter repeatedly.

You aren't trying to play just the loudest few seconds over and over again. This can damage your subwoofer after a few minutes. Make sure you have the entire track where the subwoofer kicks in and out set to Repeat.

Take the six pack and crawl into the corner that meets the longest wall. Sitting in that corner, pay attention to the sound of the subwoofer as it plays the DVD. Then, begin to move along that long wall, working your way around the room, listening to the sub's response. At some locations the sound will be boomy and rough, and at other spots the sound will be smoother and sometimes even quiet. Using a can, mark the location where the sound is smooth. Try and pick three or four spots like this where the subwoofer would fit.

When you have several spots picked out, put the subwoofer in the first location you've marked with a can. Replace the couch, reconnect all the speakers, and sit down to listen. By the law of reflections, the sound in your primary seating location now should be as smooth as the sound where the can was. Now chug the can you just moved.

If the subwoofer (along with the rest of the system) sounds great, you are done. If not, try one of the other can locations for the subwoofer and repeat the process, making sure to drink the whole can each time. Eventually you will find the best location for the subwoofer—or be so caffeinated, sugared up, or inebriated that you won't care.

—*Robert McElfresh*

HACK #53 Use Multiple Subwoofers

As the size of your theater room grows (along with your budget), you might want to add a second subwoofer to really rattle your windows. However, you'll need to take great care in placing the second subwoofer so that it helps your sound rather than hurts it.

If you get a good subwoofer [Hack #49], you're going to end up with a solid bass response from your home theater, even if you don't spend a lot of time figuring out the best location for the sub [Hack #52]. A great sub in a poor location still is going to provide a good low end for movie playback. However, when you begin to use a second subwoofer, placement of the second sub becomes critical. You actually can knock out a lot of your bass response with a second subwoofer if you're not careful.

First, forget about all those old web pages or drawings of a room with a subwoofer on different sides of a room or on either side of a couch. More often than not, these were done by people who were not audio enthusiasts.

The problem with multiple subwoofers (and many of those drawings you've seen) is wave interactions. Have you ever dribbled water from the kitchen faucet into a pan of water? Try it and notice how the ripples expand out and reflect off the sides of the pan. This is almost exactly how sound waves travel in your room from your subwoofer. Now start to dribble more water, from another glass, into a different location in the pan. The waves from the two sources of water are now interacting in a complex fashion. Instead of smooth ripples, the surface of the pan is turbulent and choppy because of the two different dribbles of water. Instead of getting smooth waves, you've got the two sources of ripples fighting each other; this is a great picture of the problem you easily can get into when using multiple subs.

Thankfully, there is a simple solution to the complexities caused by two subwoofers: put them both in the same location. If you can stack them on top of each other, that's great. If not, just place them side by side. Going back to the illustration of dripping water, now you've allowed twice as much water to drip in, but all the water is coming in the same spot; there is no conflict between waves, water, or sound.

With your subs placed, take your existing subwoofer cable and buy a Y-adapter and two short audio cables. Split the subwoofer signal near the subs and connect each woofer with the Y-cable to the main sub signal. Finally, use your SPL meter **[Hack #63]** to level-adjust the combined subwoofer stack to match your other speakers.

—*Robert McElfresh*

HACK #54 Remove Subwoofer Hum

A common subwoofer problem is hum—that low buzzing that seems to fill all the empty spaces in your movie soundtracks and annoys dogs and low-flying birds. Learn how to isolate the hum, analyze it, and remove it.

One of the most annoying problems you'll ever encounter in a good home theater setup is subwoofer hum. This most often crops up when you've got everything set up and calibrated, all your cables Velcroed together, and your racks moved back into their final resting places. Then, just as you get comfortable on the couch, something starts to hum. As annoying as this is, it's not going to go away until you locate the source of the hum and get rid of it.

Hum Created by Coaxial Cable

The number-one cause of subwoofer/speaker hum is the coaxial cable connecting your cable or satellite receiver to your provider (either through an in-ground run to a cable box or through a satellite dish). Here is how you test for this:

1. Turn your system on and get it to produce the hum by watching a movie. Pause your DVD or videotape so that the hum is all you hear.

Don't use a normal cable or satellite program for this; you're about to disconnect the cable that provides the audio and video for your cable/satellite feed.

2. Find the coaxial cable running from your cable or satellite receiver to your service provider, and while listening to the hum, unscrew the connector and disconnect the cable.

Did the hum stop or reduce by a large amount? If so, the cable you disconnected is the source of the noise. If this is the problem, you have several ways to fix the issue permanently (if not, jump ahead to the next section):

1. Call your cable or satellite company and ask for a service call. Sometimes you get a smart cable guy, and if you demonstrate the problem he

can do something upstream to ground the coaxial cable and remove or reduce the hum.

2. Buy a power strip that has F connectors as part of its surge protection. Plug the strip into the AC outlet and feed your main coaxial signal through these connectors. This ties the shielding of the coax (the source of the noise) to your AC ground and sometimes can solve the problem.

Oddly enough, this solution (surge protection) can sometimes increase the humming.

3. Go to your local RadioShack store and buy three inexpensive items: a Matching Transformer (part #15-1253), an Indoor/Outdoor Matching Transformer (#15-1140), and a Cable Coupler (#278-304). Connect your coaxial cable to the cable coupler, and then to the first matching transformer. The output is two screws for the old two-wire antenna wire. Your indoor/outdoor matching transformer has two connectors for the screws, and the other end is a coaxial connector. Hook your cable or satellite receiver into this connector and see if the hum goes away.

   Here is why this trick works: the first transformer converts your 75-ohm coax into a 300-ohm antenna connector. The second transformer converts the 300-ohm back to a 75-ohm connector. The humming, which usually is at around 60 Hz, can't pass through these conversions.

Ground Loop Hum

The second cause of hum is called a *ground loop*, and it almost always shows up right after you bring home a brand-new, self-powered subwoofer, or perhaps an external amplifier.

Take a look at all the plugs on the power cords on your home theater equipment. In most systems, the receiver (or amplifier) has a three-prong power plug, but most of your other devices have only two-prong plugs. This is not by accident; the device with the three-prong power plug is grounded. This means that device "owns" the ground. As long as no other power device has a three-prong plug, everything works well.

When you bring home a self-powered subwoofer and plug it in, though, you might notice it has a three-prong plug; this is for safety reasons. However, when you connect an RCA cable from your receiver to your subwoofer and turn everything on, you suddenly notice a loud hum.

The external amplifier in your subwoofer is now fighting with the amplifier in your receiver for possession of the ground. Both devices want to define 0.00 volts. But because the wiring in the two amplifiers to your household AC ground is different, one device is really using 0.001 volt and the other device uses something closer to 0.003 volts. The subwoofer cable connects the two, and the fighting begins.

You have to stop these two devices from trying to own the ground, or get them to not "see" each other. First, make your system produce the humming noise. Disconnect the single RCA cable between your receiver and subwoofer. Did the noise stop? If so, you have a ground loop issue.

The proper, safest way to solve this problem is to buy a special subwoofer cable with little arrows on the wire to show the signal direction (see Figure 6-1).

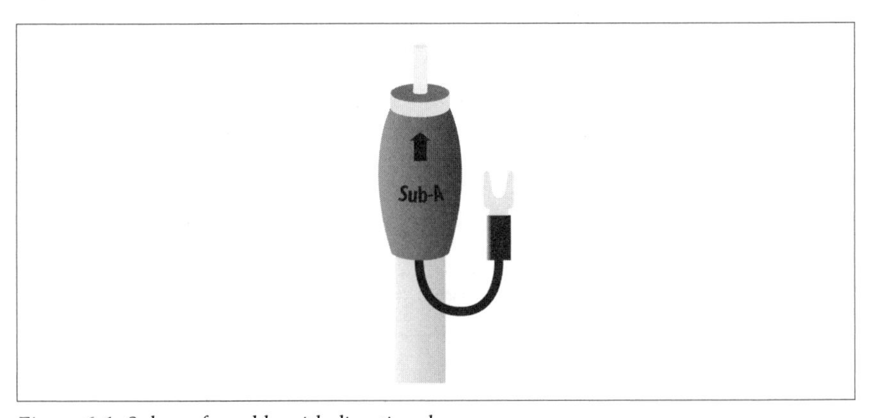

Figure 6-1. Subwoofer cable with directional arrow

Unfortunately, many people have been ridiculed when asking about these sorts of cables at their local electronics store: "But cables don't really have a direction. My expert friend at work laughed at me when I asked about this!" Yes, your friend is right. Cables don't have a direction, but these little arrows indicate that this cable will prevent or solve your ground loop problem.

Remember when I said the hum started when you connected the RCA cable? That RCA cable really contains two wires: the center wire and something called the *shield*. The center wire carries the audio signal, but the shield tries to define 0.00 volts. The shield is the wire that lets the two different components (the receiver or amplifier, and the subwoofer) see each other's ground, and causes the fight. What if you took your subwoofer cable and disconnected the shielding from just one end? Wouldn't that solve the problem? Yes, it would. This is exactly what a subwoofer cable with little arrows does. The shield is not connected at both ends. The shield has to be connected at one end, for connecting to your receiver or amp, so you should run the cable so that the arrows show the flow from the receiver to the subwoofer.

It is unsafe to use a two-prong to three-prong "cheater" plug on the subwoofer power cord to solve the hum problem. Even if the subwoofer came with a cheater plug in the box, it's REALLY not safe to do. Don't do it.

—Robert McElfresh

CHAPTER SEVEN

Connectivity

Hacks 55–59

So, you've purchased all this equipment, and now it's time to hook it all up. That's going to involve cables—lots and lots of cables. And because every component serves a largely distinct function, you'll need lots of different cables. Some carry multichannel audio; others transmit video; still others work best for single-channel audio. Getting good sound and video, though, involves more than just picking a cable that fits into the desired hole and plugging it in.

Cable Basics

Understand what makes a good cable, and understand which is the best cable for the job.

The manufacturers include cables with their equipment, but I suggest using these cables only temporarily until you can pick up something of decent quality. I usually recommend the RadioShack Gold Series cables as a good starting point, but you can work you way up to higher-quality cables from there. Before you go out and purchase these cables, though, you need to know what types of cables you require, what they're called, and exactly what they are used for. I'll be describing these cables based on the type of connector used and the actual design of the cable itself.

You'll notice that every cable shown in this section is from Better Cables (*http://www.bettercables.com*). Although you might choose to go with a lower-end cable, such as ones from RadioShack, I can't recommend Better Cables strongly enough. That company offers amazing cables, at very fair prices, and they are the only kinds of cables I buy for my own system at home.

RCA-Analog Audio

The most commonly used cables are analog audio cables, with RCA connectors on each end. These are the cables normally used to connect a tape deck, turntable, CD player, etc., to your receiver. They usually are color-coded with a red connector for the right channel and a white connector for the left channel (see Figure 7-1). There are also some multichannel sources which have an analog audio connection for each channel: left, right, center, surround left, surround right, and subwoofer.

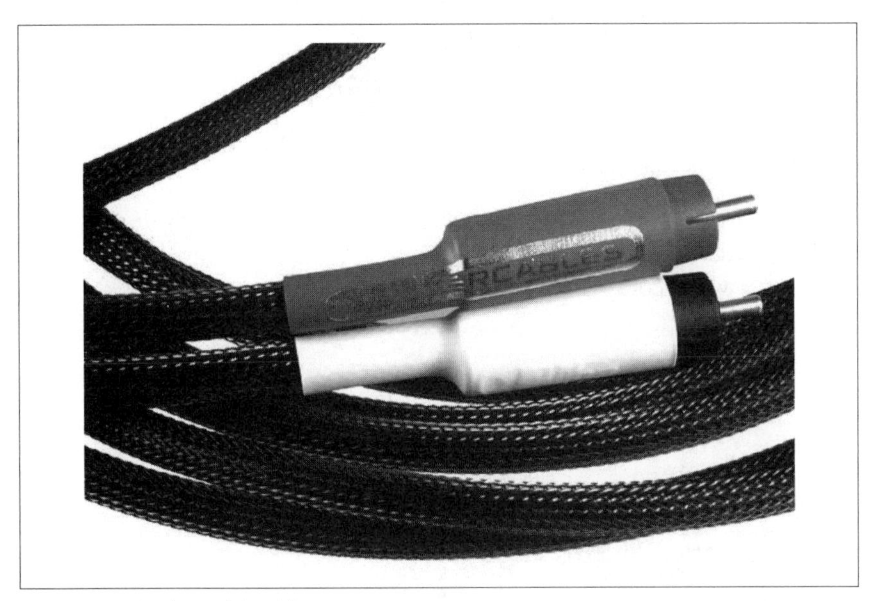

Figure 7-1. Analog RCA cables

RCA-Digital Audio

This also uses an RCA connector, but the cable itself is very different. Most DVD players have both analog and digital connections, but the digital connection is what is required to transmit Dolby Digital and DTS signals. The digital signal is meant to be carried by a 75-ohm RCA coaxial cable, which has a bandwidth of 6 MHz (see Figure 7-2).

RCA-Composite Video

As you probably have noticed by now, the little RCA connector (see Figure 7-3) gets around a bit. It is used not only for audio connections, but also for video connections. In this case, it's a single analog video connection. This cable, like the digital audio cable, should be a high-quality, 75-ohm

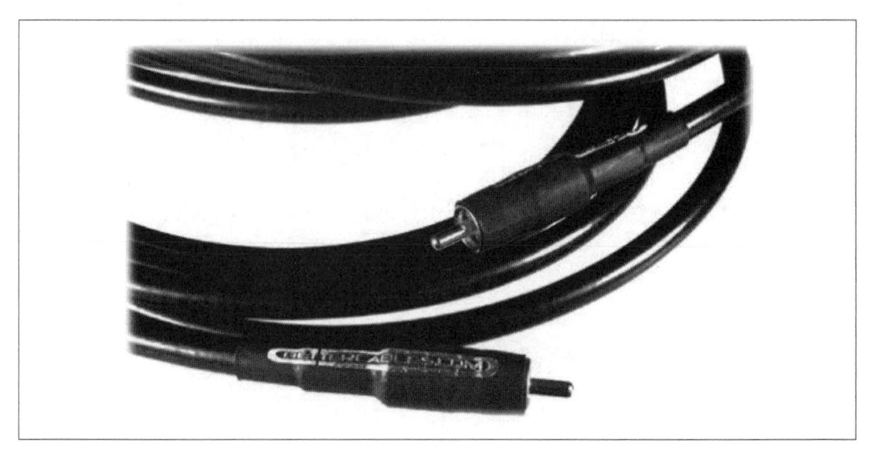

Figure 7-2. Digital coax cable

coaxial RCA cable. Still, composite video is considered to be the lowest-quality video signal. S-Video connections are a noticeable improvement over composite, while component video connections are even better still.

Figure 7-3. Composite video cable

S-Video

Another common video option is S-Video. Most midrange DVD players and TVs offer S-Video connectivity, and it is far superior to composite video. This is also the first of the newer cables that have a different connector. If you can, always use S-Video over composite (although you usually can improve by another order of magnitude if you've got the ability to use component video cables, detailed next). Figure 7-4 shows an example of this type of connector.

Figure 7-4. S-Video cable

RCA-Component Video

I'm just about finished with the RCA connector, but before I move on, I need to talk about component video signals, which are the highest-quality analog video signals. These signals are divided among several cables (see Figure 7-5) and separated into multiple elements including the red, green, and blue color information, along with chrominance and synchronization information. On most consumer gear you'll see these connections handled by three RCA connectors. They most commonly are labeled Y, Pb, and Pr. Y is the luminance, containing black and white information as well as brightness. The Pb and Pr signals carry *color difference* signals: Pb indicates the amount of blue relative to luminance, and Pr indicates the amount of red relative to luminance. Because the total brightness is supplied by the Y signal, the amount of green (which would be labeled Pg, I suppose) can be figured out from the difference between the blue and red and the total luminance. Once again, these signals should be handled by high-quality, 75-ohm video coaxial cable.

Figure 7-5. Component video cables

BNC

Although you can use BNC connectors for both composite and component video cables, you normally will find them only on professional video equipment. They are much more rugged, and actually lock into place when attached to the connection point on the equipment (see Figure 7-6). If it were up to me, these would be standard on consumer equipment because they offer a better connection than RCA connectors (used in both composite and component video cables). They also are easier to attach and remove than most of the standard coaxial F-connectors on TVs and VCRs. Keep in mind that any advantages of BNC are purely in connectivity rather than quality.

Figure 7-6. Component video cables with BNC connectors

Some audiophiles argue that a better connection makes for better signal transfer. Although this is true, the connectivity improvement of BNC is largely about convenience; it won't result in, say, the wires carrying more of the signal. I find no sound improvement in BNC versus composite or component video, although the usability is much better.

Optical

This cable is used in home theater for digital audio. It is often referred to as a *Toslink cable*, which refers to the type of connector found on each end of the cable (see Figure 7-7). This cable is used for the same application as the RCA digital cable described earlier. The major difference is that it is used to send optical signals via fiber optics. The positive side of these cables is that they aren't susceptible to radio frequency or electromagnetic interference. The negative side is that they are much more susceptible to being damaged by being kinked or bent too sharply. As a result, the RCA digital cable is my choice for a digital interconnect whenever possible.

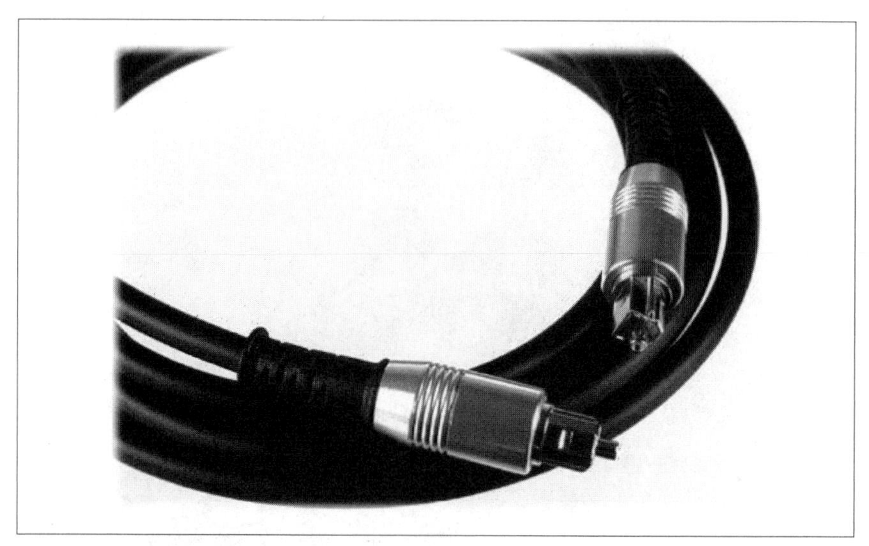

Figure 7-7. Optical connectors

XLR

Higher-end audio processors often will allow for a balanced connection between the preamplifier/processor and the outboard amplifier. In these cases, you'll need XLR cables, as shown in Figure 7-8. These are generally the very high end in terms of gear, so even midrange components won't need these.

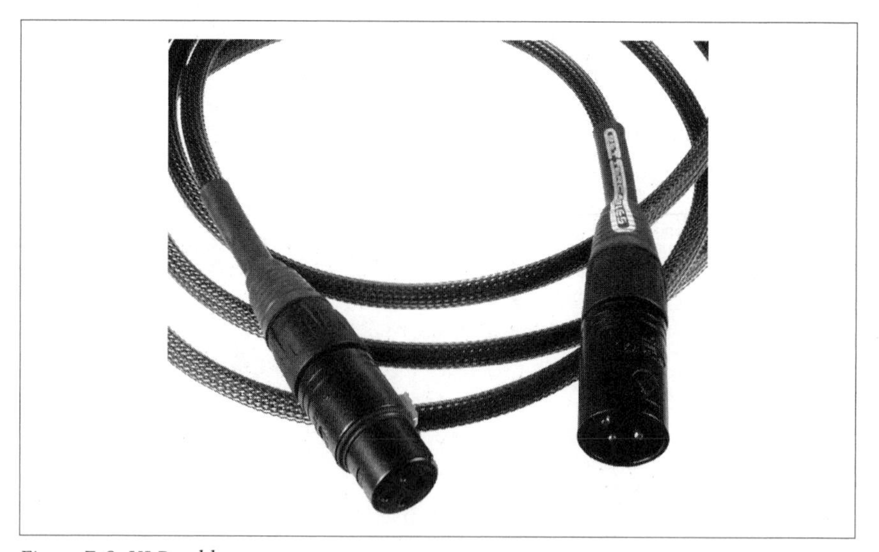

Figure 7-8. XLR cables

F-Connector

Most of you are familiar with this type of connector. It is the old screw-on coax cable connector that has been used to hook up cable TV and DBS satellite dish systems for years. In the past it was used solely for analog audio and video, but today it also can carry digital audio and video information.

—Brian Bunge

Watch Out for Entertainment Centers

There's a tendency to want to load expensive equipment into expensive containers—usually a high-dollar, all-wood entertainment center. Watch out, though, as this could cost you time, energy, and all kinds of frustration.

Furniture manufacturers leapt into the home stereo and then home theater business early, and with both feet, with a wealth of "entertainment center" offerings.

Unfortunately, these attractive pieces of furniture can cause as many problems for your home theater as they solve. I'll list some of the problems, and some of the solutions that you can undertake.

Problem #1: Overheated Equipment

Many home entertainment centers offer totally enclosed cabinets that are supposed to hold the electronics for the home theater. They even can include a solid set of doors for the front. But where does the heat get out? Amplifiers—and receivers in particular—generate a lot of heat, and totally enclosed shelves often result in an early death for your equipment due to heat stroke or stress.

Here are a few solutions to Problem #1:

- Cut large openings in the back of any closed equipment enclosures, and don't place the entertainment center right against the wall. This way hot air can escape out the back.
- Remove any solid doors for equipment cabinets so that the heat can escape out the front (of course, this might ruin the very thing you liked about the entertainment center in the first place!).

Problem #2: Impossible Wiring Access

Fully enclosed equipment spaces also make it very difficult to wire a home theater. Sometimes the entertainment center manufacturers teasingly provide a few small holes or thin slots that are completely inadequate for the number of cables that usually have to be connected.

Here's a solution to Problem #2:

- Happily, cutting large holes in the back of the closed areas, or even removing the back cover of the entertainment center, make it much easier to wire everything together, and also helps with overheating.

Problem #3: Unusable Remote Controls

Doors that close over equipment shelves can sometimes block the signals from the remote controls.

Here are some solutions to Problem #3:

- Remove the doors! Now the remote control can get to the equipment, and the equipment runs cooler, too.
- Choose an entertainment center with doors that are transparent to the remote control signal.
- Choose an entertainment center with no doors.
- Choose an entertainment center with doors that hide away inside the cabinet when opened.

Problem #4: Bad Sound from the Speakers

Putting speakers into hollow wooden boxes (the spaces in the entertainment center where the manufacturer wants you to put the speakers) can often muffle the sound, much like cupping your hands over your mouth while talking.

Here are some solutions to Problem #4:

- Mount the speakers on separate stands outside the entertainment center.
- Buy a low entertainment center that only sits underneath the television and speakers, with the speakers on short stands to place them evenly centered about the height of the center of the television screen.
- Place a good-quality, sound-absorbent material in the back of the enclosures where the speakers sit to kill the resonance of the enclosure so that the enclosure does not alter the sound of the speaker too much.
- Make sure the speaker front edges are lined up with—or even slightly in front of—the front edge of the entertainment center.

Problem #5: Bad Stereo Separation

Stereo speakers or the left and right channels of a surround sound system setup must be placed an equal distance from either side of the television screen. As a starting point, the distance between the speakers typically should be slightly more than the distance between the viewer and the television screen.

Dolby Digital–equipped home theaters might be able to lessen this distance due to the excellent channel separation between the left, right, and center channels.

Entertainment centers can force you to place speakers in locations that violate these guidelines.

Here are a few solutions to Problem #5:

- Modify any barriers or dividers in the entertainment center that prevent you from keeping speakers centered on the television.
- Mount the speakers beyond the outside edges of the entertainment center using shelf brackets or similar hardware. Make sure the front of the speakers is not behind the front edge of the cabinet.
- Mount the speakers on their own stands on either side of the entertainment center at the proper spacing.

Problem #6: Inflexible or Unmovable (or Both!)

People usually buy the entertainment center to "just fit" the television and other equipment they currently own. Then they're stuck when they want to upgrade to something with different dimensions—particularly to a larger display! Large entertainment centers with no wheels also make it a nightmare to get behind the unit to do the original wiring or to rewire it later.

Here are some solutions to Problem #6:

- Get a unit with some designed-in flexibility.
- Get a unit with good wheels or casters.

Problem #7: Inconveniently Low Access

Some entertainment centers force you to place the electronic equipment all the way down near the floor. This places the equipment closer to the dirt, splashes, vacuum cleaner activities, and (heaven help us) very small children. It's no fun having to get down on your hands and knees to push buttons, to put in a disk or videotape, or to remove a peanut-butter-and-jelly sandwich from the VCR.

Here are some solutions to Problem #7:

- Buy an entertainment center that provides at least some equipment shelves at a convenient height.
- Place the equipment that you need to reach most often at a location somewhere between waist and chest height. Place the other equipment that you don't have to mess with very often down in the lower areas of the cabinet.

Problem #8: Speaker Blockage

The doors on some entertainment centers swing out in such a way that they block the sound of the speakers from getting to the audience. Even if they don't block the sound completely, they can re-create the undesirable "hands cupped around the mouth" situation mentioned earlier. Subwoofers set next to an entertainment center also can get blocked by open doors.

Here are a few solutions to Problem #8:

- Choose an entertainment center without doors or with doors that swing out of the way.
- Remove the doors (are you detecting a theme here?).
- Place speakers so that the doors don't block them when the doors are open.
- If you use small speakers and you have doors that open out to the left and right on either side of the entertainment center, consider trying to actually mount the speakers on the inside of the doors so that when the doors swing open, the speakers are in a good position.

Conclusion

If you get the impression that I am not in love with home entertainment centers, you have got it right! I feel the best compromise for folks with conventional TVs is a sturdy, somewhat low, open set of shelves which are high enough to put the television screen at a good viewing height with the TV sitting on top of the shelves, as shown in Figure 7-9.

That said, Ginni Designs (*http://www.ginnidesigns.com/*) offers some products that address some of the problems I mentioned. If you must have a "Nice Piece of Furniture" to justify all the money you are blowing on your home theater gear, consider the Ginni Designs products as a starting point. You also should consider going with equipment racks **[Hack #7]**, or building your own TV stand **[Hack #76]**.

—David Gibbons

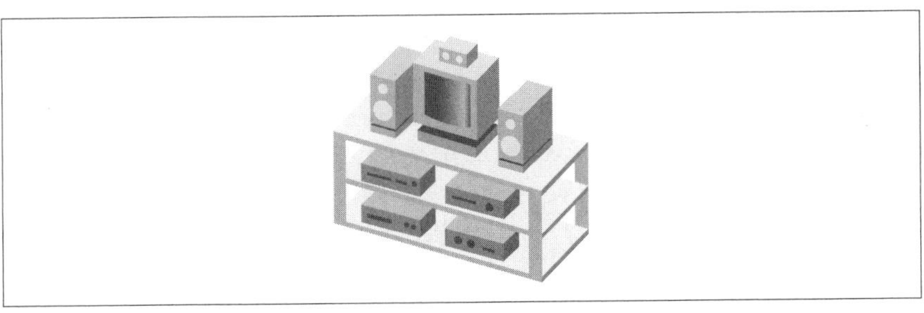

Figure 7-9. Simple entertainment center layout

HACK #57 Don't Be Swindled into Buying Overly Expensive Cables

Many salespeople will try and get you to spend thousands of dollars—just on cables! Often this is just modern-day snake oil, so be wary of what you hear.

Wiring and cabling are the plumbing of your home theater. If a plumbing salesperson came to your house and told you that all your plumbing was inadequate and had to be replaced with $1,000-per-running-foot, gold-plated, high-pressure, 6-inch-diameter piping, you would tell him to go and (insert your favorite impolite phrase here).

Don't be hustled by audio/video salespeople trying to sell you overspecified and superexpensive wiring and cabling. Here's some information to help you understand what is important and unimportant about wiring and cabling.

A Community Divided

To be fair, home theater enthusiasts are pretty divided on the subject of cables. Some insist on buying the highest-grade cables. Others opt for higher-end cables, such as those from Better Cables (*http://www.bettercables.com*). Still others actually will construct their own cables, such as for connecting speakers [Hack #79] and even power cables [Hack #80]. And, finally, some believe that all of this is hogwash, and they go with what others might call a lower-end cable. Note that several of these viewpoints are represented in this chapter. Read all the hacks and decide for yourself, but whatever you do, make an informed decision!

About Wires and Electricity

Let's start with a little bit about wires and electricity. Wires and electricity go together much like pipes and water go together. For plumbing, pipe is chosen on the basis of the pressure of the water, and how much water has to flow through the pipe. Corrosion resistance also is important. This is common sense; plus, there are lots of building codes to make sure the right type of pipe is chosen and installed in your home. The material, thickness, and diameter all need to be correct so that the water stays inside the pipe, and so that enough water can get to where it needs to go.

The same things are true of wires and electricity. Electricity is not exactly the same as water, but the same sort of ideas apply.

Water pressure=electrical voltage. A pipe has to hold the pressure of the water inside it. A wire has to handle the voltage that is applied to it.

The highest voltage in home theater system cables and wires is actually the voltage on the AC main power cords. All the other voltages are much, much lower, and super-fancy insulation is therefore not required.

Water flow=electrical current. A pipe has to be big enough to carry the desired flow of water. A wire has to be big enough to carry the desired flow of electricity.

The highest current flow in home theater system cabling and wiring is usually the speaker wire current. It is not much current, and thumb-thick copper conductors aren't required.

Water's sudden change in direction=electrical frequency/signal. Water doesn't like to stop or change direction suddenly. You can prevent this effect, called *water hammer*, by putting together the piping correctly.

The direction of flow of electrical current can be changed very rapidly. However, cables, wires, and connectors that carry that electricity must be designed to handle the rate of change of flow of electricity (*frequency*) they are carrying. If these items aren't designed and manufactured correctly, they can foul up the proper change of flow of electricity through them.

As the frequency increases, the cable and connector design and material must be able to carry the flow without problems. Almost any sort of solid connection works for audio frequencies. The frequencies coming from a broadcast cable system these days can be more than 100,000 times higher than the highest audio frequencies, and connectors and cables suited to handle those frequencies are required. Video frequencies are intermediate to the previous two, and thus cables for video frequencies need not be quite as good as the cabling required for broadcast cable wiring.

Pipes/corrosion resistance=wires and cables/corrosion resistance. Corrosion or contamination of a water pipe can cause either a blockage or a leak. Corrosion or contamination of a wire or cable can reduce the amount of flow of electricity, or mess up the proper change of flow of electricity through that wire or cable.

Pipe/watertightness=cables/shielding. A shield wrapped around a cable's center conductor keeps the signal on the cable free from interference, and also keeps the cable's signal from leaking out and interfering with other equipment. The low frequency and high power level in the speaker wires make shielding unnecessary for those wires. All other connections require shielded cables.

About electrical signals. An electrical signal is just a change in flow and/or direction of electricity, which is used to carry a message. You even can simulate this in a plumbing system:

- Water on = the message "yes"
- Water off = the message "no"

Because electricity is a lot neater and faster than water, we use it to carry signals in our home theater rather than water.

The Truth About "Wiring Upsell"

Let's go back to that plumbing salesperson who was trying to sell you a pipe-upgrade job for your house. He told you 6-inch-diameter, gold-plated pipe will never corrode, will be sure to carry all the water you'll ever need, and will never affect the taste of your water. He's telling you the truth!

So, why did you tell this salesperson to take a hike? Because you know the plumbing you have is good enough. You know it would be better to spend your plumbing upgrade money on a better bathtub or a better sink. The exact same principle applies for wiring and cabling on home theaters!

For the ordinary home theater, money spent on extra-fancy cabling and speaker wires is a waste. The salesperson in the home theater store will try to sell you wires and cables that are the electrical equivalent of the insanely expensive pipe mentioned in the earlier example. Yes, the wire he's trying to sell you will carry lots of electricity. Yes, the cable is more corrosion-resistant and will carry high-frequency signals better. He even will show you articles in fancy home theater or audio magazines that praise the wire or cable the salesperson is trying to sell you.

Ask yourself where those magazines get their advertising dollars. Ask how profitable the "high-end" wire and cable can be for the home electronics stores.

You need to remember the phrase "gold-plated pipes" when reading or listening to these pitches.

Wiring Recommendations

Enough theory; I'm going to make some recommendations on the right amount of quality you should pay for in the wiring and cables for your home theater.

Speaker wires. Type SPT lamp cord is plenty good for speaker wire. Eighteen-gauge is heavy enough, unless you have long runs—say, more than 30 feet. Then you could use 16-gauge wire. I specify this type of wire because it is very durable. You can step on it, set furniture on it, and bend it back and forth a lot before the wire will fail. You can buy good-quality, type SPT lamp cord in bulk at your local hardware or electrical store for only around 15 cents per foot.

Really cheap stuff sold as "speaker wire" is typically inflexible and will fail quickly if it is bent back and forth a few times. Further, the insulation quality of "speaker wire" is sometimes poor, and can actually contribute to wire corrosion. Stay with the type SPT cord.

Reliability can be a problem where the speaker wire attaches to the amplifier and the speaker. Putting connectors [Hack #44] on the ends of the wires can help.

You can buy expensive hook-up connectors of various types, but inexpensive crimp-type solderless spade tongue terminals are available at your local hardware store and will work well to make a good connection to the terminals on the speakers and amp. Get a matching crimp tool from the hardware store and securely crimp a connector on each wire end, taking care that no bare wire is exposed outside of the crimp connector. Having no exposed bare wire avoids short circuits.

If your equipment has the little spring-clamp-type connectors, you can cut off one of the two prongs on a spade tongue terminal and stick the remaining prong into the hole where the wire goes. However, Thomas & Betts makes a crimp terminal especially suited for connecting speaker wires: ask for the Thomas & Betts type TV14-18BL blade crimp terminal at electrical supply houses.

Banana-type connectors are a somewhat more expensive, but convenient, option where the speaker and amplifier terminals will accept them. Again, look for ordinary ones, such as those sold by Pomona Electronics, and avoid the gold-plated "high-end" versions. Whatever connections you make, ensure they are clean and tight.

Audio "line-level" signal cables. Audio frequencies don't require fancy cables either. Good-quality shielded cabling with decent RCA connectors will do the job. Gold plating is not required, although these days you might not be able to avoid it. L-com (*http://www.l-com.com*) offers excellent value in its RCA cable assemblies, found in the Video/Monitor cable section of the company's web site. RadioShack is probably the nearest local place to get the cables you need to carry the audio signals between the different pieces of gear in your system.

RadioShack is now offering "high-end" audio/video cables, too, so go for the less-expensive stuff.

Video signal cables. We are caught in an interesting situation on the subject of the cables and connectors used to carry the video signals used in home theater systems. A long time ago some unknown engineer, probably working for RCA, designed a little connector for connecting cables to (it is said) phonographs. Supposedly, this is why RCA connectors also are known as *phono connectors*. This type of connector is inexpensive, and is not a precision connector designed to carry radio frequency signals. Unfortunately, the manufacturers of television and consumer audio/video equipment somehow ended up using this inexpensive connector for carrying low-level video signals.

Gee, could it have anything to do with the connector being really inexpensive?

The phono connector obviously works. Millions of home theaters have them carrying video signals. People get decent pictures. Still, audio/video cable marketers tell us that if we spend lots of money on fancy cables to carry the video signals, our pictures will be much better.

The problem is that every piece of wire and connector between a radio frequency signal source and a radio frequency signal receiver affects the signal. When I talk about a signal source, I am talking about the actual integrated circuit buried inside your DVD player that is creating the video signal. When I talk about the signal receiver, I am talking about the integrated circuit in your receiver which takes the video signal from your DVD so that it can be passed on to the TV. The circuit board traces, jumper wires, jacks, plugs, interconnect cabling, etc., all affect the signal going from one integrated circuit to the other.

Think of a chain: the weakest link determines the strength of the chain. If you increase the strength of most of the links in the chain, the chain is not much stronger. With the RCA connector being what it is, spending big dollars on a fancy cable that connects to those connectors might not be the best use of your money.

Once again, using regular L-com or RadioShack RCA cables to carry the video between the different pieces of gear in your home theater will deliver very good pictures. If you would rather spend $300 just for one video cable (solid silver conductors, triple shielding, super-duper connectors, etc.), go for it. However, spending the $300 to help buy a new television that costs $280 more than the one you were going to buy and a plain $20 video cable probably will improve your picture far more.

Antenna/broadcast cable connections. As with the video signals, various problems make interconnect cable replacement with top-dollar cables an unlikely route to a perfect picture. Consider all the following (negative) statements:

- The quality of the signal provided by a rooftop antenna is typically so-so, at best.
- What passes for a signal coming from your local cable company is usually just that: passable.
- The F-connector that is used to connect the so-called "RF" signal to your set top box, VCR, receiver, or television is a poor conductor.

In these cases, super-duper connection cables again might not make a big improvement, except to the bank accounts of the people who make and sell those types of cables.

Even so, let's look a little at the F-connector, as it does perhaps justify considering somewhat better cable assemblies for more reliable connections. Even more than the phono connector, the F-connector is a miserable consumer-grade connector.

Cable companies favor the F-connector because it is cheap and quick to install.

Due to the low performance of the typical F-connector, the best interconnect cables in the world won't fix all the F-connector's problems. That being said, a somewhat better interconnect cable with better connectors featuring properly shaped gold-plated center pins can improve contact reliability. Of course, F-connectors mounted on televisions and VCRs have a cheap internal center contact that is prone to damage and corrosion, so even then you're still fighting your own equipment. Once again, L-com offers very good RG-6 cable assemblies fitted with better-quality F-connectors at good prices.

Here are some final thoughts on cables:

- The quality of the cable shielding, and the quality of the soldered or crimped connections that join the cables to the connectors, probably matters the most for long-term reliability and signal quality. Single-braided or foil shielding is good, particularly for the video and the antenna/broadcast cable connections.
- Unscrew the cable connector covers if you can, and inspect the actual connections inside the connector. Are crimps tight? Are no loose parts or unsoldered strands of wire floating about? Great cables cannot make up for a connector that is poorly assembled.
- Are plugs firmly seated when connected to the jacks? The plug's firm grip on the outer, shielded part of a phono connector is important for good shielding.
- A true performance nut would replace all the phono-type video jacks and the F-connectors on the back of all his home theater equipment with different connectors, which would perform excellently at the required frequencies.

  I guess it would be an interesting electronics project, even if it voided the warranty, chanced damaging the equipment, and didn't make a big improvement in the picture. Then again, I bet you could find someone who would charge you an arm and a leg to do it for you!

- If you think very expensive connectors and cables make your system sound and look great, I won't argue with you. I'm talking to the ordinary person who wants to make the most of a limited home theater budget.

Here's to spending your home entertainment dollars wisely!

—David Gibbons

Use Your Receiver for Video Switching

Run all your theater's non-component video sources through your receiver, and switch devices through the receiver.

One of the most frustrating situations in home theater is hooking up all your components and getting no picture from a particular device. It's a big shock to install a new home theater receiver and hear great sound, but see no video. Video cables are running to the receiver, one cable connects that receiver to your TV, yet you can see no picture.

The problem here is that most receivers will not switch video signals from one type of cable to another. For example, a VCR that is connected to your receiver with RCA cables will not display a picture on a TV connected to that receiver with S-Video. The receiver is incapable of converting an RCA signal (from the VCR) to S-Video (which runs to the TV). It's only in the very high-dollar ranges that this is a feature of receivers and processors.

To avoid this problem, use S-Video cables to connect all devices to your receiver. Your DVD player(s), cable and satellite receivers, and even your gaming systems should all have S-Video outputs. Some older VCRs don't have S-Video outputs, but you can buy an inexpensive composite-to-S-Video adapter from RadioShack (part #15-1242) to provide an S-Video output. The quality isn't great, but you aren't going to get good quality from a composite video signal in the first place. This solution ensures that all video coming into your receiver is in the S-Video format.

Then connect your receiver to your TV with an S-Video cable. Now all your devices will properly display on your TV, in S-Video format. Of course, you also might want to use component video (and you should, if at all possible). You can use the component switching of your receiver [Hack #59], combined with your TV's component inputs, to handle these inputs. Finally, train your family and friends to use your receiver as the brain of your system, switching audio and video sources with the receiver remote. As long as you've got proper ventilation, this will not wear out your receiver, and you'll have your audio and video tied together through the receiver.

—Robert McElfresh

Understand Component Video Switching

One of the most frustrating things in home video is running all your component video-capable devices into your video output. Get around the limitations of your processor and monitor with a little knowledge about component video bandwidth requirements.

If you've got a DVD player, HDTV cable or satellite receiver, PlayStation 2, and Xbox, you're probably running short of component video inputs. This turns out to be one of the major problems in home theater, albeit one that you can solve easily if you get past the (inaccurate) assumptions made at the local water cooler. Most home theater enthusiasts will insist that running your component video cables through the inputs on your receiver, processor, or even an A/V switch will cause the signal to deteriorate. Instead, they'll tell you, everything should be connected straight to your TV or projector. Because most display devices have only two, or possibly three, component inputs, this can be a real problem, but most people are afraid to use their processor's component video switching abilities. However, that's precisely the right way to get around a lack of inputs on your monitor.

Use the component switching on your receiver or processor for all your higher-end components, such as DVD players and HDTV receivers. Most processors will have at least two component video inputs, and some even have three these days. Then, use an A/V switch (see Table 7-1) when you need additional component inputs for gaming or additional video devices.

Table 7-1. A/V switches with component video support

Manufacturer	Unit/part number	Functionality	Price	Comments
RadioShack	15-1976	Four input, one output, manual switch	$40	Not rated for HD, but works well for folks on a budget.
RadioShack	15-1977	Four input, one output, remote-controlled switch	$60	Not rated for HD, but works well for folks on a budget.
RadioShack	15-1987	Six input, two output, remote-controlled switch	$150	Not rated for HD, but works well for folks on a budget. Ideal for gamers with multiple systems.
JVC	JSX111 AV	Three input, one output, manual switch, with S-Video jacks	$100	Not rated for HD.

Table 7-1. A/V switches with component video support (continued)

Manufacturer	Unit/part number	Functionality	Price	Comments
AV-Toolbox	AVT-5842	Four input, two output, IR-controlled switch	$90	HD-rated. Bright green color might not be complementary to other equipment.
Inday	RGB4X-R	Four input, one output, remote-controlled switch	$149	Excellent HD-rated unit.
Audio Authority	1154	Four input, one output, manual switch	$200	One of the best HD-rated units.
Zektor	HDS4	Four input, one output, IR-controlled switch with coaxial and optical inputs and conversion	$300	One of the best HD-rated units. Built like other high-end components, providing extra stability.

Many of these switches are built with the same components that go into midrange, and particularly high-end, receivers and processors. Avoid the misconception that you need a standalone box just to prove to your friends that you've spent the most money. The switches in Table 7-1 are great if you don't have enough component inputs on your receiver, but they aren't needed as a replacement for your receiver.

—*Dr. Robert A Fowkes, Robert McElfresh, and Brett McLaughlin*

Calibration

Hacks 60–74

If you've read the earlier chapters and have been trying out stuff as you go, you might know how to tweak your TV for basic HD viewing **[Hack #35]**, be comfortable with figuring out where best to place your sub **[Hack #52]**, and even pulled the screen off of your RPTV **[Hack #17]**. Hold on to your remote, though, because this chapter is going to take you to a whole new level.

To calibrate simply means to check, adjust, or determine by comparison with a standard. That might sound pretty basic, and something that you could do with a few spare minutes. In home theater parlance, though, calibration is an entire science unto itself. From getting the white on your particular TV to be "true" white, to ensuring that *Jurassic Park* sounds not how you like it, but how Crichton wanted it to be heard, to keeping the back of your TV from crushing you when adjustments are needed, calibration is the key. You'll find a huge number of hacks in this chapter, and these are culled from literally hundreds more. Take the time to work through this entire chapter, and you'll be in a movie theater rather than an old, converted garage.

Choose the Right Seating Distance

Sitting the correct distance from your television is the first step in getting the best picture from your display unit.

Although the location where you sit isn't technically a part of calibration, it's critical to enjoying a good picture. You also need to lock down the seating area before calibrating your display **[Hack #61]** or speakers **[Hack #63]**.

There are two schools of thought for determining proper seating distance:

- Base seating location on visual acuity.
- Base seating location on SMPTE (Society of Motion Picture and Television Engineers) and THX standards.

The visual acuity distances are interesting, but they change based on the source material. Because all TV programming is not HD (and even some HD programming varies between 720p and 1080i), you would have to move your seating based on what you are watching. This obviously isn't a real popular idea.

THX and SMPTE are standards used primarily in movie theaters. They calculate seating distance based on viewing angles: if the angle is too narrow, detail and perspective are lost; if the angle is too wide, it takes a lot of work to scan the entire screen. (Ever tried watching *The Lord of the Rings* on an IMAX screen? Get ready for a sore neck!)

THX places the optimal viewing angle at 36°; 26° and 30° are also popular angles, and generally are applicable for this discussion. Given these numbers and a viewing distance calculator (*http://www.myhometheater.homestead.com/viewingdistancecalculator.html*), Table 8-1 provides the recommended and maximum distances given a 16:9 TV screen.

Table 8-1. Viewing distance for 16:9 TV screens

TV size (in inches diagonal)	Recommended seating distance (feet)	Maximum seating distance (feet)	Visual acuity (based on 1080i source material)
40	4.5	6.3	5.2
45	5.0	7.1	5.9
50	5.6	7.9	6.5
55	6.1	8.7	7.2
60	6.7	9.4	7.8
65	7.3	10.2	8.5

Table 8-2 provides these same numbers for a 4:3 screen.

Table 8-2. Viewing distance for 4:3 TV screens

TV size (in inches diagonal)	Recommended seating distance (feet)	Maximum seating distance (feet)	Visual acuity (based on 1080i source material)
40	4.1	5.8	4.8
45	4.6	6.5	5.4
50	5.1	7.2	6.0
55	5.6	7.9	6.6
60	6.2	8.7	7.2
65	6.7	9.4	7.8

—Robert McElfresh

Get the Best Calibration Tools

Purchasing a few simple tools, such as calibration DVDs and a sound meter, allows you to take your home theater from an out-of-the-box experiment to an in-the-theater experience.

The ability to accurately reproduce the intended look, sound, and feel of a film is one of the major appeals of having a home theater. Respect for film and moviemaking as an art is taken seriously by hobbyists, and the intent of the filmmaker is respected above all.

When films are conceptualized and shot, a lot of energy goes into color palette choices, lighting, production design, and film stock—all decisions that help to subtly (and sometimes not so subtly) influence the viewer's mood, enhancing the emotional impact of the scene. When film is processed and printed, again, much attention is paid to maintain the intended color and visual message of the images. When a soundtrack is mixed and finalized, great care is taken to maintain relative levels of music and effects, to maximize the impact of bass and low-frequency effects (LFE), and to properly present surround information.

When the material reaches the DVD production stage, the masters are once again scrutinized to confirm that the intended color and look of the film aren't altered, and the sound is checked to ensure the channel levels and audio fidelity are maintained. Technicians and artisans work for hours on specially calibrated display and sound devices to tweak nearly every subtle hue, shade, and frequency—all to present the most accurate and proper reproduction of the original artist's vision.

When that disc gets to your home, it is only natural to want to continue this chain of correct reproduction. Without doing so, all the previous work and effort that went into maintaining the specific look and sound of the film would be wasted. By calibrating your system to the same standard the creators and technicians used, you ensure that the intended presentation is maintained in your home. You can see what they saw, exactly as they intended you to see it! Colors will be more vivid. You will see and hear more detail. Sound and video will take on a deeper and richer tone.

Types of Calibration

It is with this goal in mind that we seek to conform our home equipment to the identical standards (or as close as our budget allows) as were used in the creation of the material in the first place. We want to see what they saw. To do this, two types of calibration can be employed.

User calibration
: The consumer (or a technically adept buddy) uses basic user menus and picture/sound adjustments such as brightness, contrast, sharpness, and speaker level to conform to standards using test patterns, basic measuring devices, and filters.

Professional calibration
: A trained (and usually paid) professional uses service menus and physical adjustment/repair of the internal workings of the display device using advanced test patterns, expensive professional equipment, and a professional eye.

You always should begin with user calibration, as you can reap tremendous benefits with minimal expense, other than your time. Then, with your system running in tiptop shape, you might consider saving up your pennies and undertaking professional calibration.

Higher-end professional calibration is a great option, and an excellent investment! Organizations such as the Imaging Science Foundation (ISF), located online at *http://www.imagingscience.com/*, train technicians to calibrate and service display devices to ensure visual accuracy. Some of the better calibrators actually tour the U.S. and beyond, performing full calibration and setup.

Tools of Calibration

Whether you're performing calibration yourself or you've got a friend who knows the ropes, you're going to need to have some basic tools handy. Here's the rundown.

Calibration DVD(s)
: Calibration DVDs are standard DVDs that contain test patterns for both audio and video calibration. These discs are invaluable assets, and you'll learn which one is right for the job [Hack #62] later in this chapter. No serious home theater should be without at least one calibration DVD.

SPL meter
: SPL stands for *sound pressure level*, and an SPL meter is just as essential as a calibration DVD. An SPL meter helps you analyze the tones from calibration discs far better than possible with the naked ear, allowing finer tuning of your setup. The easiest place to obtain an SPL meter is at RadioShack (*http://www.radioshack.com*). Although RadioShack offers a fancy digital version, buy the cheaper, analog model (Catalog #33-4050). In addition to costing less, it's easier to work with. My SPL meter is shown in Figure 8-1.

Figure 8-1. SPL meter

—*Vince Maskeeper-Tennant, Brett McLaughlin*

HACK #62 Choose the Right Calibration DVD

Calibration DVDs are the most efficient and cost-effective way to optimize your audio and video components.

With the growing popularity of both DVD and home theater as a hobby, several calibration DVDs have appeared on the market, promising to improve your home theater experience by helping you with equipment setup. As many first-time buyers have realized, true enjoyment of home theater only begins with plugging in the various components.

By using a set of standard test patterns for audio and video, even a novice can adjust his equipment to conform to the recognized audio and video standards used in theaters and production facilities throughout the world. There are three common options in calibration discs, and each is covered in this hack. First, though, a word to those of you who think buying a DVD just to calibrate your system once or twice is a waste of money.

More than a Tool

First and foremost, it's silly to think you will use a calibration disc only once. I own three different discs and all of them get a spin a couple of times a year at my place—not to mention their usage for the half dozen pals I help out with basic calibration. These discs will be used every time you change equipment (and if you are into this hobby, the upgrade bug will bite as often as the wallet allows). You'll find yourself pulling it out every time you unplug or move your equipment. Don't be surprised if you throw in the disc a few hours before you have guests over for movie night, just to make sure your theater is operating at peak levels!

Beyond calibration and test patterns, these DVDs really do offer an excellent overview of video and audio technology, terminology, and pitfalls. My first big learning moment in the hobby of home theater was when I rented Digital Video Essentials from a local video store and sat down with it for the afternoon. The information contained in the original Video Essentials (or, simply, VE) walked me through how a display device works, how light level and color in the room affected my perception of the display, how Dolby audio signals were carried, and more. It covered basic wiring and concepts on how surround formats worked, and even went through an introduction of why calibration and accuracy were important—giving examples and visual aids—all before the first test pattern.

So, aside from test patterns and the raw utilitarian value of these discs, it is important to take into consideration the added value and significance of these discs as learning tools. If you're interested in home theater and home electronics, these discs will supply some excellent background info; pick up a good calibration disc and this book, and soon you'll be the expert among your friends.

Digital Video Essentials

Digital Video Essentials (DVE) is the recently released sequel to Video Essentials, which originally appeared on LaserDisc many years ago. VE was the standard for Laserdisc calibration, and VE continued to be the popular solution for DVD for the first few years of the format. It wasn't until the birth of Avia that VE got its first real competition.

VE and the new DVE are both products of Joe Kane, one of the main guys behind the ISF. DVE was released in September 2003, and is detailed online at *http://www.videoessentials.com/*. Expect to pay about $25 for DVE.

DVE pros. DVE's strongest feature is the sheer amount of material and test patterns it offers. DVE is designed to guide the novice through basic tests as well as to supply professional-level test patterns and added reference-quality film and video clips. DVE is the most recent calibration disc on the market, so it has the most modern mastering and authoring. Without getting too technical, it was created in high-definition component digital and was ported to standard-definition NTSC or PAL for the DVD release. It's the most up-to-date in terms of test patterns, information, and authoring technology.

DVE cons. Like its predecessor, DVE has been criticized for being difficult to navigate. DVE (like VE) obviously was designed with the more technical user in mind and includes a wide variety of test patterns. As a result, menu navigation and simplicity of use suffer.

I also have read that the subcalibration tones on this disc are incorrect, or at the very least are 2dB different in level from the previous VE release. I'm not familiar with why it happened, but anecdotal evidence suggests that the tones on DVE are 2dB louder than they should be, resulting in a lower level when you calibrate using this test tone. A quick solution is to simply calibrate to a 2dB higher readout on your SPL meter.

Avia

Avia is a calibration disc produced by Ovation Multimedia (*http://www.ovationmultimedia.com/*) and written by *Sound & Vision*'s (*http://www.soundandvisionmag.com/*) technical editor, David Ranada. It offers some very good explanations of home theater concepts and an expansive array of tests. Ovation's main test designer, Dr. Guy Kuo, is a respected member of the online home theater community and a regular contributor to discussions on various home theater message boards. There is a great post online from Dr. Kuo outlining the tests on Avia at *http://www.hometheaterforum.com/htforum/showthread.php?s=&threadid=28110*.

Avia was released in June 1999 and will cost you about 50 bucks.

Avia pros. Avia is the middle ground of the three calibration discs listed here. It doesn't include as many test patterns as DVE, but is easier to navigate and understand. Its strength is in its ease of use and excellent narration. Avia does a very nice job of presenting an infomercial-style overview of home theater technology and methodology.

I took this disc to the in-laws' house years ago to set up their modest home theater. Despite being in their 50s with no real techno-passion, my mother- and father-in-law both enjoyed watching Avia's explanation of audio and video concepts. They said it explained the audio and video components far better than the instruction manuals!

Avia also has some interesting test patterns not featured on other discs, and all the test patterns are easy to find in the menu system. This disc represents a nice middle ground between user friendliness and power.

Avia cons. Avia is the oldest of the three discs detailed here. As a result, it lacks a DTS audio test or any information on 6.1 audio.

Avia also has a subwoofer calibration; the main test pattern for subwoofer calibration actually puts the test signal in the main channels, and must be rerouted by bass management in your receiver. If your speakers are set to "large" [Hack #67], this test won't work for subwoofer levels in your system.

Ovation has released an updated version of Avia known as Avia Pro (*http://www.ovationmultimedia.com/aviapro.html*), but it is literally a professional package, consisting of seven discs and costing $400. As of now, the only consumer update to Avia is the Sound & Vision disc (discussed next), which is abbreviated in comparison to the original Avia.

Sound & Vision Home Theater Tune-Up

Sound & Vision Home Theater Tune-Up (S&V) is made by Ovation, the people who brought you Avia, and is housed online at *http://www.soundandvisionmag.com/*. Targeted at the home theater beginner, S&V was designed to be sold to the budget-conscious dabbler looking to run basic system tests without investing in a full $40 or $50 disc.

In the interest of keeping the disc simple, many of the tests included in Avia are removed. S&V features easy navigation and simple instructions but lacks more advanced tests and calibration tools. It's also important to note that S&V came out well after Avia, and has newer tests Avia doesn't offer, such as 6.1 test signals. Released in January 2002, S&V will cost you about $20.

S&V pros. S&V is cheaper and easier to navigate than either Avia or DVE. Some higher-end and redundant tests of Avia have been removed in favor of simplified navigation and basic materials targeted at the average user. It really is the opposite of DVE in terms of scope and audience, and as a result ends up with a much more "average user" feel than DVE offers.

Still, S&V is made by the same people who produced Avia, so the disc remains well laid out and well explained. Additionally, although there are fewer tests, all the tests that did make the cut are accurate.

S&V cons. Like most things, when you add user-friendliness, you sacrifice power. S&V is very simple; it has only the basic video and audio tests, and very limited home theater discussion and information. It really should be seen as Avia Lite (with a slight update due to a more recent release).

—*Vince Maskeeper-Tenant*

HACK #63 Calibrate Speakers with a Sound Meter

With a calibration DVD and sound meter in hand, you can play a test pattern and set levels for your speakers.

Although the on-screen instructions on calibration DVDs [Hack #62] are clearly written, sometimes questions arise as to how to calibrate the audio portion of the system using the test patterns included on these discs. This hack deals with that process in detail, adding additional detail that is left off of many discs' instructions.

If you choose to use the THX Optimizer [Hack #65] or internal test tones [Hack #66], these instructions are also applicable, although the quality of both of these test methods is suspect.

Inside your receiver menus there should be speaker-level adjustments for each channel: front left, front right, center, surround right, surround left, subwoofer, and back or rear.

Some newer receivers provide a separate level for the rear-right and rear-left speakers. You generally should treat these as a single level, unless you are substantially further from one rear speaker than from the other.

The range of allowed adjustments usually goes from –10 to +10dB, with zero being the default volume. Figure 8-2 shows a Pioneer receiver's face, with the front-left speaker level being modified.

It also is a good idea to set all treble and bass controls to 0 as well.

Set your meter to "slow" response and the weighting to the setting suggested by the calibration test you're using (both VE and Avia offer on-screen instructions on proper meter settings). Most calibration tests suggest a "C" weighting (as seen in Figure 8-3).

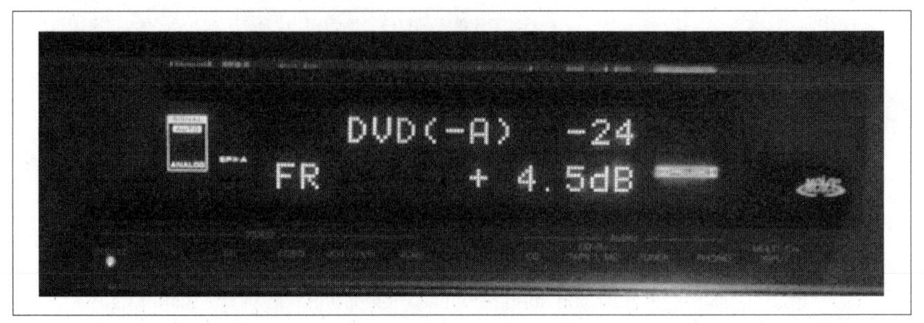

Figure 8-2. Speaker levels being modified

Figure 8-3. SPL meter weighting and response settings

Positioning the SPL Meter

Place the SPL meter at your main listening position, approximately at the height of your nose while seated, and point it forward, angled upward at about 40°; a decent rule of thumb is aiming it at the point where the ceiling meets the front wall. You also should crouch down (behind a couch if possible) so that you aren't in between the meter and any speakers. Standing up while taking sound levels results in an inaccurate reading, unless, of course, you prefer to stand through a three-hour showing of *The Return of the King*!

It is helpful to place the meter on a camera tripod; the RadioShack SPL meter even has a nice tripod mount on its back. This will help make sure you keep the meter in one spot. You'll find it is a bit tough to hold the meter, read the meter, and make receiver adjustments on the remote. You also can recruit a "calibration buddy" to help you in lieu of a tripod.

Don't move the meter during the tests. You don't need to point the meter at each speaker during the tones; in fact, you should avoid doing this. Leave the meter fixed at ear level, pointing forward and angled slightly upward.

Testing the First Speaker

Insert your test tone disc and begin the audio test for speaker calibration. Both Avia and VE have these tests, and they should be reasonably easy to find. The majority of discs start with the left speaker tone. While the left speaker tone is playing, increase the master volume on your receiver or processor until you reach a reading of 75dB on your meter.

The Avia disc has some tones that are intended to be calibrated to 75dB, while others are intended for a reading of 85dB. Be sure to read the instructions carefully to ensure which test tones are which.

Determining the sound level is simple: set the meter's dial on a number, such as 70. When the dial is set to 70, the needle is displaying sound pressure measurement relative to 70. When the needle hits the middle 0 position, the SPL of the sound you're measuring is 70dB. If it reads –4, that means it is 4 under 70, or 66dB. To measure 75dB output, set the dial on the meter to 70 and then adjust the speaker volume until the little needle hits the +5 mark on the meter display. To measure 85dB output, set the dial on the meter to 80 and then adjust the speaker volume until the little needle hits the +5 spot on the meter display.

That Stupid Tone Is Moving Too Fast!

The various calibration discs usually play a tone for a few seconds, and then move to the next speaker. This can be a bit frustrating, as that brief tone often doesn't provide you with enough time to get a good reading, make level adjustments, and prepare for the next speaker. However, the discs tend to put each speaker's test tone on a different DVD chapter; this allows you to play the tone you want, and quickly press the Repeat button on your DVD player or remote. The tone will continue to play, stop, and then, instead of moving to the next speaker, repeat. Once you've got the level locked in, you can move to the next speaker.

At this point, the first speaker should be set so that it is outputting 75dB of sound, with no level correction (the speaker level is at 0dB). This effectively determines the reference level of your receiver [Hack #64]. Once you've reached this level of volume, you are done with your master volume.

Testing the Remaining Speakers

As the tone cycles to each of the other speakers, adjust the individual speaker level in the receiver menu so that each speaker measures the proper level on the meter.

Don't continue to change the master volume on your receiver or preamplifier because this will change the output level from the first speaker tested, skewing your setup.

It might take a couple of times around the room to get everything right. Keep working at it, and be sure to leave the meter in the same place. By the time you are finished, all the channels should measure the same on your SPL meter. It's a good idea to go around the room one final time to ensure that everything is set properly. Once all speakers output the same level, you've got a well-calibrated home theater speaker setup.

—*Vince Maskeeper-Tennant, Brett McLaughlin*

Set the Receiver's Reference Level

Setting a receiver's volume to the reference level is not the mystical task it often appears to be; in fact, a receiver's reference level is whatever your theater needs it to be.

Dolby specifies an ideal playback level for its theatrical soundtracks and for the mixing environments in which those soundtracks are created. So, unlike music, movies are created with a standard playback level in mind. Based upon this intended level, dialog and effects are mixed at very specific levels to offer similar sound levels across various Dolby soundtracks and across different playback environments. In technical terms, this playback level is defined as 105dB peak level from each main speaker in the Dolby playback system. Test tones on calibration DVDs [Hack #62] are specifically defined to give you this playback level.

Before starting audio tests [Hack #63], calibration discs often instruct you to set your receiver's volume to the reference level. This is often a confusing instruction, leaving many home theater enthusiasts scratching their heads and wondering if they have done something incorrectly. All of this confusion is based on the incorrect assumption that there is a universal setting known as *reference level*.

Reference level is simply the volume at which your particular system is calibrated to play all speakers at correct volumes (75dB on some tests, 85dB on others). Because each speaker has its own level controls, this reference level can change, even in the same room. If a system had its master volume set to 0dB and all speaker levels at 0dB, and test tones played at 75dB, your receiver or preamplifier's reference level would be 0dB. However, you could turn the master volume to –2dB and increase all speaker levels to +2dB, and you would (theoretically, at least) get the same output from each speaker. In this case, your reference level would be the receiver's new volume, –2dB.

It's About Output

If this is confusing, remember that reference level is about output, not a number on a dial. Some receivers and preamplifiers have an analog control, with no numbers at all! On these systems, reference level is simply an arbitrary position on a dial. If you follow the speaker calibration steps [Hack #63], reference level will be the volume at which your front-left speaker plays a tone 75dB loud, at listener level, with no individual speaker level adjustment. That might be –10dB on your master volume, +2dB, or somewhere in the area of 11:45 on your analog dial. It's all relative, and it's related to output, not specific numbers.

A Moot Point

To a certain degree, getting too hung up on reference level is a waste of time. It's useful for calibrating speakers, but you'll hardly ever listen to movies at this volume. Most home users don't listen at Dolby's specified levels, as they are quite loud—even for larger rooms. The levels work better for an acoustically dampened room such as a movie theater, so this level is used mainly as a useful point of reference when seeking help or advice. The most important element is to get all the speakers operating on the same level in relation to one another; you can then adjust the overall volume to your comfort level.

—Brett McLaughlin, Vince Maskeeper-Tennant

Avoid Using THX Optimizer

Recent DVDs, especially those that are THX-certified, contain a THX Optimizer to calibrate your system. Although this might seem like a feature, avoid using THX Optimizer if at all possible.

One of the cooler logos to see stamped on your favorite DVD is THX-certified or THX-optimized. Most home theater newcomers love seeing the THX label and equate it with better sound and picture. However, what's on the disc and what comes out of your system are often not the same thing. With the release of *Star Wars, Episode One* on DVD, THX began including an optimizer program with THX-certified DVDs. The basic optimizer tests were supposed to serve as a calibration tool with basic audio and video test patterns for speaker level, phase, and picture settings.

Unfortunately, these tests have proved to be inconsistent and have been criticized by everyone from the average home theater dabbler to professional technicians. There are large differences between the optimizers on different DVDs, as well; *Star Wars, Episode One* appears to have a completely different optimizer than *Star Wars, Episode Two*. These discrepancies have caused some to suggest that each test pattern is catered specifically for the disc with which it is bundled, compromising the overall quality of your home theater for one disc's playback. Using a set of accepted, baseline standards **[Hack #63]** is necessary for accuracy, as no one wants to have to recalibrate for each DVD.

In the end, the THX Optimizer serves as a good starting point for those people who would never buy a calibration disc **[Hack #62]**. If nothing else, THX Optimizer makes average users understand that there are guidelines and target ideals for equipment settings. If you're serious about getting the most performance out of your home theater system, though, you should look beyond THX Optimizer (or even avoid it altogether).

If you're looking for evidence pointing to the use of Avia or DVE over Optimizer, just look at the THX web site, where THX gives information on THX Optimizer, followed immediately by links directly to Avia and DVE (*http://www.thx.com/mod/techLib/index.html*).

—*Vince Maskeeper-Tennant*

Avoid Using Internal Test Tones

Most receivers and preamplifier/processors provide test tones to measure audio levels from each speaker. These tones aren't an accurate representation of your entire system's sounds, and you should consider them to be a rough guideline at best and problematic at worst.

Most receivers have their own internal test tone generator which you can use to set speaker levels. Although this might seem a real convenience and avoid the need for a calibration DVD **[Hack #62]**, it turns out to be as problematic a solution as the THX Optimizer **[Hack #65]**. Many people attempt to set their speaker levels using these tones by ear, instead of using a sound pressure level meter **[Hack #63]**. This will not result in accurate calibration. Even if you subscribe to the value of the receiver's internal test tones, using a consistent sound measuring device is a must. However, even when using a meter, these internal tones aren't always ideal.

There are two schools of thought when it comes to using the internal tones versus disc tones:

- The internal tones are preferred. This comes from some rather important people, such as the folks at Dolby Labs.
- The tones generated by a good test disc are preferred. This is a more popular opinion among hobbyists, and the one to which I subscribe.

In my opinion, the DVD player will be the source for the actual playback of material; you don't invite your friends over to listen to test tones, you invite them to watch and listen to a DVD. Eliminating the DVD player from the calibration chain (by using receiver tones) will result in an inaccurate measurement of sound as compared to how the DVD player would produce it. Any variations introduced by the player will not be reflected in using the internal tones. It's always a better idea to use the entire intended playback chain when calibrating.

I'm not sure why Dolby Labs prefers internal tones, but one could theorize that the chips that produce these tones are based on Dolby's specifications, and thus Dolby would side with the internal tones. Although this might provide a clearer representation of the receiver in a sterile environment, it won't be accurate as far as your particular combination of components and cables is concerned.

—*Vince Maskeeper-Tennant*

Set Your Speakers to "Small" in Your Receiver Setup

Avoid setting your speakers, even if they are towers, to the "large" setting in your receiver's setup menus.

Receivers and preamp/processors typically have two settings for your speakers: "small" and "large." You need to get past what these words actually mean in English, as they are a very poor choice for this feature of a processor. This setting actually has nothing to do with the size of the speakers, and everything to do with the range of the speakers. This setting determines when low frequencies are diverted from your front speakers and into your subwoofer (the *crossover* frequency). In other words, it has a tremendous effect on the bass you'll hear in movie soundtracks.

Very few speakers should actually use the "large" setting. Even most of the big, powered towers should not be used with the "large" setting because they can't produce these low frequencies (or they produce them without power and depth). What you should be thinking is that "large" means you have a truly full-range speaker; use "small" for everything else. If your speaker can't put out more than 100dB at 20 Hz, set it to "small."

There are three main reasons for avoiding the "large" setting. The first is that crossovers aren't brick walls; they have slopes in both directions. The rule of thumb is that with typical bass management crossovers, your speaker should be flat to 1 octave below the crossover point. So, with an 80-Hz crossover point, your speaker should be flat to 40 Hz. Lots of speakers can do this. Only a few speakers are flat to 30 Hz (even though manufacturers' specs will try to tell you otherwise, there really are only a few, at least within a reasonable price range), and even fewer speakers are flat to 20 Hz (and below) at the levels a home theater will be asking for. The large setting on a receiver doesn't filter any low frequencies from a speaker to the sub. If the speaker isn't capable of the really low frequencies, they simply will be lost. Set to "small," however, these low frequencies will be filtered out and passed to the subwoofer, which is capable of reproducing them.

The second reason for using the "small" setting is that when you relieve a speaker of low bass duties, that speaker becomes a much easier load for your amp, and the midrange quality of the speaker often improves. The third reason for using the "small" setting is that bass frequencies have the greatest interaction problems with a room. Multiple sources of low bass in non-optimal places cause all sorts of sound wave problems. The best place for your main speakers is almost never the best place from which to produce low bass. Being able to produce all the bass from one spot in the room gives you the best chance of optimizing your room's bass response.

A final thing to note is you have to be wary of processors that allow you to set different crossover points for different speakers. With the exception of some very high-end processors, you should not use this feature. The vast majority of processors with this "feature" high-pass each speaker's signal at the frequency you specify, and send it to the speaker. This is good. However, to feed the sub, the processor will sum the full-range signals from all the full-range channels and the LFE channel, and then low-pass this signal at the lowest crossover point you set. So, if you have your surround crossover set to 100 Hz and your main crossover set to 40 Hz, there will be a 60-Hz hole in your surround channels' responses. This is not good. THX chose 80 Hz as its bass management crossover point for a reason; trust their research and experimentation.

Set all your speakers to an 80-Hz crossover, and let your sub and speakers do what they do best.

—*Dustin Bartlett*

To Trust or Not to Trust

You might recall that in the hack on avoiding internal test tones [Hack #66], I advised you not to take THX up on its calibration tools. Here, though, I steer you toward using THX's wisdom. This might seem confusing, but it actually outlines an important principle: specifications are great for general application, but often terrible for specific application.

When you're dealing with a specification, or a setting that affects all movie playback in all theaters, or any other widespread standard, THX works great. In the case of determining the best crossover for all subs, in all homes, to play all movies, THX can be very helpful. That's because the specification is affecting generalities; it's making sure your movies, to the degree possible with your theater, sound like your neighbor's movies, at least in terms of relative levels of effects, music, and so forth.

On the other hand, setting the acoustics of your individual room is very, very specific. In fact, your theater is going to be at least subtly different from every other theater in the world. It is here that specifications such as THX fall down, and the reason THX Optimizer [Hack #65] and internal test tones are essentially useless. No specification can accurately tune your individual home theater, and often it will make things worse.

As a rule, use specifications for general settings, but fine-tune things by hand, with your ears and a good sound meter [Hack #63].

Hack Your TV's Service Menu

The key to almost all tweaks in TVs comes through the service menu. However, getting to this menu is usually undocumented, and you'll need some help in finding it.

It's a general principle that TV manufacturers consider TV consumers fairly unintelligent. As a result of this (usually incorrect) assumption, your TV probably has tons of undocumented features and more than a few calibration menus. Lucky for you, many popular brands of TVs have their service menu access codes available on the Internet, through sites such as *http://www.keohi.com/keohihdtv/*. Even more conveniently, various web sites' information has been aggregated here, to detail just what you need, all organized by brand.

These are just the codes to get into the service menu. If you want to go on and mess with convergence, focus, and so on, consult the other hacks in this chapter, as well as online resources. Trying to document every feature of every brand would require a separate book.

Your Instructions Don't Work for My TV!

Unfortunately, in a continuing sign of distrust, manufacturers often change the codes to access their service menus in new models. What works for many of you with older TVs might not work for someone who just got a brand-new set. If you hit a wall on this, visit *http://www.keohi.com/keohihdtv/*, which I've found to have the latest variations. So, please don't panic: you just might have a newer TV than the rest of us.

Hitachi

There are actually two ways to get at the service menu, and specifically the convergence menus, on Hitachi sets.

The hard way. This is indeed the hard way to work with a Hitachi; it involves tools rather than just pushes of remote control buttons:

1. Open the front panel (this involves removing four screws—two on each side).
2. Remove the center panel (six screws).
3. Press the blue button in the bottom-left corner inside the set. This brings up the convergence grid.
4. To navigate through the grid, use the following buttons:
   - 2 for up
   - 4 for left
   - 5 for down
   - 6 for right
5. To make adjustments to any accessible point on the grid, use the joystick.
6. To toggle or switch between the colors, use:
   - 0 for red
   - Recall for green
   - Input for blue

7. When you are happy with your tweaks, hit PIP MD twice, and then hit PIP CH.
8. Click the blue button to exit the Service Menu.
9. To exit without saving any changes, simply turn off the set.

The easy way. Needless to say, opening the set isn't desirable; thankfully, there's a better option for most folks:

1. Press Magic Focus on the front of the TV once; press the same button a second time; this stops the procedure.
2. Right after the procedure stops, press the Status button. This gets you to the convergence reset screen.
3. To navigate through the grid, use the following buttons:
   - 2 for up
   - 4 for left
   - 5 for down
   - 6 for right
4. Use the Select button (or the big blue knob) to move the color around to center it on the white lines.

By default, red is the color selected. Youl need to follow Step 5 to switch colors.

5. To change to blue, hit ANT; to switch back to red, press 0.
6. When you're finished with convergence, follow these steps to save and exit:
   a. Hit PIP MODE once, and then a second time. The screen will be blank for 4 seconds, and then you will see a bunch of green dots.
   b. Hit PIP CH; this gets you back to the convergence screen. At this point, your settings are saved to memory.
   c. Now, to get the TV to use this new setting, hit PIP MODE once, and then PIP CH. This starts a fairly involved process; just let it finish before going on.
   d. After this process is through, again you are left with green dots; hit PIP CH to go back to the Convergence screen.
   e. Turn off the TV to exit.
   f. Turn on the TV and press Magic Focus to realign to the new settings.

If you press Magic Focus, and all you see on the screen are nine crosses, you did something wrong. You can't get back to the menu the same way; now you must open the case. In other words, now you get to experience the hard way!

Mitsubishi

All the setup menus on Mitsubishi TVs are accessed via the Menu button, and then four-number sequences. The two menus of most interest to Mitsubishi users are Menus 1257 and 1259. The first is convergence, and the second is geometry.

What makes things really confusing is that you then add numbers to this initial sequence to access specific functions and features. For example, Menu 1257, and then 6, takes you to the Misc menu; adding 5 at the end takes you to coarse convergence and geometry, and adding 4 takes you to fine convergence and geometry. For a lot more variations, take a look at *http://www.keohi.com/keohihdtv/brandspecific/mitsubishi/mitsu_servicecodes.html*, which has a ton of additional codes.

Panasonic

Many adjustments are available using the On Screen Display Menu, or Serviceman Adjustment Mode, and you should understand what each adjustment does and note any of their values before changing. Panasonics are actually nice because of this; for whatever reason, the Panasonic folks seem to trust their users with more direct access to tweaks.

Terminology. You need to understand the following abbreviations:

- VU: Volume Up
- VD: Volume Down
- CU: Channel Up
- CD: Channel Down
- SM: Serviceman Mode

Entering Serviceman Mode. You can enter Serviceman Mode (SM) in two ways: Remote Method and Other Method (involves opening the case):

1. Remote Method:
    a. In the SET-UP icon, set the antenna to CABLE.
    b. In the TIMER icon, set the sleep timer to 30.
    c. Exit the menus and tune to channel 124.

d. Adjust the volume to 0.

e. Press Volume Down (VD) on the control panel and a red "CHK" will appear in the upper-left corner.

f. To toggle between Serviceman and Aging Modes, press Action and VD on the control panel.

2. Other Method:

a. Short test point FA1 to cold ground FA2 (A-board: TP pin 8 to pin 3).

b. The Receiver is in Aging Mode with yellow "CHK" in the upper-left corner.

c. Press Action and VU on the control panel, and "CHK" should be in red.

Now the set is in Serviceman Mode, of which there are five separate Serviceman Modes:

- B: VCJ Sub Adjustments
- C: W.B. Adjustments
- D: Pincushion Adjustments
- S: Options Adjustments
- CHK: Normal use of VU/VU and CU/CD buttons (no adjustments can be made)

Use the Power button on the remote to switch between these modes.

Accessing and modifying the service adjustments. Once you have selected one of the four adjustment modes, use the CU and CD buttons to move between settings. Within a setting, use the VU and VD buttons to adjust the level or value.

Write down any original values before changing anything!

Sony

There are two different sets of instructions for Sony TVs: one for NTSC versions, and one for PAL versions. If you're not sure which set you have, you're probably way over your head here, so turn back now!

NTSC sets. There is a small service switch on your set. Look for a small, round opening in the rear cabinet near the video inputs, or the antenna jack. Press and hold this button while turning on the power.

Alternatively, if you have no service switch, you should try one of the following remote control sequences:

- PWR OFF, then DISP, then 5, then VOL+, then PWR ON
- PWR OFF, then DISP, then 5, then PWR ON

You have to hit each button within 1 second of the last button. This timing is critical, so it might take some practice.

This will turn on your TV in service mode, which looks a lot like Figure 8-4.

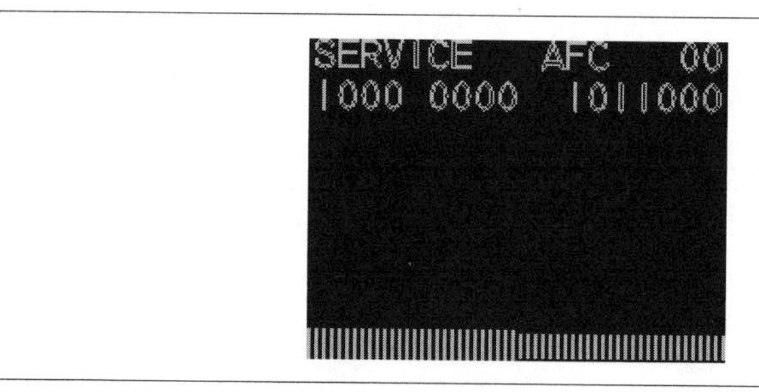

Figure 8-4. Sony service menu

Now use the following buttons to navigate:

- 1: Moves to previous menu item
- 4: Moves to next menu item
- 3: Adjusts value of selected setting up
- 6: Adjusts value of selected setting down

To save settings, press MUTE, and then ENTER. To restore all user settings to factory defaults, press 8, and then ENTER. Pressing 0 followed by ENTER will reset all values to their defaults from the last stored save.

This consults the last set of saved settings rather than factory defaults. To access factory defaults, press 8 and then ENTER.

To leave the service mode, just hit PWR OFF.

PAL sets. Be sure your TV is turned off. Hold the plus (+) and minus (–) buttons on the front panel of the TV, and turn on the power.

You must do all of this via the front panel of the TV, not the remote control.

When the TV comes on, it should show a TT on the screen, indicating it's in test mode. There are numbers you can now hit to select a specific function.

Some newer models actually use a different procedure (again, these are PAL sets, not NTSC sets):

1. Turn on the TV.
2. Enter standby mode.
3. Press INFO/DISP on the remote control.
4. Press 5 on the remote control.
5. Press Vol+ (Volume up) on the remote control.

This should get you to the same spot as the previous instructions—to the service menu, signified by a TT on your screen.

Toshiba

There are two menus of interest on Toshibas: the Service Menu and the Designer Menu.

Service Menu. Here's how to get to the Service Menu for most Toshiba sets:

1. Power up your TV.
2. Press MUTE three times on the remote.
3. Press MUTE again, and hold down the button.
4. While holding MUTE, press the MENU button on the face of the TV set (not on the remote).
5. Release MUTE and MENU. At this point, an S should appear on the bottom right of the screen.
6. Press MENU one more time, and you're in the Service Menu.

For older sets (X81 and before), press MUTE twice (not three times), and then continue with Steps 3 through 6.

To exit this menu, simply turn off the TV.

Designer Menu. Here's how to get to the Designer Menu for most Toshiba sets:

1. Follow Steps 1–6 in the Service Menu section.
2. Press the RECALL button on the remote, and hold it down.
3. Press the MENU button on the TV set front panel. A letter D should appear on the screen.
4. Hit the RECALL button on the TV's front panel (this is the second time).
5. Now you're in the Designer Menu.

To exit this menu, simply turn off the TV.

On newer sets (H80 and later), the Designer Menu has been locked out. Current workarounds introduce some pretty nasty bugs, so I won't print those methods here. Hopefully, the next printing will have a fix!

Navigation. In both of these menus, you scroll through the menus using the Channel Up and Down buttons on your remote; the Volume Up and Down buttons change settings.

—*Keohi HDTV*

HACK #69

Clean the Optics on RPTVs

With a paper towel and some know-how, you can clean your RPTV optics and drastically improve your picture.

A race car driver would never get on the track without fresh tires; an auto mechanic would never leave his tools wet, prone to rust and dirt; so why should you, home theater enthusiast, let your RPTV collect dust and end up with a distorted image? It won't take much to get into your unit and clean it up; the difference in picture quality is simply staggering! But first, some background.

What Is Dust?

Dust is thousands of tiny little refraction lenses, each with the dispersing quality of a dum-dum bullet. A coherent, otherwise highly directional ray of light hits a dust particle, and much of the light gets scattered in 100 different directions. Multiply that by the thousands of tiny little dust particles found on the typical objective lens after a year of viewing, and you have unwanted *fill-in*: the phenomenon of otherwise coherent, highly directional light rays being towed off course in 100 different directions and going where they're not supposed to go—notably, into dark areas. If all pictures were bright, with no dark areas, there would be no problem. But because perceived depth in a picture relies heavily on dark areas to separate the light areas accurately, accuracy in dark area reproduction is essential in capturing the 35mm film experience in RPTVs, and to give a sense of depth and realism to the scenes involved. Additionally, you want the projected image to retain a like-new viewing quality on a constant basis.

With haze caused by dust, areas are either light or they're dark and smeary, with no detail possible in the dark areas. It doesn't take much to compromise detail in dark areas; they're very delicate, being dark and relatively hard to see already. With a coating of dust, dark areas just look "smoked out." It looks like you're viewing something through smoky glass.

Note that I didn't say *smoked* glass; that would simply be clear, but darker. Smoky glass is glass covered with soot, resulting from thousands of tiny little dots of dust.

This phenomenon gets worse year by year, with the viewer staying acclimated to a TV picture that just seems to be "getting old." Through years' worth of gathered dust, a figure skater in pure white, with a TV camera looking up at her against the darkened skating rink canopy in the distance, will show up as having a halo-ish glow around her, which wouldn't be there if the optics were clean. Great angelic effect, but a classic indication of lousy optical dynamics!

The Effect of the Mirror

RPTVs rely heavily on a mirror [Hack #29]. Therefore, it should come as no surprise that dust on the mirror has as much effect as dust on the CRTs.

If you freeze-frame our image of a figure skater, then go off to a side angle to view the set and move your head around—back and forth and up and down—you'll find that this haziness in the dark areas will follow your head movements at half-speed, which is the image being caught up in the smokiness of the mirror behind the screen. This obviously is not the image bouncing off the mirror cleanly and invisibly, as it's supposed to do. In really bad cases, you'll actually get a slightly different view of this haze with each eye. Then the haze will be in 3D, and your eyes will be slightly confused and fatigued because of the parallax directionality involved in your two eyes seeing slightly different images because of the two different angles your two eyes see from. Obviously, the final viewed image is on a flat plane and is meant to be viewed as such. With a dirty mirror two feet directly behind the screen and visible, you're seeing the messiness in 3D, which definitely is not what 3D viewing is about.

With crystal-clear optics, the mirror is totally invisible and there's nothing but absolute black in the dark areas behind the screen's picture, where whatever is back there in the optical cavity—hopefully, nothing—shines through. With dirty optics, the mirror is visible: blatantly and visibly in the way, in those otherwise dark areas of your picture, along with everything else mentioned earlier.

This haze effect makes it impossible for any chance of *gleam* in your picture, which is defined as something small and terrifically bright against something either black or really, really dark, with a terrific amount of contrast. This is what we perceive on the screen as depth. No high, sparkling amount of contrast, or depth, is possible when there's a hazy glow around anything bright, bleeding off from the edges of the bright object and clouding up the darkness. It's the very definition of an undynamic picture. And it all results from some dust on the mirror.

Regular Maintenance Required

RPTVs are optical instruments, just like telescopes and binoculars and microscopes, and they require the level of care and feeding you'd give to any other precision optical device. Like all these other devices, RPTVs have lenses, mirrors, and an optical cavity in which they develop their magic.

However, in the case of RPTVs, high voltage is involved—it's created to drive the CRTs and thus is inescapable—and this high voltage ionizes the air, statically charging all the airborne particles, no matter how small, and thus turning the mirror(s) and lenses into powerful dust magnets.

This high voltage around the CRTs is why the optical surfaces in RPTVs need to be cleaned every year, at a bare minimum. In fact, models from Sony, Pioneer, Toshiba, and Philips need a slightly more advanced job done twice as often, as these brands have buried secondary lenses that also are exposed enough to attract the dust. Mitsubishis seal their secondary lenses so that only the upper lens's exposed surfaces need this attention, and thus need be done only yearly.

Cleaning the Optics and Mirror

To clean your optics and mirror, just pop open your unit. The best stuff I've found for wetting the surface of the lenses and the mirror is Sprayway; you can find it at Costco or specialty hardware stores. For mirrors, I use Windex or Glass Plus, in the nonammoniated versions of course. I always use nonammoniated cleaners because of the first-surface mirrors used in HD-ready TVs, which have aluminum components. You don't want to be mixing aluminum with ammonia. Allow the cleanser to penetrate and lift the first layer of grit off the surface of the lenses.

Now get some paper towels, and go to it. Yes, I did say use paper towels; I never have to worry about scratches with these. The only caveat is to not use shop towels; any additive, such as lanolin, will continue to wind up smeary, long into the night. Hours later you'll still be grinding away, trying to get your optics clean, and you simply never will, at least not with those kind of towels. You'll just scratch up your optics. Use plain old paper towels, made only of pure paper.

Avoid newspapers and Kleenex, as both leave a mess behind, defeating their entire purpose.

The way to make sure you don't scratch anything is to make sure the matting of dust is soaked and completely hydrated in your cleansing liquid before you remove anything.

Be sure to not allow any liquid to get into the multiple lenses inside the lens barrel. This liquid will evaporate, condensate, and fog up your internal lenses.

During cleaning, wipe in only one direction—either toward you or upward—during the first swipe. Don't go back and forth, over and over, grinding the particulates into the glass, or even worse, plastic, lenses. Just go in one direction, carefully but swiftly, gathering the grit toward you. You might even try rotating your towel up a bit as you go, continuing to expose only fresh paper to the optical surface. Eventually, your final swipe should get all the dirt off the lens, all while keeping it intact and unscratched.

You can use this procedure to clean the TV's mirror as well. Just be sure you keep all the swipes going in one direction, so dust is never ground into the glass. Eventually, you'll have crystal-clear lenses and an invisible mirror. Sweet!

Watch Out For...

I try never to touch the inside or outside of the fresnel and/or lenticular lenses, unless absolutely necessary. They don't need regular maintenance, and are extremely sensitive and fragile. The vertical ribs on the fronts of the lenticulars, in particular, are extremely vulnerable to residue from any liquid that touches them, no matter how professionally accepted the cleaner is.

The most I normally do in any calibration is to get the cobwebs off the inside of the fresnel and the inner optical cavity by swooshing a piece of the paper towel around in a circular motion, to gather the cobwebs around the paper towel; and perhaps gently wiping the inside of the fresnel down in a very light and gingerly fashion with a dry terry-cloth towel. Keep in mind that this is only if I can see dust on the fresnel while looking at its surface from the side.

Additionally, never touch a Mylar mirror. They don't attract dust statically like glass mirrors do, and therefore don't need cleaning in the same manner as glass mirrors. Mylar mirrors are exponentially more sensitive than glass to being scratched.

—*Robert Jones, Image Perfection*

Reduce Focus Problems on RPTVs

Most rear projection TVs have issues with focus; if you get the right side in focus, the left side of the screen blooms, and vice versa. By adjusting the lenses, you can greatly reduce this effect.

Many RPTVs exhibit focus problems. When you calibrate the picture using a good calibration DVD **[Hack #62]**, you can't get the entire screen to focus correctly. If you get the right side of the screen focused, the focus will hold for about 2/3 of the screen, but the left side blooms out; in other words, it starts to blur on the remaining 1/3 of the screen. If you try and focus this side, the same effect occurs on the right side of the screen.

This is what is called *scheimpfluge*, the angle at which the CRT hits the lens, and then how that lens hits the screen. In an RPTV, those angles will never be parallel, due to the nature of how RPTVs are designed. So, the lens is angled to compensate (not always correctly, by the way) when your TV is manufactured. However, these angles are set in stone for mass production, including the angles of the lens mounts, rather than for your specific set. As a result, even TVs from the same production line demonstrate this focusing issue to different degrees.

High-end front projectors have high-precision, spring-loaded adjustment setting screws for this purpose, providing a high-precision mechanism for changing these lens angles.

Using washers under the mounts of the screws that hold your RPTV lenses in place can make everything symmetrical. Just add some washers under the mounts and see what sort of improvement you can create; you'll have to try different thicknesses and placements, and see what works best on your set. Be sure to tape these washers in before securing the lens on top of them, so they don't fall out later and into the works down below; they're a real pain to dig out and can cause long-term problems.

The best way to figure out exactly where to place your shims is to defocus the optical so far that the grid lines are 1/2-inch to 1 inch thick, at your screen. That will tell you graphically what to do, so you won't have to try to work with a busy picture or test image. On one side of your screen, the grid-lines will be wider and thicker than on the other side when there is a scheimpfluge error. As you add shims, these lines should start to appear evenly across the screen. Get them as straight and as even as possible, and you're all set. Finally, refocus for real and you're balanced, left to right.

As an example, I had to do this on a Pioneer 510; the blue had exactly the same differing focus response, left to right. Putting in one washer each under the mounts where the screws are, at the two nearest-to-center screws holding the blue lens in, did the trick. This angled the blue lens out just a little bit more than intended, but then the factory setup was obviously designed incorrectly, which is not uncommon in RPTV production units.

You might have to recenter your image just a bit after this adjustment, but just changing the angle involved with shims doesn't really impact the direction of the image beam much. The difference it makes in getting the focus correct, on the other hand, is phenomenal.

Going back to the case of the Pioneer, the focus still wasn't perfect after this adjustment. However, the errors on each side were identical. Although this might not seem like an improvement, the blooming is gone and the focus appears drastically improved. With all that said, the ideal location for washers on your set might be the same, a little different, or completely different from on the set detailed here.

—*Robert Jones, Image Perfection*

Reduce Lens Flare on RPTVs

Lens flare can dramatically decrease the black levels on an RPTV. Learn what lens flare is, and how to knock it out of the equation on your RPTV.

Although one expects CRT-based rear projection televisions to deliver good black levels, there is a factor that compromises black level performance in these displays. This factor is something called *lens flare*. Lens flare is most commonly seen in photographs. It usually occurs when the photographer is shooting in the general direction of the sun. Bright sunlight enters the lens at an angle and bounces around inside the lens body. As it reflects off of the interior parts of the lens, some of it ends up getting on the film, usually in the form of bright circles or shapes, smears of light, or lines.

Binocular, telescope, and camera lens manufacturers all try to make the inside of their optical assemblies as dark as possible, and also add light baffles to try to waylay the misdirected light so that it doesn't spoil the image.

Rear projection televisions also suffer from lens flare. Usually the best place to see lens flare is usually on the end titles of a film. Often the titles are bright white on a black background. Look for a small single title and pause the player at that point. Look at the black areas of the screen around the title. Is there a halo, ring, or general smear of light of the same color as the title? That unwanted light most likely is due to lens flare.

If you want to see the flare at its source, pull the screen off the TV and look into the lenses while TV is displaying a bright field. Look for light being reflected off of shiny interior parts of the lenses by moving your head around so that you can look down into the lenses and see their insides.

Don't run your head into the mirror while trying to do this.

The light bouncing off of those interior edges and surfaces is the problem.

Your "black screen" black level most likely is quite good on your RPTV, especially if you have done the work to line the interior of the cabinet with light-absorbing material, and if you view your TV in a darkened room. However, one small, bright object on an otherwise black screen can cause light pollution on the supposedly black areas of the screen due to this problem. In images with a lot of bright areas mixed with some supposedly black areas, the black level of the supposedly black areas will suffer considerably.

Preparations

Find a clean, well-lit, undisturbed place to work. No kids, pets, significant others, etc., should be able to disturb you.

The following actions, if performed carelessly, will ruin your television. Further, lens flare can be reduced, but don't expect to completely eliminate it. The cathode-ray tubes, which are the light source for our systems, emit lots of light at all angles, and it is really tough to keep some small amount of it from getting out of the lens in a direction we don't want. Expect that you'll be able to make some improvement, but you won't eliminate the problem completely.

If you don't know how to get at your lens assembly, don't consider yourself qualified to attempt this procedure. If you have butterfingers, don't even think of trying to do this. Read the whole process first, and familiarize yourself with the required steps before commencing.

Get some clean, washed cotton cloths on hand. Used white T-shirts are a good choice. Clean, white cotton gloves also are good to have when doing this job. In addition, I highly recommend you obtain a source of clean, dry compressed air.

Examine the lens assembly. Determine how the lens assembly comes apart. In the case of the Delta Digital 265 lens assembly from Corning Precision Lens Inc. (this assembly is used in the Toshiba 50HDX82 RPTV, and it's what I will refer to throughout this discussion), the threaded screw which carries the wing nut used to lock the mechanical adjustments of focus must be removed. The end of the screw has been squared off. You can use a small pair of vise-grip pliers to gently turn this screw counterclockwise to remove it from the plastic assembly it mounts in.

Now the whole internal lens assembly can be rotated fully clockwise. Note that as you rotate this assembly two plastic studs are turning in a couple of spiral slots to move the lens assembly up and down in the lens frame. If the lens assembly is rotated fully counterclockwise, the studs will hit the end of the slot. Then you'll notice that a groove is running up the inside of the lens frame to allow those two studs to slip up inside the frame so that the inner lens assembly can be removed from or inserted into the frame. You might want to use a small hobby knife to put a bevel on the start of that groove to make it easier to push the lens assembly up out of the frame. Some gentle prying (and cursing) also might be necessary.

Realize that any contact with the front or rear lens while you are doing this can mean either a dirty or a damaged lens. Wearing clean cotton gloves at this point is a good idea.

Now you can disassemble the lens assembly by removing the screws holding the two halves together. Remove one half in such a way as to leave the lenses resting in the other half of the lens shell. Make a drawing at this point of exactly how the lenses fit into the lens shell. Which lens goes in which position, and which way does the lens face? Be sure about this, as you don't want to keep handling these lenses unnecessarily.

Note that the Corning lens assembly has glass and plastic lenses! Whether glass or plastic, all of them must be handled with the utmost care. *Touch only the edges of the lenses, and then only with gloved fingers!* The plastic lenses are shaped like cups, and thus you can rest them safely on a flat, clean surface with the curved side up. The center glass lens should be supported only by its edges. You should arrange some sort of cloth-lined trough of an appropriate size to support the lens.

Now that the lenses are out of the way, you can take a look at the plastic shell that held them. In the case of the Corning assembly, the shell is molded out of black, shiny plastic. You want to get rid of the shine; apply a high-quality flat black paint to all of the interior surfaces of the shell.

A search of the Web recently didn't produce any "super" flat black paints available to us ordinary mortals. 3M Corp. used to make something nice, but it's been discontinued. Argh!

I ended up using Badger brand Model Flex No. 16-119 flat black paint. This is a water-based acrylic paint of good quality. It is available at better hobby shops.

Brush or spray a thin, even coat of the paint onto all the interior surfaces of the lens shell halves. Set them aside to dry thoroughly.

Working on the Lenses

The plastic lenses in the Corning lens assembly don't have painted edges. This allows light to bounce around the inside edges of the lens and bounce back out where it should not. For painting the lenses, you should obtain a top-quality small brush at the same place where you bought your paint. The flat faces along the outer circumference of the plastic lenses, the flat outer edges of the lenses, and the outer portion of the rim of the curved face of the lenses should be painted black. As you might guess, one slip or drip could mean the purchase of a new lens assembly.

Before you start painting, put the lenses back in the shell once the paint in the shell is dry. Look through the lens assembly, particularly at the curved outer faces of the lenses. You want to identify how far in from the edge of the lens you can paint without blocking light coming through the lens assembly. I ended up painting the outer rim of the curved face of the lens on a line about 1/8 of an inch in from the circumference defined by the molded plastic retaining rings on the shell.

If you are uncertain about handling this lens painting part of the job, either skip it, or practice painting objects of the same general shape as the lenses until you feel confident. Don't load the brush heavily with paint, as this will promote paint drips running where you don't want them. This lens painting job must be done very carefully, so don't attempt it when you are rushed or distracted.

The edges of the center glass lens in the Corning assembly already were treated with some paint and with grinding. You could try painting that outer edge black, but I considered the existing treatment adequate.

Once the paint on everything is thoroughly dry, it is time to bring out the compressed air. Blow off all dust particles and dirt from the lenses and the shell assembly halves. Use appropriate lens cleaning material and liquid to remove any fingerprints or smudges. This is another place where it is easy to scratch the lenses, particularly the plastic ones. Take your time, and think about what you're doing.

Place the lenses back in the proper order in one of the shell halves. Use the compressed air again to blow the lenses clean one more time and reassemble the shell around the lenses. Inspect the shell and lens assembly for dirt, and correct any remaining problems.

Reassemble the shell assembly back into the frame, and reinstall the screw that the focus-locking wing nut rides on. Reinstall the remaining hardware on the lens assembly. Reinspect the entire lens assembly for cleanliness and any other problems. Correct as necessary and then set the lens assembly aside in a clean place.

In the Toshiba 50HDX82, the lens assembly sits atop the main CRT gun assembly. The top of the main CRT assembly just under the lens assembly consists of a cooling liquid-filled chamber with a lens molded into the top. This lens is shaped like a cup, and thus automatically gathers dust particles and dirt at the bottom, right in the main path of light going up to the lens. Use your compressed air to blow that dust out of this cup-shaped lens. The top inside rim of the cup has been blackened, but it still has a somewhat shiny finish that contributes to the lens flare.

A tremendously brave person could try to paint that blackened edge with the flat black paint we used previously on the lenses. I'm not that brave. Instead, order some black "flock paper" from Edmund Optical Supply (*http://www.edmundoptics.com*). This paper has one side that has a light-absorbing texture. Get the thin stuff without the adhesive backing. Cut a ring-shaped piece of this flock paper.

A drawing compass is very helpful for drawing circles of the right diameter on the back of the flock paper to help as a guide for the cutting.

The outer diameter of the ring should just fit into the circular depression that surrounds the cup-shaped lens. The inner diameter of the ring should be small enough to block reflections from the edges of this "cup lens" and other off-axis light without reducing the main light beam brightness too much. I found a 5mm center opening to be about right.

Use small pieces of double-sided tape attached to the back (nonflocked side) of the flock paper ring to attach it in place around the edge of the lens cup. You might want to make several trial pieces with different inside diameters for this flock paper ring before you figure out the best balance between knocking down lens flare at this point in the optical path, and reduction of screen brightness. A smaller center opening in the ring will improve the flare problem, but the main light beam can get choked off, too. Another way to estimate this ring's effect is to put the lens back on over the flock paper ring, and look back down through the lens. If the flock paper ring is not visible through the lens once the lens is in place, go smaller on the center hole size of the ring.

Once you are satisfied with your treatment of the area underneath the main lens assembly, clean it out one more time with compressed air and reinstall the lens assembly.

Now you have only two more lenses to do, unless you have a TV with a single lens, which might be the case if your rear projection TV is LCD, DLP, or some other new alphabet-soup technology instead of the old-fashioned CRTs.

Lens assemblies from other manufacturers certainly will be different in detail, but the general sequence outlined here will still apply.

Conclusion and Results

It must be noted that the television on which I did this lens treatment still exhibits lens flare; it just has been reduced due to this effort. Your mileage will vary. Please look at the image that accompanies this hack on my web site (*http://www.sonic.net/~dgibbons*) to see the difference between a treated and untreated lens assembly.

More expensive rear projection televisions can use lens assemblies that are less prone to this problem. If I were to buy another RPTV, I would look for televisions with the least possible lens flare.

Again, the easiest place to see lens flare is when a single bright object is displayed on an otherwise black screen. If the black areas become lighter when the bright object is present, lens flare should be suspected.

—*David Gibbons, Keohi HDTV*

Focus Your Front Projector

Focus on a front projection system is just as important as it is on an RPTV, but the steps to get there are completely different.

Focusing a front projection CRT is a daunting task because it involves two projection systems that operate in series: an optical and beam focus.

Problems with one type of focus make it difficult to see problems with the other. As a result, sometimes people are at a loss as to which system is the problem. Add to that the need to astigmate the electron beam and adjust lens flapping (scheimpfluge [Hack #70]), and the novice CRT setup can fall far short of the projector's optimum.

New owners confuse the two, but mechanical aim isn't the same as scheimpfluge. Mechanical aim of the CRT/lens assembly is the same as taking a telescope and physically pointing it at something. Scheimpfluge is adjusting the mounting angle of the end lens without changing where the telescope is pointed.

A good pair of binoculars that can focus at short distances is very helpful, if not essential. Do yourself a favor and obtain a pair. Also, pick up a roll of 3M easy-release blue masking tape.

Mechanical Gun Aim

If you have a projector using lens tilt rings, set them to factory specs for your projection distance before doing any aiming. First you must mechanically aim the guns properly. One way to do this is to display a white-field pattern, using a calibration DVD [Hack #62], and shift the pattern edges equally spaced relative to the phosphor edges. Look into the lenses while doing this, not at the screen. You will need to do this for both red and blue guns separately. The pattern will be widest at the CRT bottom (assuming a ceiling mount), and that is where it should be balanced from left to right to assure the raster is well clear of the phosphor edges at its closest points. Then go to the top of the projection screen and examine the left-to-right relationship of the pattern's edges and the screen edges. Swing the gun to balance the edge relationships left to right.

Actually, you can set the projected edge farthest from the gun slightly wider for the screen edge than for the nearest edge, but the difference is minimal and it's OK to make them equal.

Once you have your CRTs where you want them, lock them into position.

Next, pay attention to how the white-field pattern is positioned vertically in the green phosphor. Shift it to make the top and bottom distance from the phosphor edges equal. Adjust the projector's up/down tilt to make the projected green white-field edges equally balanced relative to the top and bottom screen edges.

Moving the CRTs

The red and blue guns of a CRT projector have mounting screws which, when loosened, allow the two outer guns to swing left and right. Sometimes the screws can go into different holes for different convergence angles, or there is a slot for the screw that allows free motion. Your installation manual should cover which screws are involved. The central green gun usually isn't adjustable left to right. This is one reason to be very accurate in getting the projector centered and square to the screen when mounting. Up-and-down mechanical aim is accomplished by altering the projector tilt.

Do a rough beam and optical focus. Mark the exact screen center with a lightly applied triangle of 3M's easy-release blue masking tape.

Don't substitute another brand of tape here.

Also, mark the center of the top, bottom, and sides of the screen frame to make geometry easier to set later. Bring up a center cross pattern. Neutralize each gun's position shift controls and then use the projector's raster centering controls or magnets (just behind the yoke, if present) to align the center cross of each gun on the screen center. Remember, you have already mechanically aimed the lens and gun assemblies, so centering the projected center crosses on the marked screen center automatically centers things on the phosphor.

Notice that I've had you use a white field instead of a center cross to do the mechanical tube aim. This technique gets the lenses and CRTs well aimed even if the raster hasn't been accurately centered on the CRT yet. Now, if you happen to know your center cross pattern is already precisely centered, you can just use the center cross to aim the tubes. The best way I know of doing that accurately is to pull off the lenses, center the cross on the phosphor while measuring with a ruler, and then remount the lenses. The white-field pattern edge comparison method allows easy, accurate physical aim without pulling the lenses, and gives a subtle plus for the red and blue guns, as I'll explain later.

I know this method seems backward, but balancing the edges of the centered white-field pattern against the edges of the phosphor, and then the projected edges relative to the screen edges, achieves precise mechanical aim in an easy manner. The advantage to this method is basically the difference between having someone mark the middle of a piece of paper without the aid of a ruler versus aligning a slightly smaller piece of paper so that it is uniformly spaced inside the larger piece of paper. The latter is much easier to do accurately.

Consider now the off-center red and blue guns. If you aim the actual center of the phosphor of those tubes to project at the center of the screen, you'll note that the phosphor usage distribution is unequal from left to right due to the throw angle. Graph out this distribution (or just take my word for it), and you will see that the farther half of the screen gets illuminated with a smaller area of phosphor. Ever notice how the side of the screen opposite the side of the gun is less well focused? This is part of the reason.

Centering the raster edges relative to the phosphor and then using those lit-up edges to guide lens aim will actually place the red and blue guns so that they are mechanically slightly off true center. The left lens ends up pointed slightly left of center and the right lens ends up slightly right of center. At first blush, this seems wrong, but this actually can be advantageous because it makes the raster usage, resolution, and illumination more uniform across the screen. As a bonus, less horizontal linearity compensation and lens flapping are needed.

Now, if you are a traditionalist and want the center of the phosphor actually aimed at the center of the screen, you can pull off the lenses, set the center cross with great precision, and then use the projected center cross position to guide mechanical aim. This is the usual way things are done, but I present an alternative approach with some advantages.

Only after the guns are physically aimed and the raster centered should you begin final optical and beam focus adjustments.

Rough Optical Focus

Display a crosshatch, again using a calibration DVD [Hack #62]. Adjust optical focus using the two knobs on your projector. The frontmost knob generally controls edge optical focus. The rearmost knob adjusts center optical focus. Dial in the center focus first. Then adjust the outer (front knob) focus while watching the corner lines of the crosshatch flare inward and outward. Try to minimize flaring using just the outer focus knob. Then go back and forth between the two optical focus knobs to get both the center in focus and the outer edge minimally flared. The use of binoculars can aid this process tremendously.

Electron Beam Astigmation

Beam astigmation fine-tunes the electron beam lens to create a uniform electron spot with minimal flaring. Poor astigmation can make electron beam focus impossible.

If you aren't a technician, skip the following electron beam astigmation step. Adjustment of CPC magnets should only be done by an advanced setup technician. Improper technique will render focus impossible. High voltages and tube neck fragility also are significant hazards during CPC magnet adjustment!

Electron beam astigmation is carried out using CPC (called *color purity control* for historical reasons having little to do with CRT projection) magnets on the tube neck, and/or electronic astigmation controls. If both CPC magnets and electronic astigmation controls are present, it is best to let the CPC magnets do most of the work and then fine-tune with the electronics. That means neutralizing the electronic astigmation controls, and then adjusting the CPC magnets.

CPC magnets are arranged as pairs of rings about the end of the tube neck near the socketed drive board. Projectors don't always have the full set of two-, four-, and six-pole CPC magnets. Sometimes the ring pairs have a small knob allowing one to adjust the angle between the two rings of a pair. More often, you will see tabs with which to manipulate the rings. The CPCs have two, four, or six magnetic poles, but don't confuse that with the number of tabs on the rings; you can't actually see the poles. By varying the angle between the two rings of a pair (moving the tabs in opposite directions), you vary the intensity of the effect. Rotating a pair about the axis of the tube neck (moving the tabs in the same direction) changes the directionality of the effect.

If all three sets of CPC magnets are present, the rearmost is the two-pole (centering). The middle is the four-pole (ovalness). And the most anterior (if present) is either nonfunctional or a six-pole correction (triangularity).

Be sure to neutralize all electronic astigmation controls prior to working with CPC magnets. Also, on projectors that lack separate electronic astigmation controls, do CPC and electronic astigmation while the highest scan frequency to be used is displayed.

This hack assumes you know how to change the electron beam focus. Don't get that confused with optical focusing.

Ovalness adjustments. The four-pole alters the ovalness of the electron beam lens. Adjust this while displaying a dot pattern while contrast is moderately high. Intentionally underfocus the electron beam, turning the dots on the pattern into uniform blobs. Adjust the four-pole magnets to make the blobs as perfectly circular at the screen center as possible. Attach a perfectly round piece of 3M blue tape on the screen to aid in judging shape (you can make this with a compass or large-hole punch). Turning the small knob, or moving the adjustment tabs in opposite directions, alters the amount of ovalness. Spinning the four-pole rings around the axis of the tube neck changes the direction of the ovalness axis.

Centering adjustments. The two-pole centers the electron beam in the electron beam lens. Adjust this while displaying a dot or crosshatch pattern while the contrast is moderately high. Intentionally overfocus the electron beam, making the dots into a flare with a bright central core. Turning the small knob, or moving the adjustment tabs in opposite directions, alters the amount of deflection. Spinning the two-pole rings around the axis of the tube neck changes the direction of deflection. Make the bright core centered in the flare.

Triangularity adjustments. The six-pole adjustment also is done with the dot pattern in underfocus. If the six-pole works (it might not do anything), it creates a triangular astigmation change. Use it to correct any residual triangularity which you could not correct using the four-pole.

Go back and forth between the two- and four-pole adjustments to get things right. As a final check, carefully watch the dots as you go from under- to overfocused. The dots should stay almost motionless as you vary the beam focus. Once your CPC magnets are set, you can fine-tune using electronic astigmation.

You will have to redo raster centering after adjusting the CPC magnets. If this is the blue gun, you'll probably want to leave the electron gun underfocused enough to make its light output measure about 20% higher than its fully focused state to improve grayscale tracking at higher light output.

Phosphor Grain Optical Focus Technique

At this point, most people have difficulty deciding whether optical, beam, or both kinds of focus problems are present. Here is a method for setting excellent optical focus without being confused by beam focus. You must have a good pair of binoculars that focus at a short distance to use this technique.

The phosphor surface of a CRT has an inherent grain pattern. Because this grain is visible and is always exactly at the plane of light generation, you can use the grain to set optical focus independent of beam focus. A small piece of 3M easy-release blue masking tape aids in keeping your eyes correctly focused on-screen. Display a bright window pattern and intentionally defocus the electron beam to make the scan lines disappear. Adjust the center optical focus while viewing the screen through binoculars. When the optical focus is correct, the inherent grain pattern of the phosphor surface suddenly snaps into view. This is nearly impossible to see with just your eyes, but binoculars make it readily evident. Once optical focus is sharp enough to see phosphor grain, refocus the electron beam finely.

Focusing Lens Cap

The phosphor grain optical focusing technique largely eliminates the need for this method, but I mention it for completeness. Some projectors come with a special lens cap having a central hole approximately 1 inch in diameter. This is intended to reduce the aperture of the optical lens and allow examination of beam focus even when optical focus is not quite correct.

Because the phosphor technique achieves good optical focus independent of beam focus, I recommend setting the optical focus using the phosphor grain technique first. This allows greater light availability while setting beam focus than reducing the lens aperture.

Final Beam and Optical Focus

At this point in the process, both beam and optical focus should be excellent, but further refinement is sometimes possible.

Use a plain white 3×5 card for finding the exact focal distance of the projector. Do this by moving the card fore and aft in front of your screen to see where a fine focus pattern is best in focus. If it is already exactly at the screen surface plane throughout the screen, you are done. If it is more than 1cm in front of or behind the screen, perform the following procedure.

Unfortunately, this will temporarily undo the hard work you just accomplished getting the phosphor grain sharp. Welcome to the back-and-forth world of calibration.

Display a fine-detail focus pattern and intentionally overfocus the optics (using the rear lens control) so that the center is best optically focused about 2cm short of the screen. This is in the direction that extends the lens barrel forward. Bringing the focal plane slightly short of the screen lets you more easily examine the focal distances throughout the screen. Check the focal distance for each screen edge by moving the 3×5 card back and forth in front of the screen. You'll be able to see, very accurately, the distance at which things are best focused on the 3×5 card. Note if the distances are uneven between left and right (indicating a horizontal lens flapping error), or top and bottom (indicating a vertical lens flapping problem). Fine-tune your lens flapping [Hack #70] to make the focal distances for the top equal to the bottom, and the left equal to the right. This is also the time to fine-tune lens flapping rings. The 3×5 card-check is so precise that you'll notice the flapping changes caused by uneven tension on the lens mounting screws!

Next, pay attention to how the left and right edge distances compare to the distance at screen center. They should be about equal between the center and the edges. If not, slightly adjust the inner and outer optical focus to bring both the edges and center to focus about 2cm in front of the screen. Notice that I don't have you check the screen corners. That's because there will almost always be a difference in the extreme corners and center, and using the left and right edges gives a good compromise which preserves the central focus where the video image is going to be the sharpest portion of a movie frame anyway.

At this point, the optics are perfectly balanced in terms of scheimpfluge and inner versus outer lens focus. The only thing left is to shift the entire optical plane to the screen surface. Make tiny movements of the rear lens focus knob while repeatedly checking the optical focus position with the 3×5 card. One thing to consider is that you probably did all this with the projector's lens hood off, and while the lenses are probably a little cooler than normal. As the lenses warm up, their index of refraction decreases and the focal plane moves slightly outward. You might need to leave the focal plane a centimeter or so short of the screen surface, so it hits exactly on when the optics are at normal operating temperature an hour or so after the lens hood and hushbox are closed.

If you use the 3×5 card for final focus, you'll note that it is so sensitive an indicator that even the slight shift of the lens while tightening the focus knobs will be detectable. Expect to have a great deal of exercise moving between screen and projector using this technique.

With optical focus perfected, set the final red and green beam focus, and then the grayscale/blue gun defocus. Recheck your raster centering, especially if the projector uses electromagnetic focus.

—Guy Kuo, Keohi HDTV

Don't Mess with Odd Screws

Ever run into a strange-looking screw, and spend hours trying to find just the right tool to get it open? Chances are that screw is for your protection: proceed with caution.

When you go to clean your TV's mirror and lenses **[Hack #69]**, you probably will be able to enter your TV from both the front and back. It's a good idea to try and go in through the front whenever possible, and if you've got a Pioneer Elite TV, you must never enter in through the back. There's a nice optical cavity, easily accessible, but you'll be digging your own grave (or rather, your TV's grave). This is a classic case of why not to mess with strange screws (in other words, screws that aren't standard Philips head or flat head).

What's the Big Deal?

On the Pioneer Elite unit on which I learned about this, there were a couple of weird-headed screws holding the back on the unit, along with several regular Philips head screws. Don't be clever and figure out how to get these screws out; they are there for your own protection! You can use a small flat-edge screwdriver to circumvent them, but you'd be shooting yourself in the foot, just as I did when I first ran into this issue.

Several Pioneer Elite models have the bottom edge of the mirror bracketed into the removable back and the top edge bracketed into the body of the TV itself. If you separate the back from the unit, the mirror slips out of the top bracket and does a nosedive straight into your fresnel screen! Once this process has begun, there is nothing to do but watch in horror. You'll find yourself standing there holding the back of the optical cavity in your hands, which has just pushed backward and into your hips for some unknown reason...and then there's the crash and things get really ugly. I've seen the mirror break, the screen break: it's a bad, bad deal.

So, if you've got a Pioneer Elite unit, don't go in through the back. Of course, if Pioneer can do something like this, anyone can. As a general rule, don't go digging into odd screws on expensive equipment, unless you're willing to pay the consequences. And on the TV front, if you can go in through the front, that might just save some real headaches, on any manufacturer's unit.

Going in Through the Front of a Pioneer Elite TV

The front of a Pioneer Elite TV's frame usually comes off via unscrewing the Philips head screws at the bottom of the frame. You get to these screws by removing the ornate plastic plates that say *Pioneer* on them; these plates are about 1 foot wide and 1 inch tall, forming the cosmetics separating the screen above from the speaker grill cloth section below. There is one plate on each side of the unit; stick your fingers under the plate and pull gently.

On some units, you have to remove the grill cloth first. That takes a little more force, but still is accomplished by just pulling by hand.

The screen frame comes off by lifting it straight up, or out at a 45° angle, and then up. This reveals the two-layer screen sandwich: the fresnel closest to the mirror and the lenticular facing the viewer. The screws to remove this sandwich are then apparent.

The upper-left corner's holder is the only one where you actually have to remove a screw. The others allow their holders to be removed by just loosening up their screws a bit and deslotting the holders from them.

Be sure not to grab the screen by its attached aluminum brace at the top of the two-layer sandwich, lying horizontally. It's not attached; it's just lying there for bracing and derippling purposes. The stack will fall out of your hands directly if you try to hold it via this piece of aluminum.

Also beware of turning this sandwich sideways once it's off. Keep it horizontal, or the very flexible lenticular will have a tendency to waffle and bend in half on you, via gravity. I once had one split itself on the edge of the frame on its way to the ground. I'm sad to say that it cost a bundle to replace.

—*Robert Jones, Image Perfection*

Annual Home Theater Tune-Up

All the calibration in the world won't do your theater any good if you don't keep it in tiptop shape. Tuning things up every 12 to 18 months will keep your theater sounding and looking good.

You might be amazed at how speaker wires come loose, connectors loosen, and dust bunnies reproduce over time. An annual tune-up keeps things working at their best. Here is a list of things to do:

- Disconnect and trim off the ends of all your speaker wires. Cut back the ends enough to expose fresh, shiny copper wire (you are getting rid of the dull-brown oxidized wire). Reconnect the speaker wires, making sure no strands of wire are sticking out.
- Dust the top and underneath of all the electronics in your rack. You can use a can of compressed air to clean out your receiver/amp, but don't do this to your DVD player. You really don't want to stir up dust inside an optical device.
- Make two labels that say "A," two that say "B," etc., and mark both ends of your power cords from all your electronics. Unplug the power cords and try to bundle the AC power cords together using Velcro strips or split-loom tubing available at RadioShack or your local electronics store.
- Disconnect all your interconnects and straighten out the tangles. While disconnected, wipe off the dust from the jacks on the back of your equipment. You can use a soft toothbrush to gently scrub the dust out of the connections.
- Disconnect all your CATV coax and satellite connectors and examine the center wire. If the center wire is dull brown, cut off the end and install new F-connectors. Examine all your F-connectors and replace any that look like they are pulling out of the coax, or that look sloppy.

When you use F-connectors, tighten the connectors by hand, and then use a wrench or pliers to go at least 1/4 turn more. The number-one cause of poor CATV signal and service calls is loose F-connectors.

- Use a laser pen to see where your speakers are pointing and adjust them.
- Use your SPL meter [Hack #63] to readjust the levels on all your speakers.

Once you have everything adjusted, write down the settings you used on your receiver. This can serve as a baseline if you ever need to start from scratch.

Taking an hour or two to perform these tasks every year will keep your theater looking good, but more importantly, it will keep it running at optimal levels.

—Robert McElfresh

CHAPTER NINE

Do It Yourself

Hacks 75–83

Home theater presents myriad reasons and options for DIY, or do-it-yourself, activities. Reasons can range from and be combinations of saving money, starting a new hobby, meeting particular performance goals, meeting particular appearance goals, the challenge, the satisfaction of the end result, or even just because you're bored on a Saturday afternoon. The formula that will determine if DIY is right for someone will be unique to each person. Actual DIY projects can range from buying a complete kit—where all you have to do is make a few wire connections and turn a few screws—to designing and building amplifiers.

I'll start with a word of caution. Building audio equipment needs to be a lifetime hobby if you intend to attempt your own designs, especially if you're trying to design a crossover for a speaker or build an amplifier. You won't save money with these activities, and it might take you years to gather all the necessary equipment and learn enough to design and properly implement a speaker or amplifier from scratch.

An exception to this is a subwoofer enclosure. With a few weeks of research, you can learn enough to build some very good subwoofer enclosures.

On the other hand, loads of designs are available on the Internet where all the hard technical stuff is done for you. All you have to do is buy the parts and assemble them. Provided you have the tools and some basic skills, these kits or sets of instructions can provide exceptional value. If you don't have the tools, the money-saving reason for DIY will go out the window, most of the time. Building amps will require the appropriate soldering and electrical measuring tools and a fair bit of electrical/electronic know-how. Not everyone will be successful at these kits. Building speaker kits, on the other hand, requires very little electrical know-how; all you really need are some basic carpentry skills. Pretty much anybody with the right tools can build an enclosure for a speaker.

This chapter provides a real cross-section of hacks, designed to interest and challenge the novice all the way to the pro. Simple designs for speaker stands are included, but so are complex instructions for combining multiple antennas into a 16-bay reception monster. You'll learn how to build your own screen and masking for a real in-home viewing experience, but you also can start with a simple TV stand. Pick something that looks interesting and achievable, and jump on in!

Build Your Own Speaker Stands

One of the most common do-it-yourself projects is the speaker stand. Almost everyone needs them, you can make them cheaply, and they'll be a great starter project.

Many home theater systems use bookshelf speakers for left and right main channels. Bookshelf speakers require stands to get them to the right height. Short of having a woodworker custom-build the stands for you, you'll probably find that one stand sits your speakers down too low, another stand raises them up too high, and the only stand that does work sports a ghastly price tag that makes you do a double take. This is a perfect place to add some do-it-yourself know-how, and get a perfect fit.

Basic Bookshelf Speaker Stands

First, measure the *footprint* of your speakers, which is just the width and depth of the bottom of the speaker. You usually want to match that pretty closely with your stand; too much excess and the stand looks like it was made for larger speakers (and we don't want that!). The stand tops shown in Figure 9-1 are 8 inches wide and 8 inches deep.

These stands are about as simple as you can get and still look quite good. Once you have the base of your speaker measured, obtain some 3/4-inch-thick, medium-density fiberboard (MDF). Cut two rectangles to your speaker measurements, and then cut two more squares slightly bigger (for example, two squares might be cut at 8×8 inches, and two more at 10×10 inches). The smaller squares are for the speaker base and the larger ones are for the stand base.

Next, cut four longer rectangles; these should be about 2 inches less deep than your speaker base. So, if you have a top plate that is 8 inches deep, these rectangles should be 6 inches deep. To obtain the height, you'll need

Figure 9-1. Basic speaker stands

to know how high you want your speaker to sit. Then, just subtract 1 inch, and cut to length.

The top and base of your stand actually total 1 1/2 inches in height, but assume 1/2 inch of settling, especially if you've got carpet. In fact, if you have really thick carpet, you might want to assume even more.

So, these might be 6 inches deep and 30 inches wide (for example). Two of these become the connectors for one stand and two for the other. At this point, if you're able, bevel the long edges of the connectors. You also should round all four edges of the tops and bases. This will really add a polished, classy look to your stands, and allow the base of the stands to sit more firmly on the floor.

With the rounded sides up on both the top and bottom plates, the two long pieces should be evenly spaced, centered, and then angled slightly (see Figure 9-1 as a reference). Now, glue and nail in place. All that's left is to paint the stands; prime first, and then lay on a couple of good coats of black, or whatever other color you prefer.

Sturdier Stands for Heavier Speakers

If you've got heavier speakers, you might want a sturdier design. This design also works well if you need your speakers quite high; the basic stand sometimes looks a little wobbly as the connector pieces get really lengthy. The basic procedure is similar, with just a few twists for stability.

First, I'll assume you've got larger speakers—mine for this job were 16×10 inches in footprint. So, my tops were a bit larger than in the basic stand. Again, round the tops. The base follows the same process; in this case, my bases were 18×12 inches, also rounded.

Instead of long connector pieces, though, I made a square "box" out of MDF (again, 3/4-inch thickness is fine). This box was half the depth of the top—in this example, 5 inches—and the height was calculated as shown in the basic stand section. To add a nice look, the edges of the box should be sanded slightly round, to look less...well...boxy. Glue and then nail this box to the base of your speaker; I set mine in the shape of a diamond for slightly more visual interest (see Figure 9-2).

Figure 9-2. Sturdier speaker stands

Now glue and nail on the top. Before you finish up and paint, though, drill a hole in the center of the top plate, which opens into the box that acts as a support. Now prime and paint. However, before putting these stands into action, fill the box support with sand. This really solidifies these stands, allow them to hold heavier speakers, and even look more substantial.

Costs Involved

Believe it or not, it cost about 30 bucks to make all four stands! Online, similar stands ran from $30 a pair (on the low end; and this is only for two) to hundreds of dollars in higher-end shops. Clearly, this is a no-brainer way to get into DIY and save some bucks. Also, don't be afraid to get creative. If you don't have a lot of tools, find materials you can work with that don't need much in the way of tools. Buy premade shelves and have the hardware store cut them to the sizes you need. Substitute threaded rode (the line you typically see attached to the anchors of small boats) or PVC pipe for the center supports. Add door or corner moldings to get a different look. Whatever you can come up with that gets the job done and saves you money will represent a successful project.

—Dustin Bartlett

Add Rollers and a Stand to Your TV

A television is a tremendous pain to move, even if it's going only six inches. With some MDF and a few casters, you can add mobility to your TV set.

If you've ever had a connection come loose from behind your huge television set, you realize what a pain it is to move the set, try and reconnect the cables, and then ease the set back into place, all without causing the wire to come loose again. A great solution to this (and all other TV movement problems) is to build a small stand for your TV, put it on wheels, and relax.

Required Materials

Here's what you need:

- Three MDF shelves (1-inch thick, cut to TV base dimensions)
- Four brass ball casters
- Four 4×7/16-inch bolts
- Four 7/16-inch nuts
- Eight 7/16-inch cut washers
- 220-grit sandpaper
- 400-grit sandpaper

- Four work clamps
- Wood glue
- Paint

Construction

The directions that follow will produce the same stand pictured in my home theater (see Figure 9-3). Of course, you can make any changes you want to produce a different stand.

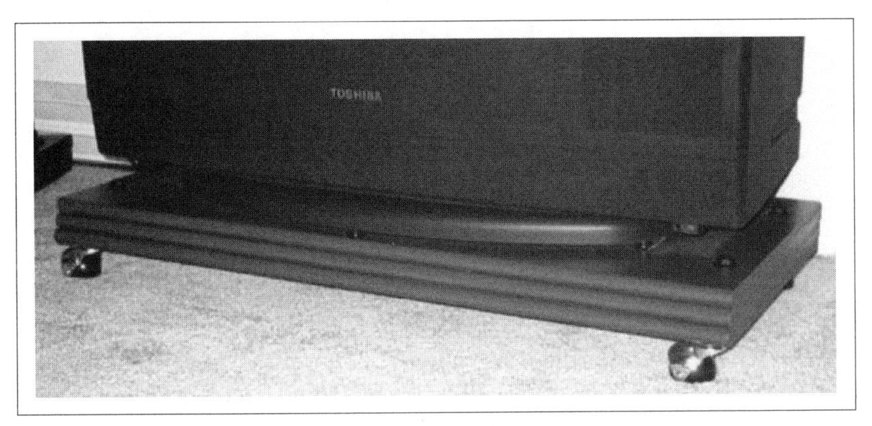

Figure 9-3. DIY TV stand

1. Most MDF shelves come with a bull-nosed edge on one side. The first thing you want to do is sand the bull-nosed edge smooth. Use the 220-grit paper first and follow up with 400-grit. This will smooth out the surface, making the paint flow better later on.
2. Take two of the MDF shelves and place one on top of the other to ensure that they are the same exact size. Place one shelf on top of the other, and then slide back the top shelf 1/4 of an inch to create a staggered arrangement.

The front edge should be the bull-nosed edge. This produces a nice "raked" detail.

When you're happy with the fit, remove the top shelf, lay five or six generous beads of wood glue up and down the length of the bottom shelf, replace the top shelf, and slide it into position. Clamp the two shelves together at each corner, and let dry for 15 to 25 minutes.

3. Repeat Step 2 for the third shelf. You will be gluing this one on top of the two shelves you already glued together. Don't forget to test the fit and measure the 1/4-inch stagger. Glue and clamp the third shelf, and let dry for 15 to 25 minutes.
4. Now, drill four holes into the stand to accommodate the four bolts.

You can leave the bolt heads poking out; I actually like the way this looks. They match the detail on the rack for my AV gear.

Because the bolts are 7/16 of an inch in diameter, you need a 1/2-inch drill bit. The extra 1/16 of an inch in size will allow easy insertion of the bolts. You basically can place the bolts anywhere on the stand that you want, but make sure you know where your casters are going so that you don't overlap their space. I chose to put the casters as close to the corners of the stand as I could for increased stability; that meant the bolts needed to be further off the corners. Pencil in the center of the intended holes, get your drill and 1/2-inch bit, and drill through the three glued shelves.
5. Turn your stand upside down and mark off where your casters and any necessary screw holes will go (this depends on the casters you chose). Once your screw holes are marked, drill them through with an appropriately sized bit. Your stand is now ready for painting.
6. Paint your stand and your hardware. I sprayed two coats of flat black onto the hardware (bolts, washers, and nuts) and three coats of brown onto the base itself. When the stand was dry, I brushed on a coat of acrylic polyurethane. This dries in an hour or so. When the assembly is dry enough, sand with 220- and then 400-grit paper to smooth out the finish. Don't worry about how it looks after you sand! Just be consistent and don't rush. Recoat with the urethane and set to dry. Do this as many times as you like. The more you sand, the smoother the finish will be. I recommend at least two sanding sessions.

Before applying each coat of paint, be sure to clean the stand with some paper towels and Windex. This removes any dust from the MDF and sanding you do. The Windex won't hurt the bare MDF, either.

7. When all is dry, install the bolts as indicated in Figure 9-3, and screw on your casters. Now all you need to do is bring the stand to your set and get a friend to help you lift that TV onto its new base. You're done!

Construct a Screen for Projection

If you've got a projection-based system, you've probably already realized that without a screen, projectors are just about useless. Good screens can cost big bucks, unless you know how to build one yourself!

With digital projectors as cheap as they are today it's a shame the screen is still so expensive. Many screens out there cost more than budget projectors do. However, you can get a decent screen from numerous online vendors for a few hundred dollars. (The larger fixed screens with fancy screen fabrics and nice masking systems you'd like to have can cost several thousand dollars.) Given my experiences with samples of the hyped screen fabrics out there my preference is definitely blackout fabric and my own frame for a total cost of about $50.

Frame It and Stretch It

Building a basic screen isn't much more difficult than building speaker stands [Hack #75]. That might seem hard to believe, but a screen is just wood and fabric. If you can build a simple wooden frame and can follow some simple canvas stretching techniques, it's a piece of cake.

Begin by getting some 1/4-inch flooring plywood. Cut the plywood to your desired screen size (note that this technique limits you to a 48-inch-tall screen). You'll also need some 3/4-inch pine shelving material that you should cut into 3 to 4 inch wide strips, sized to make a frame that will mount on the plywood.

I chose to construct the frame using simple but very strong fishplate joints. Figure 9-4 shows 1/4-inch cuts into the pine on each side, with a piece of plywood for connecting the two edges.

Once you've glued the joint, use a staple gun to sturdy things up (see Figure 9-5).

Now, you can mount this entire frame on your plywood, as seen in Figure 9-6. The plywood makes this frame very rigid and provides numerous options for hanging the screen. I've had success cutting a couple of small square handholds in the plywood. This makes the screen easier to handle and provides a perfect fit for 1/4-inch mirror hooks. I'll leave it up to you to determine the best way to mount the mirror hooks to your wall, or let you come up with another method that works for you.

Before you start stretching the fabric I also recommend sanding down the outside edges of your frame to round them off. This reduces the chance of ripping the fabric you stretch over the frame. If you have a router, I also

Figure 9-4. Fishplate joint with pine and plywood

Figure 9-5. Completed joint

Figure 9-6. Frame with borders and bracing

recommend rounding off the inner edge of the frame to prevent it from pressing against the fabric. If you don't have a router, sanding will have to do.Now you need to add fabric. Plain white blackout fabric will work, and you can get this at any fabric store. If the salesperson doesn't know what you are talking about, ask for the fabric that backs curtains to prevent light from passing through. Lay out the fabric on a flat surface, cloth side down, and then place the frame on top; make sure you've got at least one inch of excess fabric on all four sides of the frame, and if you've got significantly more than that, cut it off and discard. Also, smooth out any wrinkles.

Fold one side over the frame, and tack the fabric to the frame in the center of the side you're working on. Do not add any other tacks at this point. Now stand the frame up so that the side tacked in is on the bottom (against the floor). You should be able to stretch the fabric opposite the tack over the frame, until there is a slight crease in the fabric. This tells you things are just tight enough. Tack in the fabric.

Many will suggest using a staple gun at this point. I prefer using a few tacks, and then coming back with the staple gun; if I mis-tack something, a tack is a lot easier to pull out of fabric than a staple.

Now you can move to one of the two remaining untacked sides, and repeat. Then, tack in the final side; at this point, you should have a sort of diamond formed from the creases, running from one tack to the next. Now, you're ready to get out your staple gun. Start in the center of one side, slowly move out a few inches, and staple. Then repeat, in the opposite direction. So, you're always moving out in one direction, stapling, and then moving out in the other direction, keeping the tension even and the creases moving toward the corners.

When you get to the corners tuck one side under and fold over, and you're all set (see Figure 9-7).

Figure 9-7. Completed screen with staples in

Finish up with the stapling, and you should have a great, taut fabric, as seen in Figure 9-8.

> Although a regular staple gun will work, a pneumatic staple gun will make the task much easier; besides, what better excuse for getting a new power tool could you ask for?

This is but one method for building a screen. I found it easy and effective. I've seen examples online of screens that ranged from simply hanging a piece of Parkland plastic on your wall, to frames made out of PVC, all the way to

Figure 9-8. Completed screen

guys with the skills to weld an aluminum frame. There's no single right solution to most problems and that's the beauty of DIY. Copy exactly what someone else did, come up with your own unique solution, or do something in the middle. With DIY, you're the client, engineer, worker, and boss, all in one.

—*Dustin Bartlett*

Mask Your Screen

Whether you built your own screen or purchased one new, adding masks to the top, bottom, and sides will create a true movie-style experience in your home.

Having your screen bordered by a black frame is a must. Not only does it make the picture look better, but also it makes the screen look better when the lights are on. The simplest masking is the fixed variety. All you have to do is build another simple frame, minus the plywood backing this time, sized to fit around the screen you built earlier. Wrap it in a matte black fabric and use some I-brackets to attach it to the screen.

If you're a true-blue movie fanatic, fixed masking just won't cut it; you'll need the variable kind. If you've already read about aspect ratios [Hack #13], you know that movies come in several ratios; most notably, though, you'll need to deal with 2.35:1, 1.85:1, and 1.33:1 (more colloquially known as 4:3). If you've got just a plain white screen [Hack #77] hanging on the wall, you're probably going to have leftover space in all of these formats.

It's actually possible to build a screen specifically for one format, and ensure that movies in that format fill the entire screen matching up to fixed masking. Most people do this. Of course, movies in other formats won't match up to a fixed masking system, and that's hardly acceptable to any real DIYer.

Constant Area Viewing

There are three types of adjustable masking systems: constant width, constant height, and constant area. Most people run a constant width, some run constant height, and a few run constant area. In my opinion, constant area is by far the best option.

Constant width. A constant width setup doesn't require you to manipulate the picture in any way. You don't need to use the zoom lens or a home theater PC's aspect ratio control abilities. All you do is build a screen in a ratio that matches the native aspect ratio of your projector (either 4:3 or 16:9) and cover up the black bars on the top and bottom of the screen that show up with certain aspect ratios. You can do this with removable boards of various sizes or with horizontal curtains on rollers. The roller method does require you to counterweight the bottom mask, though. Although this setup is easy, it has two major drawbacks:

- You can mask all aspect ratios in this manner only if your projector's native aspect ratio is 4:3. This system doesn't work with 16:9 projectors and 4:3 material.
- The 4:3 images will be the largest and the 2.35:1 images will be the smallest. I don't know about you, but I prefer the opposite to be true.

Constant height. A constant height setup does require the use of a zoom lens and a home theater PC. With a constant height setup you use the zoom lens and a home theater PC to shrink or expand the image so that its height stays constant in all aspect ratios. Vertical curtains are used to mask off the unused portions of your screen on the sides. Most people build a 2.35:1 screen for this type of setup. The big advantage here is that 2.35:1 movies look the biggest and 4:3 movies look the smallest. The big disadvantage is that most projectors' zoom lenses can't manage this alone. So, you either have to be able to move the projector back and forth in the room, or you need a home theater PC that will allow you to shrink and expand the image at the expense of panel resolution.

Constant area. The third option is constant area. With this system you adjust the image so that each aspect ratio takes up the same surface area. This means both the height and width of your screen will change between the various aspect ratios. Its biggest advantage is that it really does give the illusion that your screen is always the same size; 4:3 images don't lose their impact like they do with a constant height setup, and 2.35:1 material doesn't lose its impact like it does with a constant width setup. Like constant height, however, a constant area screen isn't easy to achieve, especially if you try to achieve it across all aspect ratios. To properly achieve it over all aspect ratios you need four-way adjustable masking, a good zoom lens, and a good lens shift feature or a home theater PC.

I really like the idea of a constant area screen, but I couldn't pull it off across all aspect ratios (at least not with the room and cost constraints I had). So, I did what I felt was the next best thing: I went for constant area between 2.35:1 and 1.78:1 and then constant height between 1.78:1 and 1.33:1. I did this with a three-way adjustable manual masking system and by taking advantage of how a zoom lens works on most projectors. You can come very close to a constant area between 2.35:1 and 1.85 with just a zoom lens, and you can get all the way there if your projector's lens shift, DVD player, or home theater PC will let you shift the 2.35:1 image up a few inches.

This works because most projectors, when ceiling-mounted, have a zoom lens that expands down much faster than it does up. So, I built a 2:1 aspect ratio screen and mounted the projector so that when it was zoomed all the way out, its 1.78:1 image filled the entire height of the screen. In this setting I had the top adjustable mask fully retracted and I used the two adjustable vertical masks to cover the unused side portions of my 2:1 screen. Then when I display 2.35:1 material, I zoom the projector in until the bottom of the 2.35:1 image matches up with my fixed bottom mask. To get a true constant area here you'll have to zoom until the image fills the entire width of a 2:1 screen (2.08:1 if you want to get picky), which will result in some of the image ending up on the bottom mask. Then you'll need to use your projector's lens shift, your DVD player, or a home theater PC to shift the image up off the bottom mask. Next I adjust the top mask to match the top of the image and fully retract the vertical side masks. For 4:3 material I zoom back out, retract the horizontal mask, and bring the vertical curtains in further. This results in the 4:3 image appearing smaller, but who really cares about 4:3 material anyway?

Constructing the Masking

The masking is quite simple. Cut a strip of 3/4-inch laminated pine (or other appropriate wood of your choice) to the width of your screen, plus a few inches (so, perhaps, 100 inches for a 96-inch-wide screen). Then affix black masking cloth, or blackout cloth, to it. This serves as the bottom mask. Your top mask will be a valance that will house a curtain rod for the vertical masks and a roller for the horizontal mask (more about this later). The valance and the fixed bottom mask will be attached using two strips of pine such that they will form a slot your screen will fit into. With the bottom mask acting as a ledge, you can get away with using Velcro to mount the screen into the masking system. One advantage of this system is you can keep the screen out of harm's way while you mount the masking to the wall. Once the masking is mounted and the screw, drills, and other dangers to a screen are gone, you can bring the screen in and drop it in place.

Now, back to the valance. This involves framing out the sides and top, coming out from the wall several inches (see Figure 9-9), also with pine. Within this valance, attach a curtain rod (see the figure again for a visual aid). This curtain rod will connect to and control the vertical masking.

Figure 9-9. Valance, curtain rod, and top assembly

Attach some blackout curtains to this curtain rod for use as your side masking. I've sewn a piece of straight steel (light, but inflexible) into the inside edges of the curtains to ensure they hang straight and form nice straight vertical masks.

The last step is to add the horizontal masking (which reduces the effective height of the screen); this is where things get a little tricky. The horizontal mask is wound on a roller that feeds out underneath the top curtain rod (you can see this in Figure 9-9, as well as Figure 9-10).

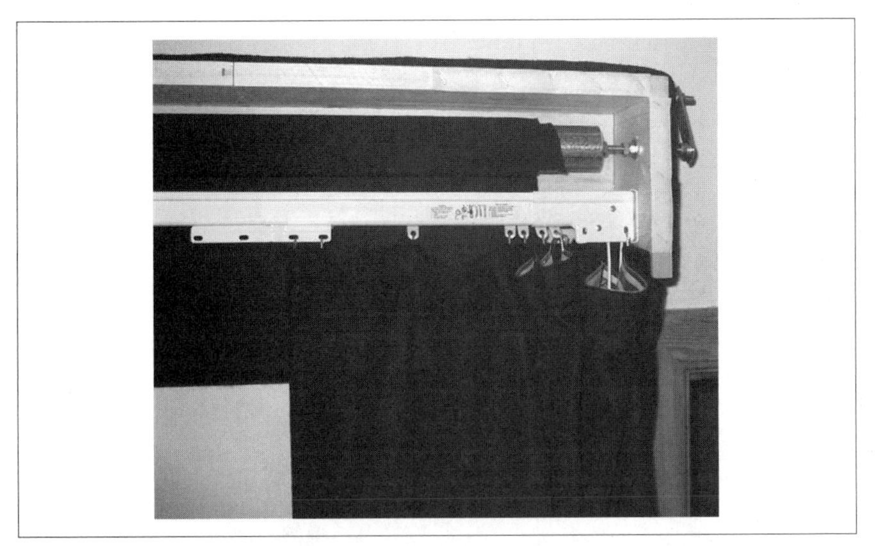

Figure 9-10. Horizontal masking assembly

This roller is made of 1 1/4-inch Electrical Metallic Tubing (EMT) pipe. I've hammered into the ends of the pipe a 1/2-inch piece of circular plywood, with holes drilled in them to accept 1/4-inch bolts and T-nuts on the inside. The T-nuts are there to provide something very solid for the bolts to screw into. This essentially creates a roller that I can easily set up to rotate. On the side that will not have the cranking mechanism (the left in my theater), thread a 1/4-inch bolt (one that isn't threaded all the way to the top) through a hole you've drilled in the outside of the valance, then into the plywood in the end of the roller, and finally into the T-nut. Don't forget to put a washer, lock washer, and nut on the bolt before you thread it into the plywood in the end of the pipe. You'll use this to lock the bolt to the roller, making it so that they turn together.

The right side will need a little more hardware. Here I recommend using a 1/4-inch stove bolt (a bolt that is threaded all the way to the top). Again you'll need to drill a hole in the valance to accept this bolt. On this end, though, because the stove bolt is threaded all the way, I mounted a T-nut that was too large for a 1/4-inch stove bolt to the outside of the valance to prevent the bolt from wearing away the wood. Use the same means as the noncrank side to lock this bolt to the roller so that they turn together. On the inside of the valance, put two nuts (turn them together tightly so that they won't move) and a washer. These two nuts and a washer serve two purposes. The first is to prevent the roller from sliding too much from side to side. The second is to provide a support for a wing nut on the outside of the valance to tighten against and stop the roller from turning, without popping out the wooden plugs in the rollers in the process. Finally, rig a simple hand crank to the stove bolt on the outside of the valance (see Figure 9-11).

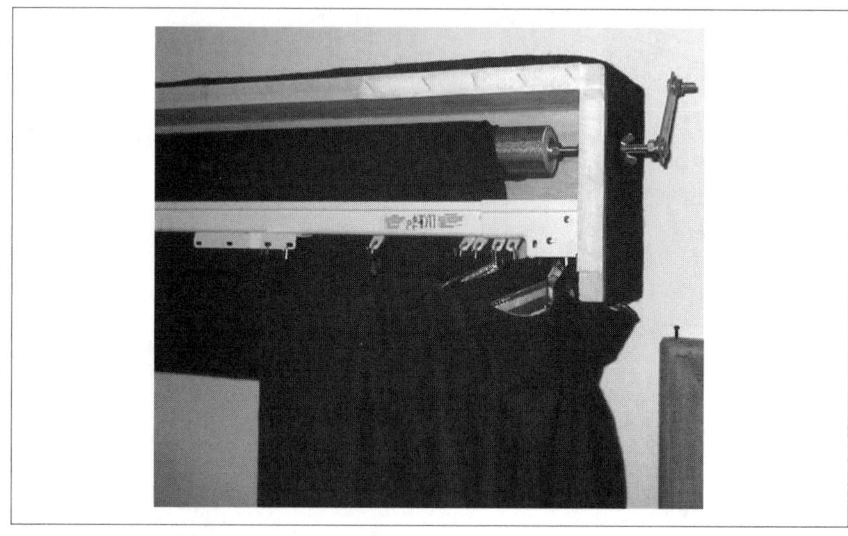

Figure 9-11. Crank assembly

Next, wind black masking onto this pipe, and sew another steel piece into the bottom of the masking (I used 3/4-inch EMT pipe for this purpose). This will ensure the masking hangs evenly with a straight edge, and winds down when needed.

Granted, this is no remote controlled, motorized, four-way-adjustable masking system that uses magnetic stop points (there are plans on the Internet for masking of that type). It was, however, an effective solution that fit my wants and budget, which was less than $200. Even pictures of just the screen with the lights on show the rather drastic improvement adjustable masking can provide. See Figures 9-12 and 9-13.

—Dustin Bartlett

Figure 9-12. Masking setup for 2.35:1 format

Figure 9-13. Masking setup for 4:3 ratio

HACK #79 Construct Speaker Cables Using CAT 5

Speaker cables are one of the most overlooked items in your system. Building your own cables allows you to get maximum performance at a minimal price.

This all started back in 1995 when I read a letter published in *Speaker-Builder* extolling the virtues of using Category ("CAT") 5 network cable as high-end speaker cable. I bought a 1,000-foot spool and spent more than a year tinkering around with different configurations to best optimize the use of this cable for high-end audio. After countless hours (and many blisters on my fingers), I found this construction to be the most ideal. In direct side-by-side comparisons I found this cable to be sonically better than the very best in high-end cables, such as Audioquest Midnight and Straightwire Maestros, and even equal to or better in some areas than the Kimber 8TC. The completed cables are shown in Figure 9-14.

Figure 9-14. Completed CAT 5 speaker cables

Materials

All you need for this project is some CAT 5 cable. I recommend either Anixter (*http://www.anixter.com*) or MECI (*http://www.meci.com*) as a supply source. You should order 50% more cable than you plan to use. In other words, if you need a 100-foot cable run, order 150 feet of cable.

You'll also need spade connectors or banana connectors (it's a personal preference as to which you like better). You can find some decent spades (Part #091-395, #093-546, #093-547, #093-550, or #093-551) or bananas (Part #091-1165 or #091-1260) at RadioShack.

You also might want to use the new Eichmann Bayonet banana plugs or Furutech solderless spade connectors, found at VH Audio (*http://venhaus1.com/VH_Audio_Test.html*).

Construction

Cut 14 pieces of the CAT 5 to a length of 6 feet (I needed about 5 1/2 feet, and allowed for some shrinkage of length due to braiding). Remove the blue

PVC jacket (you might have to do it a foot at a time), and remove the twisted pairs.

A good trick to remove the jacket is to use the fiber rip cord, which you will find within the PVC jacket along with the wires.

Simply secure the cable at one end and pull the rip cord through the PVC, and you should be in business. At this point, leave the pairs twisted. You should now have 56 pairs of wire (112 wires). You won't need two of the 56 twisted pairs; this project needs only 54 pairs. Next comes the fun, and time-consuming, part.

Take three twisted pairs and secure with a bench vise (or any other method that provides tension on one end of the cable). I actually used a heavy-duty staple gun and stapled the three pairs to a vertical wooden doorjamb. Next, braid the three twisted pairs. Starting with the left pair placed between the center and right pair, and then the right pair placed between the left and center pair, continue back and forth until you have completed the entire length. Try to braid the pairs as tightly as is reasonably possible to yield the lowest possible inductance. However, do not go overboard with the tightness on these initial braids. I found that there should be about a 1/16-inch gap (at its widest) between twisted pairs. A good idea is to practice on a short length first to get your technique down. You do not have to braid the last 5 inches or so, as you will need to strip the ends of the wire later on and you will need to have these "pigtails" to span the distance between binding posts.

Start the next batch of three twisted pairs and continue until all the wire has been braided. Now you should have 18 braids containing three twisted pairs each (six total wires per braid). Now take three lengths of your braided beauties, and braid these together reasonably tightly. Now you should have six braided lengths containing nine twisted pairs (18 wires). At this point you should be an expert at braiding, and will welcome the fact that you have only two more braids to go.

Next—you guessed it—take three of the six lengths and braid those together. Now braid the last three together. You should now have two lengths of 27 twisted pairs of wire (54 wires). One length is for your right speaker channel, and the other is for your left speaker channel.

Your next step also will be time-consuming, and an absolute bear on your fingers, but trust me, it will be well worth it. Each twisted pair within each braid will have one solid color-coded wire and one striped color-coded wire.

Separate and group together all the solid wires and then do the same for the striped wires. You might have differences in the lengths of these wires; just go ahead and cut them so that they are all about the same length (in other words, the ends of the wires all match up). Now strip about 1 inch of insulation from the ends of all the solid wires, and twist them all together tightly. Do the same for the striped wires, and then continue with the other ends of your cables.

At this point all you need to do is check to make sure you really have all the solids with solids and striped with striped by checking with a multimeter or continuity tester.

If you don't check it with a multimeter you might ruin your equipment if one of your pairs is mixed up and causes a short.

Now all you have to do to complete your speaker cables is to terminate them with your spade or banana connectors.

Make sure you note whether the solids or striped wires are positive polarity, and be consistent with your other cable channel, or you will have your speakers running out of phase with each other. This will be easy to hear because your bass response will suffer dramatically.

That's it, you're all done. Now you have very low-inductance speaker cables that are equivalent to 10-gauge! It took me an entire weekend to construct these cables, but they're worth it. If you want, give them a while to break in before giving a seriously critical listen (but see [Hack #25]—Ed.).

Some Notes on the Design

Several factors influenced my design, and in particular, why I chose this approach. This is the high-end techno-babble, so if you're not into cable construction, you probably won't be interested in much of this.

Low inductance

It has been demonstrated that low inductance is a desirable quality to have in a speaker cable due to the strong relationship between inductance and signal risetime. My source for this information was an article in the winter 1995 issue of *Audio Ideas Guide*, written by a retired Bell Labs engineer named James H. Hayward. Simply put, he concluded that the lower the inductance, the faster the risetime when using a cable in the amplifier/speaker interface. By using twisted pairs (with each wire within the pair used as opposite polarity), I can keep the inductance to a minimum.

Symmetrical field interactions
By using the braiding technique described, asymmetrical field interactions are significantly reduced, as no pair (or wire) rides on either the inside or outside of the cable more than any other pair/wire. I learned this theory from reading an interview with Roger Skoff of XLO, published in *Stereophile* a few years back.

Quality materials
Although it would be more ideal to use higher-purity, oxygen-free copper with long grain structure (or better yet, OCC copper), the solid bare copper in the CAT 5 should be sufficient for the purpose of making an inexpensive DIY speaker cable which will rival many higher-priced commercial cables. The Teflon insulation has a low dielectric coefficient and is considered to be one of the best dielectrics available. I would like to thank Jon Risch, who has done extensive research in the area of sonic attributes of different cable materials, and has proven to be a valuable reference throughout my cable projects.

Individually insulated 24-gauge conductors
"Skin effect" should be negligible through 20 kHz, due to the 24-gauge conductor size. I am not a fan of stranded copper cables due to the oxidation that builds up between strands. Copper oxide is a semiconductor, and can adversely affect the signal quality. For this reason, individually insulated conductors are ideal.

—*Chris VenHaus, VH Audio*

Home-Grow Your Power Cables

Although building your own power cables won't really save you money, it can increase the efficiency of your system, at far less than professional prices.

One of the less known, but well-respected, terms in home theater is *clean power*. That's usually the term reserved for high-end power cables that are getting the maximum amount of power to your components, in the most efficient way. Of course, the cables that come with most components don't function at this level, and it's incredibly expensive to buy power cables that do. For the electrically inclined, though, the DIY route provides a great alternative: make your own power cables!

There are a variety of approaches to power cables; several different designs are supplied here, each optimized for different applications. Additionally, a fair bit of electrical knowledge is assumed. If the instructions confuse you, you're probably not to the point where you should be playing with power cables anyway!

What You'll Need

The parts list is identical for both flavors of power cables.

- VH Audio 12 AWG (Shielded) Teflon cable
- Wattgate or Furutech IEC plugs
- Marinco, Wattgate, or Furutech 3-prong AC plugs
- Expandable nylon sleeving
- Teflon-insulated 12 AWG stranded copper wire
- 3:1 heat shrink tubing

You can purchase all of these components online from VH Audio at *http://venhaus1.com/VH_Audio_Test.html*. If there are additional components for a specific design, they will be listed in that section of the hack.

For Grounded Digital Components

This approach works best for your grounded digital components. The design is illustrated in Figure 9-15 .

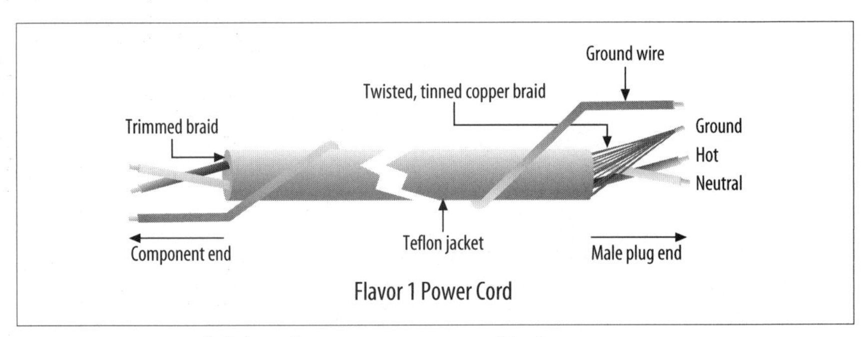

Figure 9-15. Grounded digital component power cable design

Here's what you need to do:

1. Cut the shielded cable to the desired (finished) length.
2. Strip about 2 inches of the outer Teflon jacket off the ends. Be careful not to nick the conductors!
3. Carefully comb out the braided shield on one end of the cable, in effect "unbraiding" it. Then twist the braid together tightly, to form your shielded drain wire.
4. On the same end of the cable, trim both conductors and the shield drain wire back so that they are an appropriate length to fit nicely into the male plug without excess wire.

5. Strip the wire enough to connect the conductors (both hot and neutral) to the appropriate spades on the male plug.
6. Secure the (formerly braided) shield drain wire and the 12 AWG safety ground wire to the ground on plug.
7. On the other side of the cable, completely remove the braided shield up to the beginning of the insulation on that end.
8. Determine the manner in which the conductor pairs are twisted within the cable (either clockwise or counterclockwise).
9. Starting from the male end, spiral the 12 AWG safety ground wire around the cable jacket in the opposite direction as the twisted conductors within the cable. Do this until you get to the female connector. The spiral ratio should be about one complete turn every 4 inches.
10. Secure the spiraled wire about 4 inches from each end with heat shrink.
11. Starting at the end with the male plug, tightly secure the spiraled ground wire by using 2-inch pieces of the heat shrink every 6 inches or so. The object here is to tightly secure the ground wire so that it is not flopping around when the cable is bent. Continue until you get to the end of the cable's Teflon outer jacket (you should remove the rubber band or tape before you secure the last piece of heat shrink).
12. If you plan on using flexible nylon sleeving, now is the time to cut an appropriate length and slide it over the cable. You can use additional heat shrink to join the plugs to the sleeving by sliding it onto the cable.
13. Eyeball the length you will need for the conductors to fit into the IEC plug without excess, and connect the conductors (hot and neutral) to the appropriate spade. Also connect the safety ground at this time. Do not connect the braided shield here—you should have already trimmed back the shield in Step 7.
14. After using a continuity tester to make sure you didn't mess up, plug these babies in!

For Grounded Analog Components and Amplifiers

I found the following design, shown in Figure 9-16, worked best with my analog components and most amplifiers.

1. Cut the cable to your desired finished length.
2. Cut the air hose about 6 inches shorter than the cable length in Step 1.
3. Feed the wire through the air hose until about 3 inches protrudes from each end of the hose.

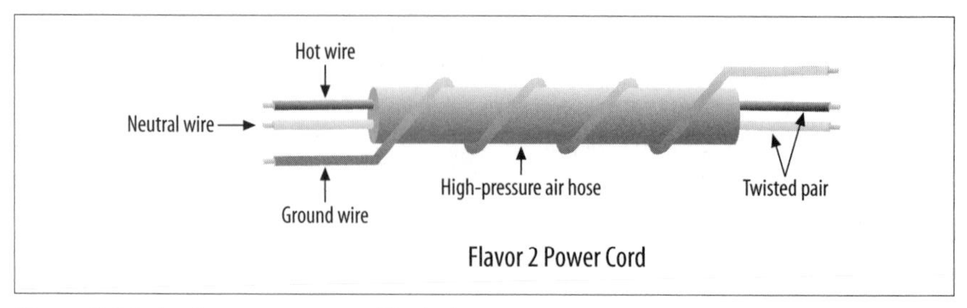

Figure 9-16. Analog component cable design

4. Strip an ample amount of insulation from each wire on the cable, and connect to the male plug at the appropriate spades (black is usually "hot" and white is "neutral").
5. Connect the 12 AWG Teflon ground to the ground spade on the male plug.
6. Determine the manner in which the conductor pairs are twisted within the cable (clockwise or counterclockwise).
7. Starting from the male end, spiral the 12 AWG safety ground wire around the outside of the tubing in the opposite direction of the twisted conductors within the cable. Do this until you get to the end of the tubing. The spiral ratio should be about one complete spiral per 4 to 6 inches.
8. Secure the spiraled wire to the hose with a thick rubber band or tape temporarily (until just before you secure the last piece of heat shrink).
9. Starting at the end with the male plug, tightly secure the spiraled ground wire by using 1 1/2-inch pieces of the heat shrink every 6 inches or so. The object here is to tightly secure the ground wire so that it is not flopping around when the cable is bent. Continue until you get to the end of the tubing (you should remove the rubber band or tape before you secure the last piece of heat shrink).
10. If you plan on using flexible nylon sleeving, now is the time to cut an appropriate length and slide it over the cable. You also can use the additional heat shrink to join the plugs with the sleeving/hose by sliding onto the cable.
11. Eyeball the length you will need for the conductors and the safety ground to fit into the IEC plug and trim the wires back.
12. Strip the wires and terminate to appropriate terminals on the IEC plug.

Additional parts. You'll need one extra item:

- 3/8-inch I.D. by 5/8-inch O.D. Synthetic Rubber Hose, also available at VH Audio

Another less costly alternative is to use a 3/8-inch I.D by 5/8-inch O.D. high-pressure air hose available from Home Depot. You'll pay about $10 for a 25-foot roll; you're looking for the orange stuff. The cheaper hose won't affect sonics, but it makes for a less flexible cord, which might cause some frustration when working with shorter cords.

Additional notes. Use the unshielded version of the VH Audio wire. This lowers the capacitive coupling even further between hot and safety ground. Additionally, ensure that there is more than 1/8-inch spacing between the safety ground and inner conductors by using the heavy-walled, high-pressure air hose as a spacer. Combined with the shield removal and the counter-spiraled ground, I believe this design can achieve the absolute lowest capacitive coupling between the inner conductors and safety ground. The end result of this design has a diameter of about 5/8 of an inch and looks really serious, especially with the nylon sleeving over it (see Figure 9-17). Your friends will ask, "Why do you have garden hoses attached to your stereo components?" Just smile and tell them you built it all by hand!

Figure 9-17. Completed power cable

For Ungrounded Components

This design is intended for components that don't need a ground; the design is shown in Figure 9-18.

To make an ungrounded cable, simply follow the steps in the "For Grounded Digital Components" section, and don't add a safety ground

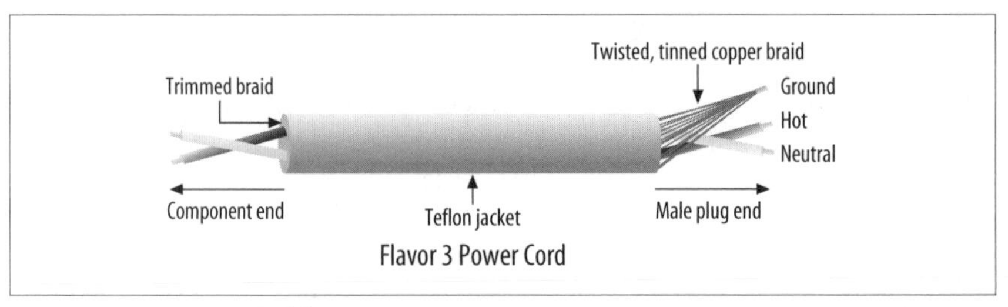

Figure 9-18. Ungrounded design

(Step 9). Just use the raw VH Audio shielded cable (no need to use or wind a 12 AWG safety ground).

> You still must comb and then twist the braiding at the source end (the male plug) and connect it to ground.

Some General Notes on Design

In this design, it is very important to place a separate safety ground outside the twisted pair conductors to decrease the capacitive coupling between the hot lead and the safety ground (as opposed to using a safety ground within the hot/neutral conductor bundle). This can minimize the chance of leakage current causing bad "gremlins" to invade your component, which is, of course, a good thing.

I also found it was best to connect the braid/foil shield only to the ground, and not to the component itself. Once again, this is to help keep the "bugaboos" moving in the right direction—away from your component and toward ground. I believe this braid/foil shield also has the benefit of further buffering interactions between the hot lead and safety ground. By spiraling the safety ground in the opposite direction of the twisted conductor pair, we further mitigate the effects between the hot and safety ground.

Something I found to be very interesting was that no single construction method sounded best on all components. The components I used seemed to prefer one method or another. Please be sure to follow the right recipe for your component to get full performance from these power cords.

Lastly, after extensive listening tests, it seems these cables need to "cook" for about 100–200 hours to reach optimum performance (but see [Hack #25]—Ed.).

—*Chris VenHaus, VH Audio*

Build a 16-Bay UHF Antenna

If you're in a weak signal area, you might need more than an off-the-shelf antenna can provide. By stacking two eight-bay antennas, you can really pull in broadcasting from further away.

Short of dropping hundreds of bucks (and maybe more, these days), the best antenna you can buy for basic reception is an eight-bay. Although that's not bad, it's certainly not going to pick up stations from 90 or 100 miles away. And, of course you're thinking: who needs reception from the next city? The answer, though, might be you! If you don't have a good selection of HD local channels in your city [Hack #28], a good antenna often can pick up stations from a nearby city. And because HD is an "either you got it or you don't" situation, if you have a strong antenna, you might be able to watch the Cowboys in HD after all.

Gang Up

When two identical antennas are mounted together, or *ganged*, pointed in the same direction, and wired together properly, there is a theoretical possibility of a 3dB improvement. That is, twice the signal power is delivered to the TV compared to what a single antenna would do. In practice, 2.5dB is readily achieved, as 0.5dB is typically lost in the combining mechanism. But if the two antennas are pointed in different directions (toward different stations), a 3.5dB penalty for each antenna is the likely result.

Further, these statements remain true regardless of whether the antennas have shared or separate amplifiers. For a shared amplifier, if the antennas point in different directions, half the power each antenna takes in reflects off the combiner and is rebroadcast out the antennas.

For dual amplifiers, when the antennas are pointed the same way, this signal is increased by 6dB, but the noise is increased by 3dB, so the overall improvement is still 3dB. When the antennas are pointed in different directions (still talking about a dual-amplifier situation), the 3dB noise increase causes a signal/noise ratio loss of 3dB for both stations. Dual amplifiers can eliminate the combiner loss, but only if the amplifiers are closely gain-matched.

Ganging nonidentical antennas together isn't recommended. They need to produce equal voltages, and adjusting out the phase difference might not be possible for all stations.

Ganging a pair of Channel Master 4228 8-Bay antennas gives you the best UHF antenna a consumer can achieve, with reasonable ease of installation.

For the Geeks

There are two approaches to finding the total power when two signals are added together. If the signals are on different frequencies, just add the powers normally. But if the signals are on the same frequency, you must add the voltages, taking into account the phase. For a 75-ohm system, the increase in power is the square of the increase in voltage. When a voltage component from one antenna reaches the combiner, it is reduced by 0.707 directed toward the amplifier, with the difference representing power reflected back toward both antennas. The second antenna adds another 0.707 so that 1.414 is directed toward the amplifier, and the reflected currents subtract to zero; 1.414 squared is 2, which ultimately results in a 3dB power gain. Did you get all that?

Mount Types

Your only major decision is deciding between a *side-by-side* mount or a *one-over-the-other* mount.

Side-by-side mounting. The elevation view of the radiation pattern, shown in Figure 9-19, is the same as for a single 4228 antenna.

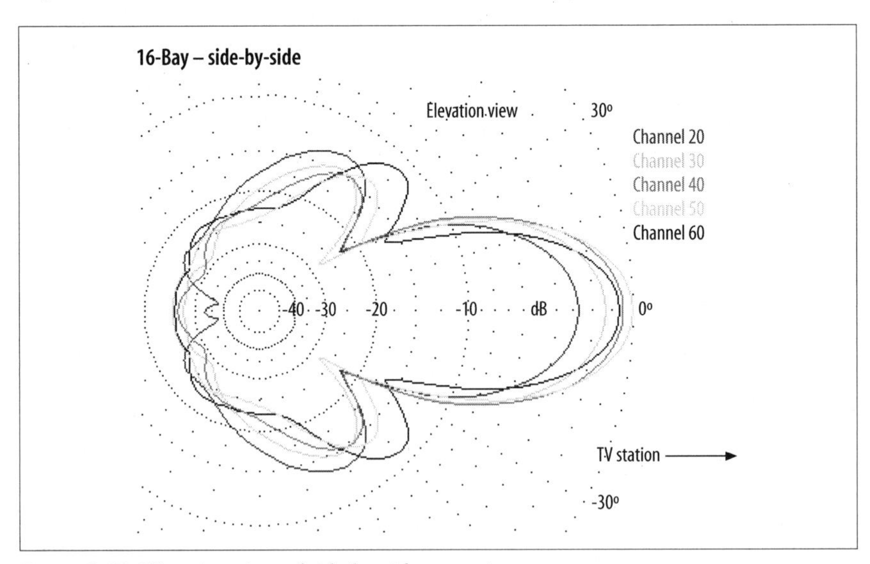

Figure 9-19. Elevation view of side-by-side mounting

In the view from overhead (see Figure 9-20), the 16-bay antenna is 2.1 times more directional.

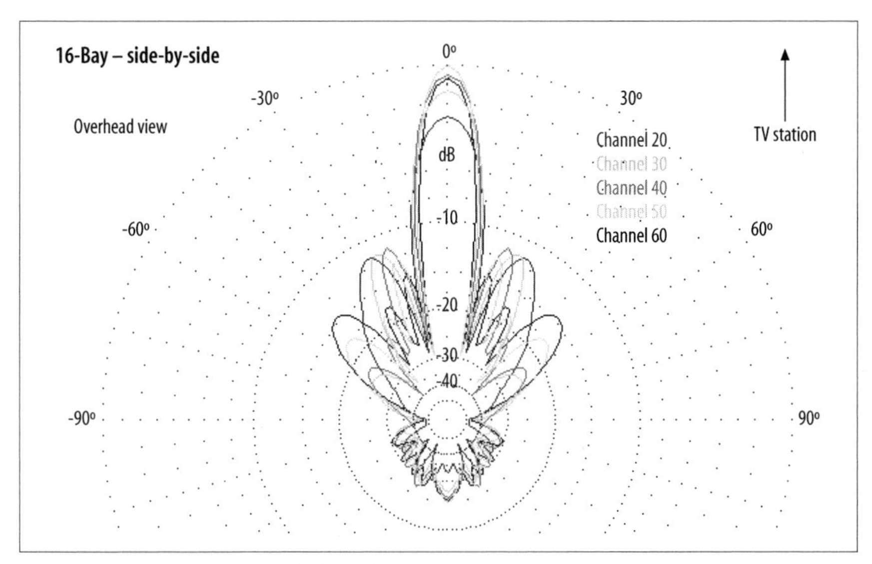

Figure 9-20. Aerial view of side-by-side mounting

This configuration can be good or bad; there is no better antenna than this setup for eliminating ghosts that arrive from near the front. However, if your antenna is going to have to rotate, the Channel Master rotor will have a hard time nailing a specific direction, at least when it has to move to that direction from another.

Hopefully, all your transmitting antennas are in the same direction, either because they are on the same tower or because the city sending the signal is so far away.

This is precisely why this antenna is ideal for picking up channels from a neighboring city; direction issues go away.

When a rotor is required, the one-over-the-other mount is usually wiser.

One-over-the-other mounting. In most situations, a one-over-the-other is the wiser choice for a 16-bay mounting. The radiation pattern viewed from above (see Figure 9-21) is the same as for a single 4228.

But in the elevation view, the 16-bay is 2.2 times more directional (as shown in Figure 9-22). This is enough to require taking the horizon elevation into account; the antenna should be tilted up to point at the horizon, and perhaps 1 degree higher.

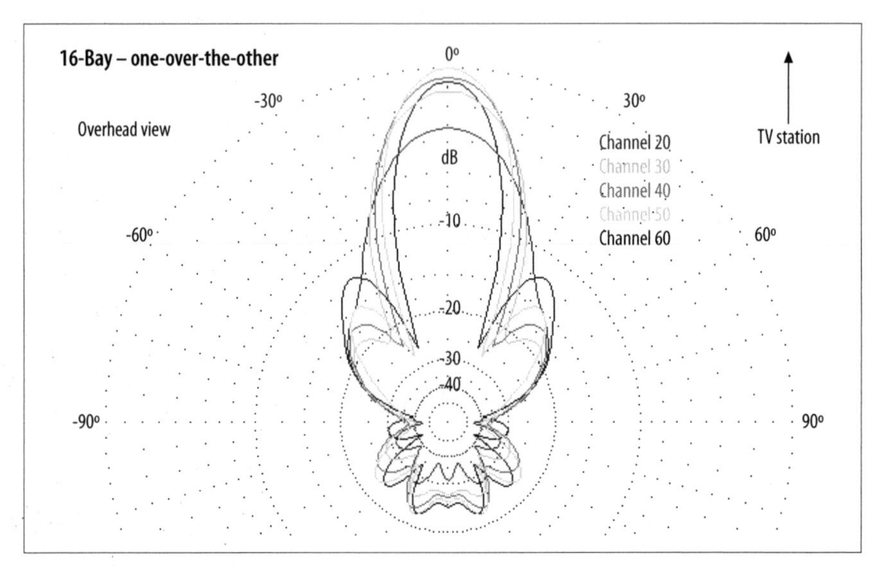

Figure 9-21. Aerial view of one-over-the-other mounting

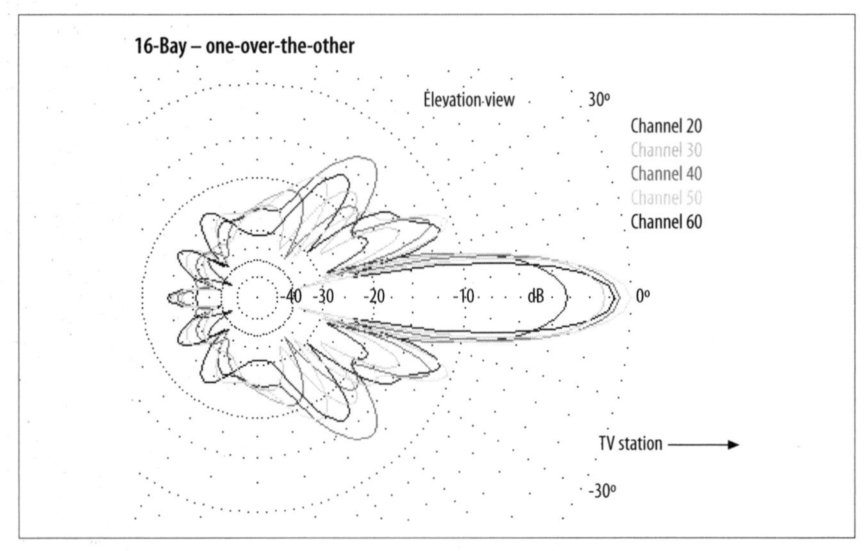

Figure 9-22. Elevation view of one-over-the-other mounting

Some authors and installers will recommend that a motorized tilter be used because the angle of the incoming signal can change from day to day, and the angle does change. However, high-angle days are strong-signal days, and the loss of a dB won't matter in those situations. I recommend a tilter only when a rotor must point the antenna in different directions with different horizon angles.

The simplest mounting technique requires a single heavy angle iron, 65 to 70 inches long. Attaching it just below its midpoint to the top of the mast will keep the assembly from being too front-heavy.

Mounting the Antennas

At 15 pounds, the 4228 is a heavy antenna. Putting up two of them requires a 1 1/2-inch, 16-gauge metal mast.

A RadioShack mast will bend with the breeze. Don't even bother.

The total weight of the antennas, mast, mounting irons, etc., will exceed 40 pounds. Trying to erect this beast by yourself, especially on a sloped roof, is something akin to suicide, even without wind. You need help; you need a large helping of good judgment; and you need a rope around your waist so that you don't fall off the roof when the whole thing tips over. Some antenna adjustments will likely be necessary too, so don't think you can put it up once and be done with it.

The good news is that there really aren't any special tricks to the setup. Mount the mast, and then mount the antennas on the mast with the supplied antenna hardware. Figure 9-23 shows a side view, and Figure 9-24 is a front view, of two antennas in a one-over-the-other mounting pattern.

Connecting the Antennas Together

The two antennas must be *phase-matched*. This means that the two signals must arrive at the combiner in phase.

Plus or minus 15° isn't a noticeable error, but anything larger should be avoided.

You achieve this phase matching by maintaining symmetry in the feed system. In other words, the wires for each antenna should be identical in type and length, although the actual length is not critical. You could have two 10-foot cables, or two 50-foot cables; just make sure both cables are of the same length.

If a ground reflection causes one antenna to be phased ahead of the other, you should adjust this out by repositioning one antenna. The easiest way to do this is by finding a new horizon tilt angle. Simply adjust the tilt while

Figure 9-23. Mounted, ganged antennas

watching the signal strength. Different stations could require different angles, but that is rare.

There is also a chance that you will mix up the polarities such that the two antennas subtract instead of add signal. This will result in two forward lobes, reduced in size, with a null straight out the front (see Figure 9-25). After the antenna is fully hooked up, you should rotate the antenna to check for this pattern. If this appears to be occurring, reverse the connections on one of the antennas. The antennas come with a balun that has a "China" stamp on one side. I believe this stamp is the key to getting addition on the first try; just make sure connections are identical on both antennas.

Possible Problems

My neighborhood has hot spots and cold spots—places where the signal strength is particularly strong or particularly weak. This is a consequence of *overlapping fields*. Because I'm 40 miles from San Francisco and behind some hills, DTV reception is possible only when my antenna is positioned in a hot spot. These hot spots are 10 to 16 feet apart for any channel, and are in different places for different channels. If you have hot spots for any one

Figure 9-24. Front view of ganged antennas

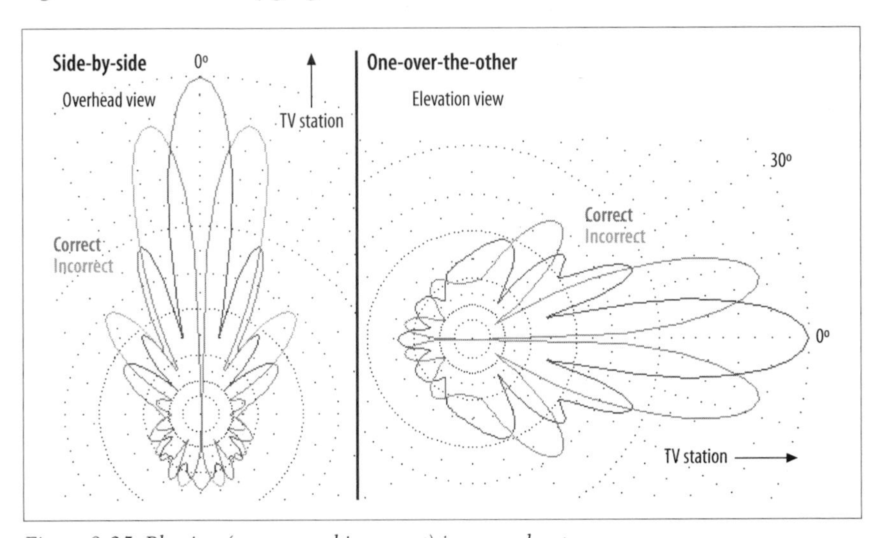

Figure 9-25. Phasing (correct and incorrect) in ganged antennas

UHF channel, you will have hot spots for all UHF channels (in the same direction). The distance between hot spots is determined by the frequency, the distance to the horizon, and the geometry of the ridgeline at the horizon. In fact, it's that ridgeline that is producing the overlapping fields.

If one 4228 is in a stronger field, part of its signal will be retransmitted out of the weaker antenna. This loss can equal the little bit of gain you had hoped to gain from the weaker antenna. In these cases, you likely will find that two 4228s are no better than one. This retransmission is avoided only when both 4228s are in equal fields.

When dual amplifiers are used, the reasons are different, but the result is the same.

At my house, the change in field strength in just 3 feet is enough to wipe out most of the hoped-for 3dB gain when the antennas are mounted side by side. Fortunately, hot spots are not generally spherical. Rather, they tend to extend upward and forward more than they extend laterally. So, the one-over-the-other configuration is much more likely to work in a neighborhood with hot spots.

But, of course, there is an exception to that general rule. If the 16-bay antenna is close to the ground, and the ground is bare extending toward the desired station, an *efficient ground reflection* is likely. This is another case of overlapping fields. But in this case, the hot spots are mainly arrayed vertically; they are likely close together vertically, but farther apart laterally. In this situation, the side by side is the configuration more likely to work. The incoming wave is angled downward by only a couple of degrees, and so the ground reflection occurs on ground that extends perhaps hundreds of feet toward the station. If this ground is paved, dirt, water, or a grass lawn, the reflection is efficient and will produce extremely weak cold spots. If it is covered with weeds, shrubbery, trees, or somebody's house, the reflection is scattered too randomly to have any effect on UHF reception.

If you want to explore the locations of your hot spots, a Silver Sensor antenna on a 10-foot pole is a good method. This antenna is small enough to fit in any hot spot, and probably strong enough for a digital-lock in a hot spot for your strongest station. If possible, work at about the elevation where you plan your permanent antenna to be mounted. You will need a monitor positioned there so that you can see the signal strength from the receiver. The distance between the hot spots will be roughly the same for all channels.

If your hot spots are too small both vertically and laterally, a 16-bay might be out of the question. Your option, then, is to put each 4228 in its own hot spot. But this works for only one channel.

You might curse your bad luck if you find you have hot and cold spots, but you would be looking at it wrong. In fact, your neighborhood is concentrating the signal for you. An antenna in a hot spot can be at least 3dB smaller than it would need to be in a "flat" neighborhood. Now, if only the hot spots never moved. But, that is another story…

—*Kenneth L. Nist*

Ground Your Outdoor Antenna

Whether you build an antenna on your own, install a prebuilt antenna, or have someone else build it, grounding is a critical part of installation.

Offering advice to not ground outside antennas is irresponsible, contrary to all known laws of physics, civil law, and city codes pertaining to static charge buildup, and downright criminal!

Types of Grounding

There are three types of grounds. In order of difficulty to achieve, they are (from most difficult to least difficult):

- Static grounding
- Radio frequency (RF) grounding
- Electrical grounding

The methods and practices used to achieve each of these are different. Static grounding is the most difficult to do with guaranteed results, while electrical grounding is easy and can be considered to work perfectly in most cases. RF and electrical grounding are not important for static discharge (lightning) safety, so I've left them out with regard to outdoor antennas.

The general rule on outside antennas is that they need to be grounded. There is literally no known government agency that recommends antennas not be grounded for safety. Indeed, many local and city codes require grounding.

The probability of direct and secondary static electricity (lightning) strikes increases with the buildup of static charge at points of conductivity, and of course the metal mast or pole of an outdoor antenna is perfect for doing just that. Static electricity is built up during a thunderstorm as wind blows over the metal structures. This static charge builds and becomes an attractor to the opposite charge of static buildup in the storm clouds. By draining off the static charge continuously, via a ground, you reduce the probability of strike because the potential difference of charge is reduced.

Studies conducted on the Empire State Building showed that the probability of a strike was in direct relation to the quantity of static buildup on conductive structures. Conductive structures with no ground path were at the highest risk, while structures that were intensely grounded over several contact points were at the lowest risk.

Grounding the Antenna

On an antenna, you can ground the mast, the boom, the dish, and the director and reflectors of the antenna by contact metal bonding to a ground wire. Then run that wire directly into the earth via a deep ground rod. However, you cannot directly ground the driven element or active element of the antenna. All you can do is make a reasonable attempt at grounding it via a special *coax grounding block* to reduce static charge buildup and reduce probability of a hit. This block is a device designed to bleed off high-voltage spikes that reach dangerous levels, keeping them from damaging your receiving equipment. The block doesn't directly short out the center conductor to the ground because this would kill the signal, but rather, allows a small gap that will—on a continuous basis—bleed off the building static charge before it reaches dangerous and damaging levels. Using one of these grounding blocks located just before the coax feed wire enters your building is what is recommended to effectively reduce the probability of small static electricity damage to your receiver. Still, a grounding block won't protect against a direct lightning hit. Additionally, both the grounding block on the coax and a direct ground wire to the mast of the antenna should be used.

Risks of Damage

In a simple static charge buildup, the minor hits can be silent killers. These tiny hits will be damaging to your receivers' RF frontend. It will most likely short out sensitive diodes in the receivers, rendering them useless. In the next worse case, you take a secondary hit; this usually occurs when the direct hit struck a tree or utility pole nearby. In this case, you might see some signs of obvious visible damage, such as the wiring in your house catching on fire, or your TV set getting fried right before your eyes. This happens far less often than the hidden damage hit. Finally, there is the rarest type of hit, which is a direct lightning bolt strike to your antenna or house. This usually will cause major fire damage to your dwelling and contents. Fortunately, these hits are rare except in places such as open farmlands where the house structure is the only corona point sticking up out of the flat ground for miles around. These structures serve as a big static attractor to a thundercloud. In areas where you are surrounded by trees and other structures, your odds of a direct hit are greatly reduced, but you are still at risk for the secondary hit and the silent static killer.

With all this in mind, realize that grounding will not protect you if you take a direct, or even a secondary, hit. What grounding does do is reduce the probability of getting hit in the first place; it also provides much better protection against the damage caused by silent hits. Because this is the most common type of hit, that's a real advantage to have in your system. The idea of the grounding is to reduce the probability of getting hit in the first place, and to continuously bleed off small static charges to prevent the silent killer to your equipment.

—*Keohi HDTV*

HACK #83 Build a Lens Hood for Your RPTV

In the unending quest to remove any interior light from your RPTV, a lens hood can provide a fairly easy approach to getting the best picture possible from those CRTs.

One the easiest ways to have inferior picture on a rear-projection television is simply to take it out of the box and turn it on. These televisions—even when bought from high-end boutique stores [Hack #4]—are set up at the factory (and the showroom) to appear bright and showy [Hack #9]. That's not the key to a great viewing experience in your home theater, though.

In Chapter 8, you learned about getting rid of lens flare [Hack #71]. That's important, but involves a lot of work, including some pretty intense work on the CRTs and glass lenses, which is dangerous if you're unsure of yourself. For those of you a little intimidated—or just more enamored with wood, nails, and the DIY approach—there's an easy way to get 80% of the lens flare in your TV eliminated, with a lot less work.

Taking the Last Part First

Realize first that lens flare, and almost all artifacts that are generated within your set, result from light bouncing around and reflecting and refracting where it ought not. Since the basis of your picture is light, reflected off of a mirror, this might seem impossible to correct. However, you can work on eliminating light from reflecting and getting back into the lens system. In other words, you just want to allow light from the lenses, to the mirror, and onto the screen. Anywhere else and light within the TV is your enemy.

Figure 9-26 shows the end result of what we're going for. This will sit in your TV around the lenses and ensure that no light from the lenses goes anywhere but straight to the mirror. The building process isn't always clear as to what the final product is, so you may want to refer back to this figure as you go along.

Figure 9-26. Completed lens hood

Mocking Up

First, you need to build a mockup. You can experiment with measurements, make design changes, and perform other spur-of-the-moment adjustments without worrying about screwing up expensive material or throwing away several hours of work. Start with some corrugated cardboard; I prefer white because it is easy to see in the darkness of your RPTV's innards. You'll need five different pieces; I'll note my measurements, which are tailored to my Toshiba TW56X81. You can probably start with similar measurements, and work from there:

Piece 1
: This will serve as the base of your lens hood. Mine is 14.5 inches by 22.75 inches. For stability, this should be as big as will fit in the cavity where your CRT lenses sit. You'll then need to cut out a hole, which allows your lens assembly to squeeze through. My hole is 15.75 inches by 5.625 inches. This hole should be centered from left to right, so the hood sits centered in your TV's innards.

Piece 2
: This is actually the cutout from Piece 1; therefore, mine was 15.75 inches by 5.625 inches. It will become the "front" of the lens hood hole.

Piece 3
: There are two of these, both set up to act as the "sides" of the hole of the hood. On my setup, these measured 3.75 inches by 7.62 inches. Note that these are a good bit longer than the hole itself.

Piece 4

This is the "back" of the lens hood hole. Mine is 4.2 inches by 15.7 inches. This piece should be cut to the same width as the hole in Piece 1.

Figure 9-27 shows these pieces, laid out nicely and cut to dimension.

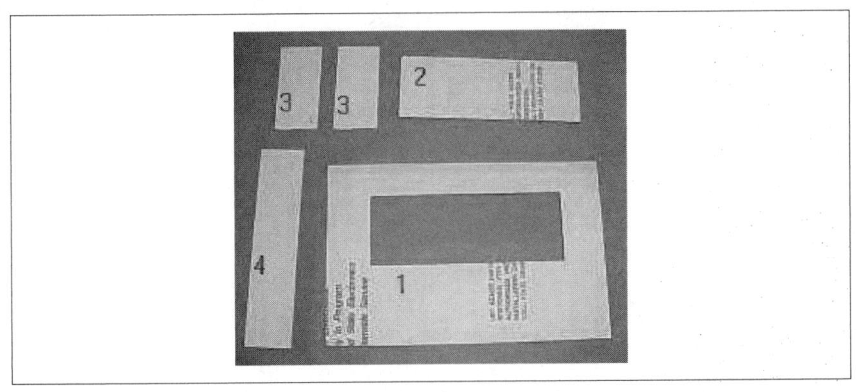

Figure 9-27. Individual pieces of the lens hood

It's pretty easy to then put these together. Rather than bore you with a page of instructions, I'll let Figure 9-28 do the talking.

Figure 9-28. Assembled lens hood mockup

Notice how Piece 3, on both sides, protrudes beyond the hole. You may want to draw some angled lines to make sure you get it lined up, and Piece 2 is correctly attached. Figure 9-29 shows Piece 3 with angled lines drawn in.

Sit this into your TV, and see how it works.

Now, you should make any needed corrections. If the base is too big, if the hole for the lens assembly is too small, or if the sides are too wide, you can make adjustments (it's just a mockup, after all!). I had to cut arcs in the

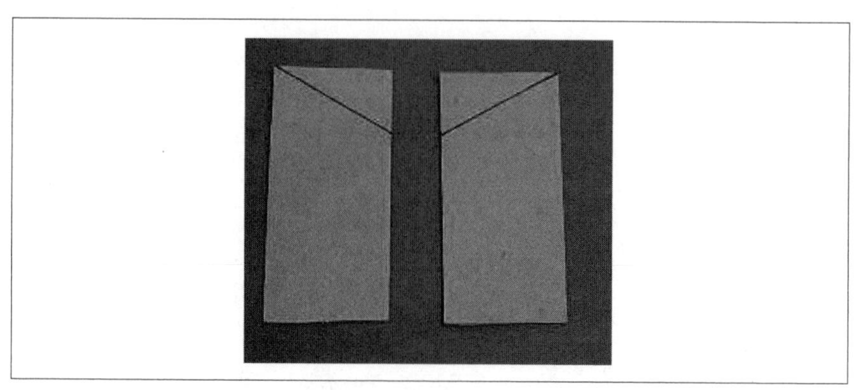

Figure 9-29. Piece 3 with angle lines drawn in

sides to avoid part of my lens being blocked. This is exactly why you mockup before working with the better material.

Actual Construction

Once you've got things figured out, reconstruct the entire hood, using natural colored cardboard. Then, cover this with black Duvetyne fabric. I used 3M 77 spray adhesive to permanently fix the Duvetyne to the hood assembly.

I recommend against using white cardboard, as some minimal light will shine through the Duvetyne, catch the white cardboard, and create flares.

Once you've got your hood, place it inside your TV, close up, and enjoy the results. I noticed a real improvement: my contrast increased, and "halos" around bright objects against dark backgrounds was dramatically lessened. And, all without voiding my warranty!

—*Tim Procuniar*

CHAPTER TEN

Remote Controls

Hacks 84–89

Imagine the joy of inviting over your closest friends to check out your new home theater system. You've got the expensive sconce lights dimmed, the curtains drawn over your projector screen, and all your gear elegantly tucked away into cabinets that match the wood veneers of your speakers. Your 7.1-channel processor is glowing, and speakers are all around. Your friends sit down, ready to see and hear something special...

... and they're greeted with the sounds of you fumbling around in a drawer. They begin to snicker as you pull out first one remote, and then another, and then yet another. Ten minutes later, you've managed to get the curtains open and the sound working, but instead of *Stargate*, the picture on the screen is of a little cartoon girl with her monkey friend, apparently saying something about a backpack. You've been completely embarrassed by the smallest, but arguably the most important, part of your home theater system—the remote control.

No matter how sophisticated your home theater system is, your remote has to be more so. In fact, as this little anecdote illustrates, your remote better be able to handle everything your system has to offer, without a hitch and without needing a stand because it's huge and unwieldy. Welcome to the world of remote controls.

Add a Programmable Remote

Once you have your basic home theater components, you need to buy a good programmable universal remote. It's worth more than luxury seating, theater lighting, and great cabinets combined.

There's nothing as annoying as walking into a $20,000 home theater, sitting down in a plush leather chair, and noticing six or seven remote controls strewn about the room; it's one of the ultimate turnoffs in home theater today. To avoid just such a situation, purchase a good programmable remote control for your home theater. But be warned: these units aren't super-cheap, and often, so-called bargains turn into full-fledged disasters.

First, avoid the $20 "universal remote" at all costs. The typical universal remote (which is not the same as a programmable remote) has hundreds of brand codes, for all types of different devices, built into it. Although this might sound like what you want, realize that it is limited by the specific codes input into it when it left the factory. These become out-of-date very quickly, and often don't contain codes for higher-end brands anyway. A much better idea is to buy a *programmable remote*, which allows you to program your existing remote control codes into it. This means that as long as you have a remote or infrared (IR) code, a good programmable remote can learn it and can operate devices in your theater. This is the ultimate insurance that your remote won't go out of style, even as you upgrade and change equipment.

As proof of this, I've been into home theater for almost 10 years now, and I still own and use my very first programmable remote, the Marantz RC5000. Although I bought a color unit for my theater, my Marantz still happily runs both my bedroom and living room setups.

Before diving into specifics, you need to know a few important terms.

Macros
: Macros are sets of commands strung together under one heading (or, in the case of remotes, one button). A typical macro would be `Watch TV`, which might change a receiver input, select an aspect ratio on the TV, and power up the satellite receiver.

Activity-based remotes
: Activity-based remotes are macro-based, allowing for one-touch operation of complex functions. Additionally, these remotes respond to a specific activity by displaying options relevant to that activity; so, `Watch TV` might result in channel buttons being displayed, and DVD and VCR controls being disabled or hidden.

Touch-screen remotes
: Touch-screen remotes operate based on the user touching a graphical icon rather than pushing a physical button. These often are preferred because you can customize the "buttons" of a touch-screen remote with fancy icons and animations to interact with and illustrate what is going on in the home theater system.

Although many good programmable remotes are available, home theater enthusiasts have settled on a few select models as the standards. You'll find that high-end devices even come with preprogrammed setups and codes for these specific remotes, so you'd do well to go mainstream instead of blazing a trail and trying to be different in this area. Determining which of these remotes to get becomes largely a matter of budget, as Table 10-1 details.

Table 10-1. Choosing a remote control

Price range	Manufacturer	Model	Description
Up to $250	Intrigue Technologies	Harmony Remote SST-659	A powerful Internet-programmable remote that is activity-based.
$250–$500	Philips	Pronto NG TSU3000	The latest version of the de facto standard: fully programmable and touch-screen-enabled, with macro support and editable via ProntoEdit on the home PC.
$500–$1,000	Philips	Pronto TSU6000/ TSU7000	These touch-screen models are the full-color, enhanced versions of the TSU3000 and are considered the best remotes available for less than $1,000.
$1,000+	Crestron	Take your pick!	Crestron goes beyond simple remote controls and provides a complete touch-screen device capable of controlling everything in your house (and then some)!

Marantz typically produces rebranded versions of the Philips TSU remotes; for example, the RC5000 and RC5000i are rebranded versions of Philips' TSU1000 line of remotes. There is no functional difference, and these devices are interchangeable.

Most home theater users settle on either the monochrome or color Philips TSU models because they are a great blend between power and value, and ProntoEdit is a killer software application [Hack #85]. If you've got a tight budget, the Harmony Remote (*http://www.logitech.com/harmony*) is a great buy, and for those who insist on the very best, the Crestron certainly won't disappoint.

Program Your Remote with ProntoEdit

Download and set up the ProntoEdit software for programming your Marantz and Philips remote controls.

If you've got a programmable remote control, don't even bother with the instructions detailing how you can program the remote by hand. That process is incredibly time-consuming and frustrating, and doesn't make available even half of the remote's true capabilities. Instead, head on over to the Mecca of remote control web sites: Remote Central (*http://www.remotecentral.com*). There, you can find files and utilities for every conceivable remote control, and especially for the Marantz and Philips units. The program you're looking for to get the most out of those models is called ProntoEdit, and you can find it by clicking the Files link in the top menu bar of the Remote Central web site. (Drill down into Phillips Pronto & Pronto Pro, then into Original PC Software, and select Philips Pronto Series.)

Mac users are out of luck; as of this writing, there are no Mac-friendly versions of ProntoEdit, and Remote Central provides only PC-compatible software.

When the Files area comes up, select your brand of remote control. The Philips Pronto and ProntoPro as well as the Marantz RC5000 lines use the same program. If you've got a newer model, such as the Philips TSU3000, there is a separate link; you'll also find a number of other remote controls represented, along with utilities and setup files. Still, if you've got a unit that works with ProntoEdit, you won't want to look elsewhere. On that page, scroll down and select Original PC Software, select your brand and model number, and download the latest version of ProntoEdit. As an example of what to expect, Figure 10-1 shows the final screen before downloading ProntoEdit for a Marantz RC5000i.

Most of these remotes come with software on a CD. Don't bother with the CD; the software is often out of date, and you're always better off going straight to Remote Central.

Once you download the software, you can install it and fire it up. Select the version of the remote that you have to avoid memory problems when programming and uploading files.

You'll also need the remote-to-serial-port connection cable that came with your remote (shown in Figure 10-2).

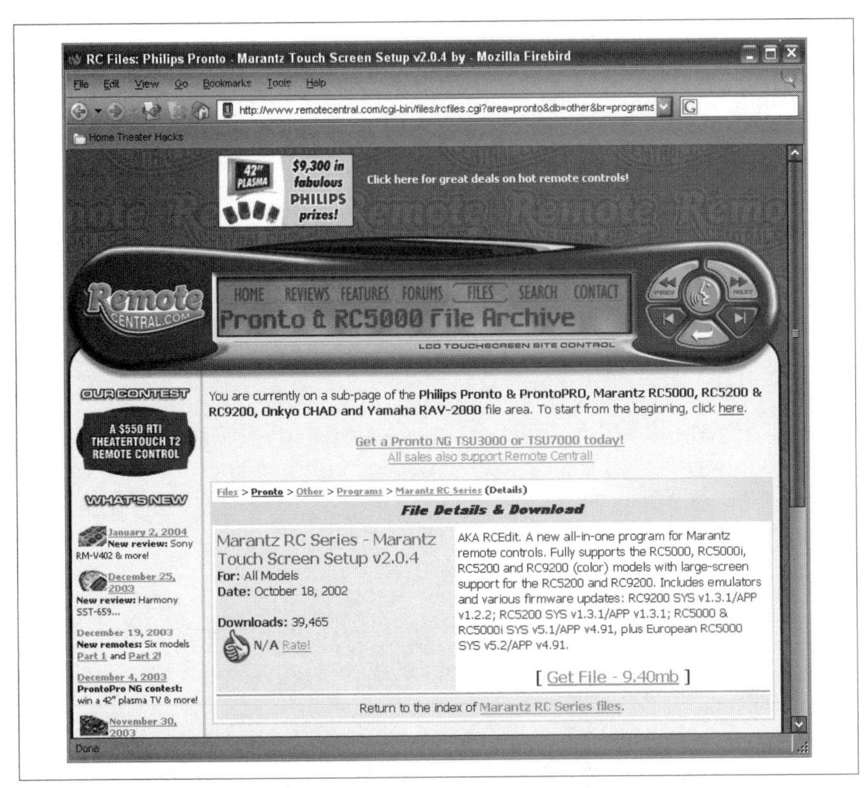

Figure 10-1. Downloading ProntoEdit for a Marantz RC5000i

Figure 10-2. The serial connection cable for a Marantz RC5000i

Connect one end of this cable to your remote and the other to your computer. Now you can upload and download configurations, as well as update your remote control's firmware.

Updating Your Firmware

Keeping your remote control's firmware up to date is an important part of home theater maintenance. You'll find that with the latest software, your remote behaves better, and often will even have a longer battery life than with older software.ProntoEdit (and RCEdit, Marantz's branded version of the same program) makes this update easy via an automated update (and download) process.

Start up the editor, select Tools → Update Firmware, and then choose your remote's model. You'll get a progress screen, and then you can click to start the update (shown in Figure 10-3). You also can choose whether you want to keep your old configuration file (you generally will).

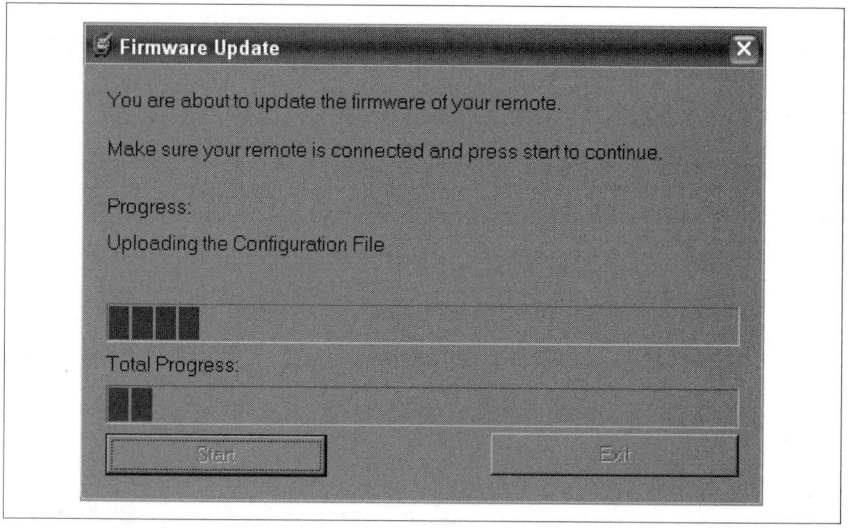

Figure 10-3. Updating remote control firmware

ProntoEdit is even nice enough to tell you if you're trying to upload the wrong firmware for your device.

Create Custom Graphics for Color Remotes

Most high-end color remotes allow you to create custom graphics and icons. These can turn a decent-looking setup into a killer one and really spruce up your control system.

Once you've gotten a good programmable remote such as the ProntoPro [Hack #84], you'll soon find that working with the default icon set is really limiting. If your remote interfaces with your computer [Hack #85], though, you can remedy this easily with a decent graphics program and some creativity. Most of these programs allow you to create custom graphics, load them into the program, and then use those graphics in your remote control's layouts.

That said, the purpose of this hack is not to tell you how to design your Philips ProntoPro or aid you in coming up with an original layout, but rather, to assist you in obtaining graphics of the best possible quality, regardless of your artistic abilities. There's no reason why your remote shouldn't look the best it can! Most graphic artists or web developers will already know many of these tricks, but for anyone who is new to bitmap editing or who has not yet designed a graphical remote control, there are some important techniques that you might not be aware of.

This hack focuses specifically on the ProntoPro in terms of working with the software. However, the guidelines for icon design are useful for any color remote (and probably black-and-white remotes as well).

ProntoEdit uses CCF files for its layouts (e.g., *theater.ccf*) so, when you see CCF, it refers to an entire layout complete with commands, graphics, and macros.

The graphics I used in my CCF file I created as vector artwork in CorelDRAW 10, then converted to ProntoEdit-ready bitmaps via Corel-PHOTO-PAINT 10. Although I have experience with these programs, almost any vector and bitmap-editing software have similar capabilities—and limitations.

Choose Colors That Work

Naturally, ProntoPro isn't as graphically capable as your desktop PC. This is a given with the obviously limited black-and-white Pronto, but might need clarification on the color ProntoPro: the word *color* doesn't actually imply a PC-quality interface. First, most computer desktops are set to run in 64,000 or 16.7 million colors, which results in photorealistic images. The Pronto-Pro can display 256 colors—a far cry from millions, or even tens of thousands, of colors. Two hundred and fifty six colors might still sound like a lot, until you work out exactly how many shades of red, yellow, blue, orange, green, purple, black, etc., are possible.

Thus, when designing, it's always a good idea to first test out how your graphics will look when converted to a limited color set. The color bars in Figure 10-4 demonstrate this issue. On the left side of each bar is a 24-bit, 16.7-million-color gradation, as your computer would normally display, and on the right side of each bar is an 8-bit, 256-color image as the ProntoPro would display. You can see how much rougher the bars are on the right. What this amounts to is a limitation to how much fine detail the remote can reproduce faithfully. For instance, green might not be the best color choice if your design involves a lot of small, detailed buttons because the end result would be fairly crude. Finally, never start off working in 8-bit color; for best results always use 24-bit color and convert down to 8-bit later.

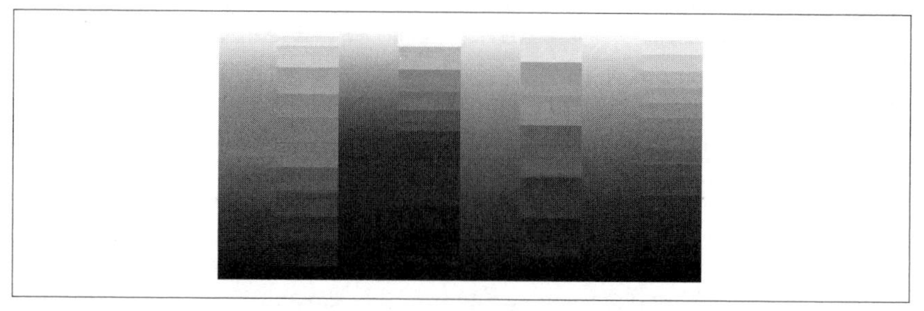

Figure 10-4. Comparing 24-bit color to 8-bit color

Another consideration in color choice involves what ProntoPro's screen can actually show. The "passive matrix" LCD in most color remotes doesn't reproduce colors exactly like your computer. What looks great in ProntoProEdit might not look quite as good on the actual remote. Remember that colors on the remote will be more washed out. Dark colors will tend to fade out to black, while very light colors will disappear into white. For best results, always use strongly contrasting colors to differentiate your buttons from the background and the text labels from your buttons. Avoid using similarly toned colors; for instance, don't place medium-red buttons on a medium-blue background.

Smooth Out Images with Antialiasing

Simply put, *antialiasing* is a method of smoothing out the edges of images, eliminating what are commonly referred to as *jaggies*. This is a technique similar to Windows' built-in *font smoothing*, which makes screen fonts easier to read. For example, instead of creating a black object solely out of black, the edges of the object are filled in with different shades of gray (see

Figure 10-5). So, a small circle appears as a smooth, even curve rather than a jagged stack of Lego bricks.

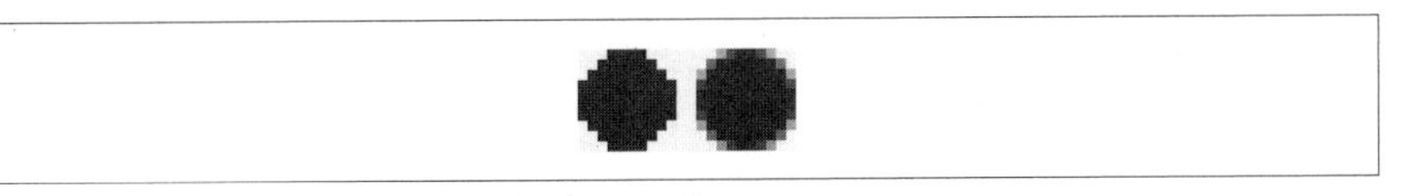

Figure 10-5. Blue dot without and with antialiasing

You can use this professional technique for a lot more than just sharply contrasting objects. For instance, look at the two examples in Figure 10-6. The one on the left was created without antialiasing, while the one on the right was created with antialiasing. You'll notice that the right-side graphic appears to have greater detail and is more pleasant to look at. Thus, when importing or creating graphics for your remote, always ensure that the "antialiasing" or "smoothing" option is enabled. This will give your remote's user interface a truly refined look.

Figure 10-6. Adding antialiasing to a button graphic

If such an option isn't available, you might be able to emulate this function by resizing a larger image to a smaller one; keep reading for more on this approach.

Think Big, Finish Small

Although this might seem like a strange idea, it's always better to start off with an image larger than you need. To be specific, if you start off with an image at least four times bigger than the finished piece will be, the end result will be that much better.

Over the years, I've found that creating graphics at a ratio of 1:1 (actual size)—although resulting in decent-quality images—just isn't quite as good as starting off larger and then resizing to a smaller size afterward. I imported all the images in my ProntoPro layout from vector artwork four or eight times as large as I wanted them, then shrunk them down to the proper dimensions. This allows the software's resizing algorithms, which generally are better than the import or creation algorithms, to refine larger amounts of detail and come up with the best-looking image.

Compare the two examples in Figure 10-7 closely. The one on the left was originally imported at 250 pixels wide with antialiasing enabled, while the one on the right was created at 1,000 pixels wide, then resampled down to 250 pixels with antialiasing. Look closely at the television, the text, and the vertical lines immediately below the text.

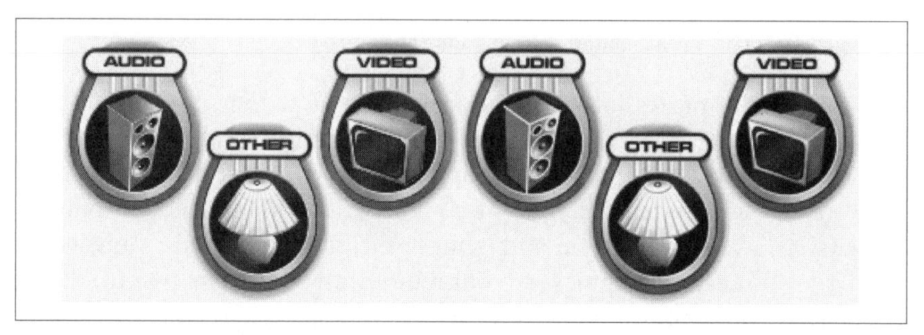

Figure 10-7. Converting larger images to smaller ones

Even if you're not starting off with a vector-based file and instead are working solely in a bitmap-editing program, you still can obtain better results, especially from special-effects filters, if you work on a larger file.

Convert to the Right Colors

As already mentioned, ProntoPro is capable of displaying 256 colors (8-bit). However, those colors can't be any 256 of your choosing; you must work with the remote's preset palette. Although specifications claim that the remote uses a 215-color "web-safe" palette, this isn't completely accurate. The first 215 colors of ProntoPro's palette are indeed made up of 215 "web-safe" colors; however, an additional 41 colors are specified—mostly pinks, blues, and grays. So, because this doesn't match up with any default palette, this presents a tricky issue: exactly how can you take advantage of these "hidden" colors?

If you merely convert an image to the 215-color "web-safe" palette, you'll be losing out on a lot of color—particularly gray shades. Some users have found that the default "Windows system" palette that many programs include uses the "web-safe" palette and adds an additional number of gray shades; however this still doesn't make use of everything that your remote is capable of. To find out what colors are valid, load up the default configuration in ProntoEdit and start the ProntoProEmulator. Take a screen capture by pressing Print Screen, then start up your favorite bitmap-editing program and load the capture. Most programs will allow you to take the color palette stored in that file and save it to disk for use later in converting 24-bit "true-color" images.

Take a look at the three examples in Figure 10-8. The one on the left was converted to the "web-safe" palette. The graphic in the center uses the "system" palette, while the one on the right uses the true "ProntoPro" palette. The enlarged sections show the difference that some additional gray shades can make. One of the reasons my layout uses a lot of blues and grays is because that's what ProntoPro's color palette leans toward. Note that most of these additional colors aren't mentioned anywhere in ProntoProEdit, and can be discovered only by the method outlined earlier.

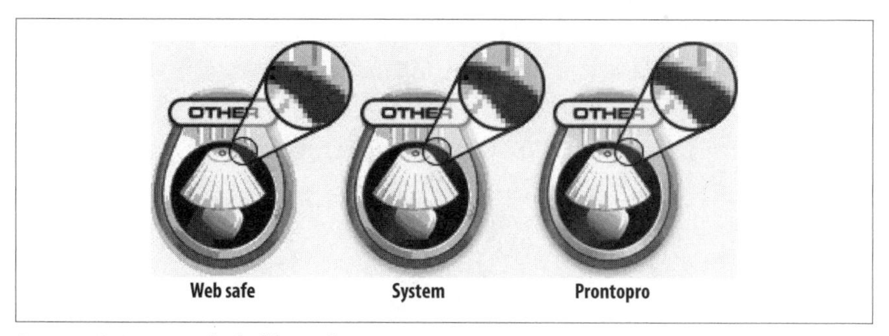

Figure 10-8. Using the hidden colors in ProntoPro

Dithering Is a Must

Something that goes hand in hand with using the right color palette is a technique called *dithering*. Dithering is a method of alternating a limited number of colors to emulate an even greater number. Remember those strong, stepped colors in Figure 10-4? You can "smooth these out," as it were, by using dithering. Although quite a few automated dithering techniques have been created over the years, I can only suggest using one of the randomized versions (often called *error diffusion*) to convert your image from 24 to 8 bits.

Don't use the ordered variation, as this creates visible patterns that don't look very good.

Observe the differences between the examples in Figure 10-9. Both images use the ProntoPro color palette and antialiasing. The image on the left uses no dithering, while the image on the right employs diffused dithering. You might think the differences are negligible, but it's these small touches that take "good" to "professional."

The green bar in Figure 10-10 demonstrates the difference between 24-bit true color (left bar), ProntoPro normal (middle bar), and ProntoPro dithered

Figure 10-9. Comparing a nondithered to a dithered image

(right bar). Although ProntoEdit can do some image conversion, it does a poor job of color selection and dithering. You always are better off doing this beforehand, even for the black-and-white Pronto remotes.

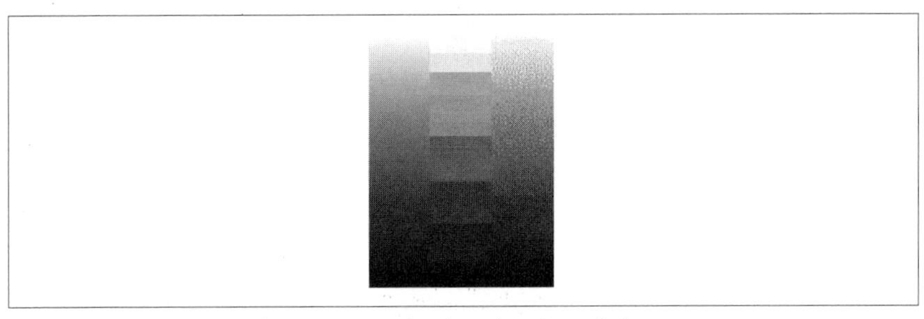

Figure 10-10. Various color variations, with and without dithering

Tweak, Hack, Play

Even if you employ all of these techniques, there's often still a need for manual editing. Some dithered pixels might be in places you don't want them to be; a small bit of text might not be quite readable; you might find that part of an image looks good with dithering while the rest doesn't. A little bit of hand editing can go a long way toward a perfect layout. Figure 10-11 shows two examples. One uses all the suggestions made earlier, while the other uses none. Quite a difference!

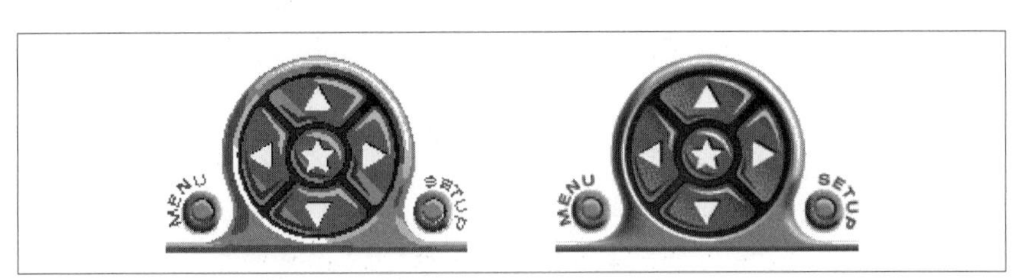

Figure 10-11. Nontweaked and tweaked images

—*Daniel Tonks, Remote Central*

Learn IR Codes the Smart Way

When you just can't get your programmable remote to learn that certain code from the manufacturer's remote, you might have to do things manually. There are tons of tricks to get to harder IR codes.

If you've ever tried using your programmable remote [Hack #84] to learn infrared codes off of myriad remotes, chances are you've hit upon a few troublesome buttons. There are those that your programmable remote just wouldn't learn properly, and those that, once learned, don't seem to work as they should. Although there might be some very good reasons why a particular IR code can't be taught, there are a few special learning techniques you can try before pulling out your hair.

Press or Hold?

The most accepted method of teaching another remote an infrared code is by holding down the original button until the programmable remote finishes capturing the code. In fact, this is how the process usually is described in manuals. Continually sending a signal allows the remote to sense its autorepeating portion and store it once, properly.

There are some situations where a particular remote requires a different learning technique; if you experience problems, be sure to read your remote's original documentation.

Even when following directions, often one or two buttons from a particular remote will simply refuse to learn. It's even been reported that codes for entire brands can experience problems. In these cases, it's worthwhile to try pressing the button just once, briefly. You might want to experiment by pressing for different lengths of time, but generally a good solid push is all that's required.

The reason this "single-push" technique is not normally recommended is because learning remotes are generally unable to sense repeating codes when buttons aren't held down (repeating codes are needed for commands such as Volume Up and Fast Forward to work properly). Plus, the remote might store the signal one and a half or even two times, wasting valuable memory. But when you simply must learn a particular button, this technique can be helpful.

Proper Spacing

The sensitivity of your learning remote's *eye* (the IR sensor) can play a large part in determining how far apart your remotes should be from each other when learning commands. Though most manuals recommend 4 to 6 inches, certain original remotes can send a particularly strong signal that can overwhelm the target remote's circuitry. By placing the remotes further apart, often more than 1 foot, you provide a weaker, easier signal to learn. Conversely, remotes with faint IR beams might need to be placed as little as 1 or 2 inches away. Finally, you can try placing the remotes at a 45° angle if you continue to have problems.

Replace the Batteries

Yes, your original remote might appear to still work...but if you're having trouble learning from it, its signal strength might be too weak or erratic to be picked up properly. Simply installing brand-new batteries can solve your difficulties. If you use NiCads, you might find you need to recharge them more frequently than you would have to replace alkalines because NiCads tend to lose their charge at a fairly fast rate, even when sitting idle.

Weak batteries also generate poor-quality signals with fluctuating code structuring and frequency. The original remote might still work with your devices, but these unreliable signals can cause serious issues when captured and then reproduced.

Lights Out!

Attempting to learn codes in a dark or near-dark room can help with some difficult remotes or buttons. In particular, some fluorescent light fixtures emit a fair amount of IR interference at 50 or 60 Hz that can confuse your learning remote. Incandescent bulbs and plasma televisions also can throw off great quantities of infrared light that can overwhelm sensitive learning remotes.

Software or Hardware?

If you own a Philips Pronto, Home Theater Master MX-700, or other computer-programmable remote control, you have the option to program your remote through a personal computer. Although this is usually the best option, you might find certain codes that the software reports as properly learned don't, in fact, work during testing. The software might even filter or process codes in a way dissimilar to the hardware, causing certain protocols to fail completely.

Something that you might want to try in situations such as these (particularly with the Pronto) is to use the learning function of only the hardware. Unplug the remote from your PC, and learn everything through the built-in menus. In many cases, codes that don't work through the software are captured perfectly. Alternatively, if you are having a similar problem with one of these remotes but aren't using the PC software, you might want to try it.

If your remote's computer software has a Test button to transmit an infrared signal through a connected remote, don't rely on this to test learned signals. Instead, to test codes, download your complete configuration to the remote and use the buttons on the remote, not your software simulator.

Right Side Up

Can't figure out why not even one command will learn? Something that has stumped more than a few users is that the learning eye on certain popular remote controls is not at the front of the unit, but is instead on the bottom.

Two-Way Confusion

Two-way IR equipment is gear that sends infrared signals back to the remote when activated. This is common in components that have higher-end displays and want to indicate state to the remote. If you use two-way IR equipment, such as many high-end receivers or DVD jukeboxes, you might find that infrared signals sent from the device back to the remote can interfere when attempting to learn codes. To provide a clean IR environment, you instead should learn codes in another room, out of sight of the original equipment.

IR Reflections

Sometimes, learning signals on a glossy or reflective table surface can produce IR reflections that confuse the learning remote. Instead, try learning codes on a matte or soft surface, or try placing the remote you are learning to slightly off the table's edge, and then hand-holding the remote you are learning from.

Duck Blind

In the rare case that none of the previous tips works, you might want to try another, more complicated, method. When placing both remotes head-to-head, place a book or magazine in between them. Make sure the book is of sufficient size to block the infrared signal completely. Then, start pressing the button you want to learn on the original remote and continue to hold it down. Set the target remote to learn mode and, while continuing to press, remove the magazine. This blocks off the initial burst of code and, in situations where that burst confuses the learning remote, can sometimes allow a remote to pick up the remaining important parts.

This method also is handy when trying to learn a particular segment of code from a "hardcoded macro" button. For instance, many Sony receivers have input source selection buttons that perform various steps besides changing the receiver's inputs; they also can turn on the TV and change its video input. The problem with this is that the code becomes too long to be learned, and so the whole thing is rejected. The idea is to remove the magazine and put it back so that only the short portion of code you desire is seen by the learning remote. This might take many tries before you can claim success, but it could be worth it. This also works with some "system on" and "system off" buttons, and can work with other mysteriously long infrared sequences.

The Flutter Effect

Is nothing working? Then you might want to try "fluttering" the code. To do this, set your learning remote ready to capture, and then quickly and repeatedly press the source button as many times as you can in one second, effectively sending brief "bursts" of the command to the remote. You might want to experiment with exactly how quickly you flutter the button.

Flyboy

Here's one final method you can try: set your learning remote ready to capture, but this time start with the source remote a distance away—perhaps 6 to 8 feet. Start pressing the source button and then, while continuing to hold down the button, quickly move the original remote closer to your learning remote. Although this is effectively similar to the duck blind technique (see the section "Duck Blind"), it has helped in rare situations. You might want to try combining this with the flutter effect (see the section "The Flutter Effect").

Out of Memory?

Another explanation as to why your remote won't learn a particular code could be that its memory is full. Some remotes have the capability to store only a few codes—15 or 20—though most should be able to hold many more. Low-memory remotes include the Marantz RC2000, Sony RM-AV2000, almost anything from One For All or RadioShack, and many "learning" remotes that ship with televisions or receivers.

To save space, try to use preprogrammed codes first, and use the learning function only to fill in holes for other unsupported devices. Plus, remember to use the "Hold" method detailed in the section "Press or Hold?" because quite a bit of space can be wasted if you unnecessarily use the "Press" method. If all else fails, try resetting the remote and starting over from scratch, being careful not to make too many mistakes that can fragment memory on older, less advanced models, and reduce their overall capacity.

—Daniel Tonks, Remote Central

Work around Proprietary Remote Systems

Sometimes you'll find that even the trickiest tricks won't get your remote codes to work correctly. In these cases, you might have a proprietary system that just won't play nicely. Learn how to work around even these problem children of the remote control world.

In the rare cases where you have a component that doesn't respond to any other IR trickery **[Hack #87]**, you might have stumbled onto one of the so-called *proprietary systems* of the remote control world. These are products from manufacturers who, for one reason or another, just don't play nicely with programmable remotes **[Hack #84]**. This is pretty annoying, so here's what to look out for and (in most cases) how to work around the problem.

Sony VisionTouch

Folks with an older Sony receiver who have just purchased a brand-new learning remote might ask, "Why won't my new universal remote learn any codes from my receiver?"

The very first thing you should check is your receiver. See if the word *Vision-Touch* is printed on the faceplate, if the model number ends with G, or if it came with the Air-Egg remote. If any of these items is true, chances are you are the proud owner of one of Sony's infamous VisionTouch (also known as CAV) controlled systems. The short answer to your problem is that almost no learning remote can work with a VisionTouch receiver (nice surprise, huh?).

A longer explanation is that the VisionTouch system uses a 455-kHz carrier frequency which uses a completely unique IR protocol that is unlearnable by almost every universal remote to date, including the Philips Pronto and Sony's own models. In fact, you can't use "normal" Sony remotes or IR codes to control the receiver; even the Sony RM-AV3000 won't work.

Affected models include the STR-DA90ESG, STR-DE805G, STR-DE815G, STR-DE905G, STR-DE1015G, STR-D760Z, STR-G1ES, STR-G3, STR-GA9ESG, TA-VE800G, and TA-VE810.

If you own one of these models, your options are:

- Buy a One For All remote and mail it in to the manufacturer for its new June 2003 preprogrammed code set.
- Continue using the original remote alongside a universal for the rest of life of the system.
- Buy a new receiver.
- Buy a Philips Pronto, build a special IR demodulator cable, and download the VisionTouch CCF file.
- Spend the money on an RTI TheaterTouch T2, which is the absolute only remote that can learn and transmit VisionTouch codes directly.

Sony's newer receivers advertised with "two-way communications" are fully compatible with regular learning remotes. Sony has dropped the entire VisionTouch system from its lineup, probably due in part to these aftermarket incompatibilities.

High-Frequency IR Systems

Several brands of equipment feature IR systems that operate at frequencies much higher than normal. Though the overwhelming majority of IR remotes operate in the 30– to 40-kHz range, several brands and components don't. Equipment from Bang & Olufsen, some Kenwood equipment, and some lighting control systems use a 455-kHz frequency. More unusual and difficult to work with are the Pioneer and Pioneer Elite components built around 1997 that use a 1.125-MHz carrier frequency.

The only remote that can begin to cover all those frequencies is the Marantz RC2000 MkII, featuring IR coverage up to 1.125 MHz (the original RC2000 will not work). Pioneer owners also can use regular 40-Hz codes from older and newer equipment.

The Philips Pronto models [Hack #85] are capable of learning up to only about 56 kHz, although they have the ability to send at much higher frequencies. However, Philips has devised a way in which its remotes can capture certain high-frequency Bang & Olufsen and Kenwood codes through a trick whereby the remote "guesses" what the code should be via information received at a low frequency, and then matches it up to a high-frequency database version. Codes not previously decoded by Philips can't be learned as well, however.

If you need only 455-kHz coverage for Bang & Olufsen devices, most of Sony's newer remotes (such as the RM-AV3000, RM-AV2100, RM-VL900, and RM-VL700) are capable of capturing up to 500 kHz.

Pace Cable Boxes

Having trouble getting your Pace cable box to operate with your snazzy new learning universal remote control? Well, you might have one of Pace's infamous IRDA boxes. For its 1000 and 2000 series digital cable boxes, Pace designed a remote control using a protocol not actually designed for remote control use. The IRDA variant it used was intended for high-speed data transfer over short distances (such as from a laptop to a printer) instead of using the slow-speed/long-distance standard for remote controls. The IRDA standard specifically includes a format for remote controls; Pace just decided to ignore it.

Trying to learn such codes is much like asking an AM radio to receive FM signals—it simply isn't possible. At this time, the only remote controls that can offer even partial functionality are certain models from One For All, and even then you'll most likely have to ship the remote to the factory for these codes to be added. Apparently, TiVo also is releasing a converter for control of these devices. More recent—and older—Pace models aren't affected by this problematic protocol.

Parity Bits

A somewhat common problem is when a device (such as a cable box) will accept a learned code once, but not twice in a row. For instance, you can enter the channel 12, but not 33. This is not the fault of your new remote, but rather, a very-hard-to-work-with design employed by your equipment.

In these situations, your original remote tacks on a *parity bit* to the end of each code. So, the first time it sends the code (say, a 3), it follows up that code with a 0. The next time it sends the code (another 3), it ends the code with 1. So, your equipment wants to get the code for 3 followed by a 0, and then the code for 3 followed by a 1, to recognize 33.

The problem is that a learning remote can learn or send the signal only one way: the way it learned it. That will be either 3 with the 0 at the end, or 3 with the 1 at the end, never both. Your equipment, unfortunately, will not accept the repeated code unless that code ends with a new parity bit or you send a different code to clear the memory buffer.

The Philips Pronto is the only remote I am aware of that can learn codes with alternating parity bits in the method required for several (but not all) brands of equipment. Using this approach, your button for 3 would send the 3 code followed by another special code to clear the buffer.

But what should that code be? Anything the equipment senses as a real code but doesn't affect operation. So, if you can find a "do-nothing" code, you're in business. However, it might be next to impossible to find such a code for many devices. So, even with the Pronto, you've got a problem that might not have a solution.

RF Satellites

For anyone attempting to consolidate all their remotes into a single unit, satellite systems can pose a potentially major problem: radio frequency (or RF) through-the-wall controls. As satellite receivers are the only mainstream audio/visual devices manufactured today using this technology, IR universal remotes don't include RF support.

Satellite receivers come with either an RF-only remote, or an RF and IR remote. In the former case, you should contact your manufacturer to determine if your model still has IR capability (most do today), in which case you can purchase a new remote or capture the codes from another preprogrammed model. For the latter, you might have to make a choice between RF and IR; most receivers can't operate with both activated at the same time. You can find this adjustment in the receiver's setup screens and you probably will have to toggle it to IR before the remote will transmit IR signals.

Tricking Sony DSS receivers. A trick is available for the Sony SAT-A2, A3, and A4 receivers that will allow use of both RF and IR at the same time:

1. Ensure you're already in RF mode.
2. Press MENU, 9 (SETUP), 1 (REMOTE).
3. Select REMOTE CODES, which will bring up a list of various devices.
4. Now select TV CODES to bring up the list of TV codes.
5. Press EXIT.

Your DSS receiver now should be in a mode where it will respond to both IR and RF codes at the same time. However, you might encounter two problems with this setup. First, if you use the RF remote in the same room as the receiver, it might respond to both RF and IR signals at the same time, as though you pressed the button twice, although this doesn't occur very often. Second, if you have another similar DSS receiver, you can't use this trick, as it always appears to default to the same RF security code (001), meaning that both RF remotes end up controlling both receivers.

If your receiver doesn't include infrared support, you will need to continue controlling it via the original remote (you can up-fit older RF-only DISH receivers with IR).

Other RF Equipment

All other RF equipment, such as most Bose systems, can't be controlled by any universal remote currently in existence. Don't be confused by remotes advertising RF capability; this is not for the control of RF devices, but rather, for controlling IR components from other rooms. The remote sends an RF signal to a base station in the same room as your equipment, which then rebroadcasts it as infrared. There is no way to consolidate control of an RF-only device into a third-party remote, learning or otherwise.

X10 Automation

If you're currently using a wireless remote to control your X10 home automation system, it's most likely transmitting via RF signals to a transceiver. The transceiver plugs into the wall and resends commands as actual X10 signals through the house wiring. If you want to use a universal remote to control your system you will need to purchase an IR-to-X10 transceiver, which takes IR signals from your remote and rebroadcasts them directly as X10.

The only such economical device that I'm aware of is the IR543, shown in Figure 10-12. It is a small, black console unit with white buttons on the top for manual control of up to eight devices on a particular house code. Note that the console can control only one house code at a time and that the IR codes remain the same no matter which code it is set to, making it nearly impossible to work with more than one house code at a time.

Figure 10-12. X10 transceiver

Note that a more expensive model, the IR543AH, is available which will operate all house codes.

Many inexpensive preprogrammed remotes include compatible X10 codes. Pronto users can download a complete X10 CCF file for control of a full 16 modules. Owners of other learning remotes might want to purchase an inexpensive X10 IR remote direct from their X10 dealer to capture codes. Though more expensive and complicated home automation systems exist—some, such as those by JDS Technologies, also can work via IR remotes—the price of the IR543 can't be beat.

Remember that X10 is not an RF-based technology. You can control it via a radio frequency remote, just as you can with an infrared one. X10 actually sends commands through the electrical wiring in your house. This means that any X10 transmitter plugs into the wall, including the IR543 IR-to-X10 controller mentioned earlier, as well as RF-to-X10 and manual pushbutton-to-X10 consoles.

European owners of the Philips Pronto NG RU950 remote controls are in a special situation. Because the RF frequencies of the RU950 and X10 RF transceivers are identical, it's possible to use the built-in RF section of the RU950 to operate those transceivers (thus avoiding the purchase of an IR transceiver). An appropriate PCF file with codes is available on Remote Central (*http://www.remotecentral.com*). This is absolutely the only exception to the "can't control RF devices" rule, and it will absolutely not work with the TSU3000 in North America.

—Daniel Tonks, Remote Central

HACK #89 Disable NetCommand for Faster Response

For users with a NetCommand remote control and Mitsubishi TV, certain functions can be extremely slow. Disabling NetCommand can help out, and it's easy if you know the right code sequence.

I recently had to recalibrate a Mitsubishi TV set on which I had already recentered the picture [Hack #31] once before. That first time, it took forever to get the fine convergence done; however, it was time-consuming because I had to wait for the TV to respond to the remote, not because it was an overly complex task.

When I came back to calibrate the set again, I realized the TV's remote control had NetCommand (*http://www.mitsubishi-tv.com/netcommand.html*), which is, of course, enabled by default. NetCommand, unique to Mitsubishi TV sets, attempts to control all your devices with a graphical user interface, all through the TV. Although that sounds pretty cool, it results in most commands taking a while to execute, as one command might attempt to perform multiple functions, on multiple devices, throughout your system.

When you're calibrating your TV, or working with one particular unit, this extra functionality doesn't help you; in fact, it can drastically slow you down (like my calibration attempts the first time around). The easy fix is to just turn off NetCommand. Go into another room, so your TV won't pick up commands, and follow this sequence:

1. Press and hold the Power button.
2. While holding down Power, press 0, then 0 again, and then 0 once more.
3. Release the Power button.

This turns off NetCommand. In my case, I was able to go back into the TV room and cruise through calibration in no time. To turn NetCommand back on, again leave the room and use this sequence:

1. Press and hold the Power button.
2. While holding down Power, press 9, then 3, and then 5.
3. Release the Power button.

You'll be amazed at how much time this will save you!

—*Robert Jones, Image Perfection*

CHAPTER ELEVEN

HTPC

Hacks 90–93

Although modern home theaters are beginning to take advantage of the latest developments in technology, perhaps nothing is so "modern" as the home theater PC. This is exactly what it sounds like—a personal computer driving your home theater. Instead of having a DVD player, CD player, HDTV set top box, and a slew of other components, an HTPC can handle all these tasks, and often better than the components can on their own!

Here are some common functions of an HTPC, especially on a high-resolution, progressive scan display (such as an HDTV display).

DVD player
: HTPCs feature progressive scan DVD players, able to scale to any resolution, with aspect ratio control capability! You'll spend at least $1,000 for any standalone component with even close to these capabilities.

Line doubler/scaler
: Using a computer as a line doubler or scaler for analog video sources means you can get higher resolutions of even poor source material, such as VHS tapes or standard-definition cable broadcasts. Although even a scaled normal broadcast won't compete with HD, it's a lot better than unscaled content.

CD player
: High-quality, 100% digital setups are possible, to make your computer (and theater) sound like an audiophile's dream CD player.

MP3 jukebox
: You can store all your music, categorized and organized, in one place. This includes audio CDs, downloaded music, and anything else you can think of. Plus, you can get killer jukebox frontends to easily navigate and play music.

Digital photo viewer
: An HTPC can be the ultimate home video display. You even can use it to transform digital photographs into fancy slideshows.

Video game console
: Imagine *Star Wars Racer* at 1024×768 at 60 frames per second, versus N64's 320×240 at 30 frames per second! This blows your average console out of the water!

Internet terminal
: Some people like to switch to the big screen for some Internet surfing. Normally an HTPC is not your primary Internet browser, but it can be a lot of fun to surf some web sites on the big screen!

Movie previewer
: Did you know that some cable modem trailers at QuickTime.com are now 640 pixels wide and play in better-than-VHS quality? See these trailers in near-DVD quality even before they show in the local movie theaters!

Some people use an HTPC to perform only one or two of the preceding tasks; others take advantage of all of these features. Although you don't have to use your HTPC for DVD playback (for example), you'll get superior results, often at greatly reduced costs.

New readers might be interested to know that some new computer video and sound cards provide a lot of flexibility, such as the ability to do custom screen resolutions in 1-pixel increments, making it possible to use an HTPC at high resolution on an HDTV set or a 16:9 projection display without distorted images.

New software developments—including dTV/DScaler from *http://www.dscaler.com*—make it possible to add an inexpensive TV tuner card to a PC. With this low-cost card, you can get stunning picture quality from analog video sources connected to the back of the computer. DScaler software replaces the TV viewing software normally included with the TV tuner card. Some people have reported picture quality being competitive to multithousand-dollar line scalers.

I could go on and on about HTPCs, but I'd rather dive into some cool hacks. To be clear, this chapter isn't a definitive guide to HTPC. If you're just getting into HTPC, start with a good site such as HTPC News (*http://www.htpcnews.com*), which will ground you in the basics. In particular, the guide at *http://www.htpcnews.com/main.php?id=htpcmidbuyersguide_1* will help you build a great HTPC for about $1,000. See also *Building the Perfect PC* (O'Reilly Media Inc.) for an excellent guide to building PCs that also includes step-by-step instructions for building an HTPC. Then you can come back here for specific recommendations and hacks that will help you make the most of your new system.

HACK #90 Choose the Right Display Resolution for Analog-Input HDTVs

When it comes to HTPC, there are literally 10, 15, even more possible resolutions. Because you're in control of the machine, you need to ensure you're displaying things optimally.

This section applies if your HDTV set has analog high-definition component inputs or high-definition RGBHV inputs rather than digital inputs such as DVI or FireWire. Most HDTV sets don't have a VGA input.

If your set has a digital connection (DVI, HDMI, FireWire, etc.), you're already set for computer resolutions. Just plug in your computer and you're ready to go.

Let me begin by saying that you can almost always safely show images in 640×480, at 60 Hz, from any PC, on any HDTV set. If this is all you need, you don't need to read anything else; of course, that's hardly high-end HD, so I'll assume you went to HTPC for the "more is better" philosophy.

You can output almost anything to an HDTV set, as long as the horizontal scan rate is 31.5 or 33.75 kHz (or 45 kHz if your HDTV can do 720p), and the vertical refresh rate is 60 Hz (30 Hz for interlaced).

Most HDTV sets can safely support 480p, 540p, 960i, and 1080i via analog high-definition inputs. If you are lucky enough to have an HDTV set that supports 720p as well, you also can do 1440i. Horizontal resolution is not important here, as the HDTV set can safely support any horizontal resolution (within certain restrictions, as will be explained later). Here's how these resolutions translate to the scan rate mentioned earlier:

- 480p and 960i have a 31.5-kHz horizontal scan rate.
- 540p and 1080i have a 33.75-kHz horizontal scan rate.
- 720p and 1440i have a 45-kHz horizontal scan rate.

The letter *p* stands for progressive scan (noninterlaced); the letter *i* stands for interlaced.

You can do interlaced resolutions that have twice the number of scan lines as a progressive scan resolution. You can think of interlaced resolutions as

What Is Horizontal Scan Rate?

Scan rate is the number of scan lines generated per second. As an example, 480p is actually 480 visible scan lines and 45 invisible scan lines (vertical blanking interval, or the *sync* interval), for a total of 525 scan lines generated per 480p refresh. There are 60 refreshes per second at 60 Hz. This means 525 scan lines, times 60, equals 31,500. There you go; that's a 31.5-kHz horizontal scan rate!

being the same thing as progressive scan resolutions, except that every other refresh (these individual refreshes are called *fields*) has the whole image offset downward by half a scan line. So, in a 1080i setting, the second 540p refresh "fills in the gaps" between the scan lines of the previous 540p refresh! So, basically, two 540p fields combine to make a single 1080i image. At the same vertical refresh rate (per field), 540p and 1080i have the same horizontal and vertical scan rate, which means it is safe to do 540p on all HDTV sets that support 1080i, as long as almost exactly the same number of scan lines are generated per second for both 540p and 1080i. The same goes for the other vertical resolutions.

A few early models of HDTVs can do only 1080i; in other words, neither 480p nor 720p is supported. In this case, don't attempt to do 480p or 960i with these HDTVs, unless you put them inside 1080i timings (letterboxing 480p inside 540p is a recommended workaround).

Some HDTV sets are improperly advertised as being able to do 720p. Be very careful to make sure this is done by the HDTV set [Hack #29] and not by the set top box [Hack #30]! Some HDTV sets can do 720p only through a set top box (which converts 720p to 1080i). You shouldn't attempt to try to use 720p timings in these cases. Instead, use 480p/540p/960i/1080i timings with such HDTVs.

Horizontal resolution doesn't matter; you can display any horizontal resolution on an HDTV set. Also, it is possible to display lower vertical resolutions by letterboxing the computer resolution inside the middle of a higher computer resolution! For instance, you can do 1024×768i in the middle of a 1920×1080i signal.

Attempting to use a horizontal or vertical scan rate beyond the specifications of an analog HDTV set can potentially damage older analog HDTV sets. Most HDTV sets include circuitry to blank the screen if the signal is incompatible for displaying on HDTV sets. However, care must be taken with analog HDTV displays that might not function properly with nonstandard modes. In the event of a scrambled, dim, or distorted image (out of sync), immediately turn off the HDTV as a precaution. You can test HDTV modes on a multisync CRT computer monitor before connecting directly to an HDTV display to reduce the chances of this happening.

Interlaced Versus Progressive Display Formats

Interlacing must be done for 960i and 1080i display resolutions. However, interlacing is not good for computer text; this causes a lot of flickering! Also, most people prefer 480p or 540p to 960i or 1080i when it comes to playing back DVDs on an HDTV set. Similarly, those who are lucky enough to have true 720p HDTV sets probably will prefer to use 720p for DVD playback over any of these resolutions, as well as 1440i.

On some models of HDTV with excellent 1080i image quality, video might actually look better at 1080i than at 480p or 720p. However, on other models of HDTV displays, 480p or 720p can look a lot better. This can vary depending on the usage of the HTPC—whether you use mostly text or mostly video. You might want to test different modes to determine the most suitable modes for your system. The quality of 1080i from a computer varies greatly depending on the type of graphics card and the graphics driver software.

Also, some older NVIDIA GeForce cards don't do interlaced resolutions properly; they simply vibrate even scan lines up and down rather than interlace between both odd and even scan lines. This means you've lost half of your vertical resolution, and the even scan lines are simply vertically pixel-doubled. There should have been more vertical resolution during stationary computer desktop images (if the GeForce interlace was working properly). This might or might not be fixed in newer GeForce drivers or newer versions of PowerStrip [Hack #91]. Currently, most people don't try to do 960i or 1080i, especially for GeForce video card setups.

To clear up any confusion, HDTV always uses a 60-Hz vertical refresh. This is the refresh rate I refer to throughout this hack. For 480p, 540p, and 720p, there is no confusion; it's always 60 Hz.

However, for 1080i (and 960i or 1440i) there is some confusion. Some people say 1080i is 30 Hz and other people say 1080i is 60 Hz. Both groups of people are correct in different ways, when you consider 1080i is all of the following:

- 30 full frames per second
- 60 fields per second (odd scan lines or even scan lines)
- 60 vertical blanking intervals per second (60 sync signals per second)

This first description—which is technically accurate—is where the number 30 comes from. You might remember the old days when interlaced 1024×768i displays on IBM 8514-compatible computer monitors were referred to as 43.5 Hz or 87 Hz. Of course, only one refresh rate for 1024×768i is used on IBM 8514/A, but both 43.5 and 87 can be correct, depending on precisely what is being measured.

Most forum members prefer to refer to HDTV as always being 60 Hz. However, PowerStrip needs to be told to do 30 Hz when doing 960i or 1080i, even though there are actually 60 vertical blanking intervals per second. Bear this in mind when adding interlaced resolutions to PowerStrip.

To keep your life simple, don't worry about interlaced resolutions at first. Focus only on adding progressive scan computer resolutions, such as 540p, as they are easier to set up than interlaced resolutions. Then you can test out interlaced resolutions later to determine if the image looks better.

—*Mark Rejhon*

HACK #91 Add Custom Resolutions with PowerStrip

Breaking the 640×480 display resolution barrier is easier, and more flexible, if you add PowerStrip to your HTPC setup.

Once you've figured out what resolutions you want [Hack #90], you'll almost certainly need some help in actually getting your display to cooperate. Although preconfigured HTPCs might allow you to use custom resolutions right away, more often you'll want software help in getting beyond 640×480. Ever dream of playing *DOOM3* at a sharp 1920×1080i [Hack #93] from your HTPC? Or playing DVDs upconverted to 960×540p? This is when life get very interesting (or complicated)!

You can achieve this and much more via the software known as PowerStrip. PowerStrip is a power-user software program available from EnTech (*http://www.entechtaiwan.com*). It allows you to tweak your computer video card to the maximum, set up custom resolutions, and even fiddle with the image in those resolutions.

You can download a shareware version of PowerStrip from the EnTech web site (see the next section). As of this writing, the latest version available for download is Version 3.52.

Getting PowerStrip

It's a snap to obtain PowerStrip:

1. Download the latest free trial version of PowerStrip from EnTech's web site (*http://www.entechtaiwan.com*).
2. Install the program by running the downloaded executable.

Once you've got PowerStrip, you're ready to create some custom resolutions for display.

The trial version is free to test out and works with the instructions listed within these sections. That said, you really should purchase PowerStrip; it costs only a few dollars. In either case, it is essential that you obtain the latest version of PowerStrip from the web site.

Adding Custom Resolutions

Rather than talking about PowerStrip in concept, I'll show you how to set up a specific custom resolution. As an exercise, let's begin by creating 960×540p. This will guide you through all the essential steps of using PowerStrip.

These instructions are compatible with present versions of PowerStrip, Version 3.x, and might be different with newer versions of PowerStrip.

1. First, hook up an ordinary 15-inch CRT computer monitor to your computer (bigger monitors are also OK).

Don't hook up your computer to your HDTV yet! Do your testing on a computer monitor first, before connecting your HTPC to an HDTV set.

2. Make sure PowerStrip is installed, and that you see a PowerStrip icon in your Windows system tray next to your clock (this means PowerStrip is running).

3. Right-click the PowerStrip icon and then select Display Profiles → Configure.
4. Click the Advanced Timings Options... button.

You might be startled by a complicated-looking screen that pops up. Don't worry; don't touch anything in this screen yet.

5. Click the Custom Resolutions... button.
6. Underneath New Resolution, enter the desired custom resolution. For this exercise, enter **960** in the first box (horizontal) and **540** in the second box (vertical).

If you enter a number greater than 540 in the second box, you must enable the Interlaced checkbox if your HDTV doesn't support 720p. In the case of HDTVs that support 720p, you must enable the Interlaced checkbox if entering a number greater than 720.

7. Enter **60** as the Refresh Rate under Vertical.

If you enabled the Interlaced checkbox, use **30** instead of **60**.

8. Click Add new resolution.
9. If your video driver supports the custom resolution, you will be prompted: "The display driver has accepted the new resolution. Do you want to try to switch to the new resolution at this time?" In other cases, the driver will not be able to add the resolution right away, and you will be prompted to reboot the computer at this time. If this happens, reboot the computer and then repeat Steps 1 through 4 before continuing on with the next step.
10. Click Yes.
11. Click Close. You're back to the Advanced Timings options.
12. Look at the Total number under Vertical Geometry.

There are two Total numbers; make sure you are looking at the one under Vertical, and not under Horizontal.

13. The Vertical Total must match 525, 563, or 750. In rare cases, deviations by 1 scan line might be required, such as 562 instead of 563 when using 540p. Use the smallest possible vertical total that is at least 5% bigger than your desired vertical computer resolution. This means the following settings should be applied:
    - 480p: Vertical Total must equal 525 lines (on 480p-capable HDTVs)
    - 540p: Vertical Total must equal 563 lines (on 1080i-capable HDTVs)
    - 720p: Vertical Total must equal 750 lines (on 720p-capable HDTVs)
    - 960i: Vertical Total must equal 525 lines (on 480p-capable HDTVs)
    - 1080i: Vertical Total must equal 563 lines (on 1080i-capable HDTVs)
    - 1440i: Vertical Total must equal 751 lines (on 720p-capable HDTVs)

    For many video cards, interlaced resolutions work best when configured with an odd number of scan lines. This forces the GeForce card, for example, to do an even number of scan lines for one field and an odd number of scan lines for the next field. Thus, the one-line difference between 720p (750 vertical total) and 1440i (751 vertical total).

    Increase or decrease the Vertical Total until it matches one of the preceding numbers. For the 960×540 exercise, make sure the Vertical Total is 563.

    Be sure to use the vertical resolution to determine the total, not the horizontal resolution.

    You aren't restricted to computer resolutions matching 480/540/960/1080, as long as the Vertical Total is 525 or 563. Basically, you will be letterboxing a computer resolution inside one of the higher resolutions supported by your HDTV set; for example, you could view 1024×768i centered in the middle of a 1920×1080i frame. In this case, you need a Vertical Total of 563.

14. Now look for Refresh rate (Vertical) and Scan rate (Horizontal). Increase or decrease these values until you get as close as possible to the following:
    - A vertical refresh rate of 60 Hz (or 30 Hz for interlaced).
    - A horizontal scan rate of 31.5 kHz or 33.75 kHz (or 48 kHz for 720p if your TV supports it)

This means that:

- 480p and 960i should use a horizontal scan rate as close to 31.5 kHz as possible.
- 540p and 1080i should use a horizontal scan rate as close to 33.75 kHz as possible.
- 720p and 1440i should use a horizontal scan rate as close to 45 kHz as possible.

For the 960×540 exercise, aim for a 33.75-kHz horizontal scan rate.

You'll notice as you change these settings that the adjustments often will jump by counts of 0.3.

The amount of these jumps depends on your graphics card (GeForce, Radeon, etc.).

Every time you tweak one of these adjustments, it affects the other; the two are interrelated. Just keep alternating between adjusting horizontal scan rate and vertical scan rate in an attempt to get as close as possible to the preceding values.

Your goal should be to get less than 0.1 away from the target, so try the following:

- A vertical refresh rate between 59.9 and 60.1 (progressive) or between 29.9 and 30.1 (interlaced)
- A horizontal scan rate between 31.4 and 31.6 for 480p or 960i
- A horizontal scan rate between 33.65 and 33.85 for 540p or 1080i
- A horizontal scan rate between 44.9 and 45.1 for 720p or 1440i (if your TV supports it)

Otherwise, your HDTV might not be able to display anything! Most HDTV sets have very fussy and tight tolerances.

15. Click OK at the bottom of the Advanced Timings options.
16. Click Save as... near the bottom of the PowerStrip Display dialog box under Profiles.
17. Enter the new name for the custom resolution (for example, **960x540**).
18. Click OK.
19. Finally, disconnect the computer from the computer monitor and connect your HTPC to the HDTV set (don't turn off the computer when doing this).
20. Turn on the HDTV set.

Cross your fingers: your Windows Desktop should display itself after having adjusted to the custom resolution. If the image looks stable, go to the next section and tweak it further.

Tweak PowerStrip on an HDTV Set

While the computer is connected to the HDTV set, you can go back to make adjustments by right-clicking the PowerStrip icon, selecting Display Profiles → Configure, and then clicking the Advanced Timings Options... button. Don'tt make adjustments or click anything (except the Cancel button) without being familiar with what you are clicking; this advanced screen is very dangerous. A single click on the wrong number or button and your HDTV screen might blank out or screw up automatically. In this case, unplug the HTPC immediately (or turn off the HDTV set first, if it's faster to do so). Then reconnect your HTPC to your computer monitor. This is what you should do whenever something is wrong with the image on the HDTV set.

You might run into the following common problems with computer images on an HDTV set. Here are included solutions, in the order that you should follow.

The computer image is too tall for the HDTV set. When your image is too tall or too wide, you've got an overscan problem [Hack #92]. Fixing this is somewhat cumbersome and requires creating a new custom computer resolution (all over again). If the image doesn't look very bad, you might find that it's not worth the effort for a negligible effect.

The computer image doesn't appear or is badly distorted. Unplug your HTPC from the HDTV set immediately! Connect your computer monitor back to the HTPC. Then you will need to follow Steps 12 and 13 of the previous section, "Adding Custom Resolutions", to make sure the VGA signal from your HTPC is as close as possible to HDTV specifications. If you can, try to get even closer than you were before, to a 31.5/33.75-kHz horizontal scan rate (or 45 kHz for 720p HDTVs). If you fail, there are a few last resorts you can implement while following Steps 12 and 13 of the previous section:

1. Try adjusting the overscan [Hack #92]. Chances are that the horizontal retrace is well beyond spec of your HDTV set; this can happen when the computer image is too wide or too narrow.
2. Try changing the horizontal scan rate by a fraction of 1%; if you used a value close to 33.8 kHz in your custom resolution, try a value closer to 33.7 kHz this time around.

Don't go more than 1% less than or greater than 31.5/33.75 kHz (and 45 kHz for 720p HDTVs).

3. Try adding 1 to 8 pixels total to the horizontal front porch or back porch. (Many graphics cards can be adjusted in only 8-pixel steps at a time.)

The computer image is flickering. If you are doing interlaced resolutions, such as 960i or 1080i, this is normal. There is nothing you can do about the flickering. Just make sure you've followed the steps in the section "Adding Custom Resolutions" correctly (especially Steps 12 and 13). If you are using a noninterlaced resolution, something is wrong; you should unplug the computer and replug your computer monitor into the HTPC.

The Windows taskbar is 100% chopped off at the bottom of the screen. You can resize the Windows taskbar by dragging its edge upward, to make it thicker. If you can't see the Windows taskbar to do this, unplug the computer from the HDTV set and connect to your computer monitor. After you're done, reconnect to the HDTV set. The problem should have resolved itself; if not, you've got a pretty serious overscan problem [Hack #92].

The computer image is not centered. Under Position and Size, click the up or down arrow—and the left and right arrow—to move your computer image until it is centered and has an equal amount of overscan on all sides.

Ignore the other buttons available to you, and ignore the text that says "Pick up and move."

You normally can keep the computer connected to your HDTV set while making this adjustment.

—*Mark Rejhon*

HACK #92 Adjust the Overscan on Your HDTV Display

Fixing the horizontal and vertical overscan on an HDTV's display isn't trivial; if you're not careful, you can ruin a carefully planned custom resolution.

Overscan problems result from having too much picture. Vertical overscan will have picture bleeding off the top and bottom edges of your display, and horizontal overscan will result in lost picture on the sides. Just the opposite is true for underscan; you don't want a 16:9 picture to appear with black bars, for example. Fixing both categories of problems is easy, once you know exactly what sequence of steps to follow.

Newer versions of PowerStrip **[Hack #91]** include a checkbox called Lock Frequency in the Advanced Timings options. This greatly simplifies the ability to adjust overscan in many cases. Enable this checkbox, and then it becomes much safer to adjust the horizontal image size using the horizontal size adjustment buttons in the Advanced Timings options. If you have this option, you can try this method first and skip the section "Adjusting Horizontal Overscan/Underscan" if the results are more than satisfactory.

Adjusting Horizontal Overscan/Underscan

Here's what you need to do to resolve horizontal overscan/underscan problems:

1. First, unplug the computer from the HDTV set, and plug it into your computer monitor.

Resist the temptation to resize the computer image while the computer is connected to the HDTV set. This is important; the procedure of resizing the image will change the horizontal and vertical scan rate. With most HDTV sets, you will experience frustration if you try to resize the image without testing on a regular computer monitor first.

2. Right-click the PowerStrip **[Hack #91]** icon and then select Display Profiles → Configure. Then click the Advanced Timings Options... button.
3. Under Position and Size click the sideways double arrow buttons (‹› or the ›‹) to grow or shrink the horizontal image size. If your image was too wide on the HDTV, click the ›‹ button. If the image was too narrow, click the ‹› button.

Don't adjust vertical image size; doing so will mess up the timings on most HDTV displays.

4. Because your computer is not currently connected to the HDTV set, you need to guess how many times you need to click the horizontal resize buttons. The horizontal size of the image on your computer monitor doesn't always equal the horizontal size of the image on the HDTV set.

One click usually will resize by about 1% at a time; never use more than 10 clicks at a time.

5. Because resizing often affects the horizontal scan rate and the vertical refresh rate, you will need to readjust them to compensate. Repeat Steps 12 and 13 in the hack on custom resolutions [Hack #91] to make sure the horizontal scan rate is as close as possible to 31.5/33.75 (or 45 for 720p) and that the vertical refresh rate is as close as possible to 60 for progressive scan (30 for interlaced).
6. Finally, disconnect the computer from the computer monitor and connect to the HDTV set (don't turn off the computer when doing this).
7. If the image changed size but is still too wide or too narrow, repeat Steps 1 through 5 in this exercise, clicking the resizing buttons as necessary.

Adjusting Vertical Overscan/Underscan

First and foremost, you should try to avoid tweaking your vertical resolution. This tweaking creates undesirable side effects:

- Fixing overscan requires decreasing vertical computer resolution.
- Fixing underscan requires increasing vertical computer resolution.

If you are in a situation where you must work with the vertical overscan, here's how to do it:

1. Keep your computer connected to your HDTV set.
2. Try to estimate how much total overscan there is. An easy measuring guide is the thickness of the Windows taskbar, while it is resized to its minimum thickness. As a base, know that the taskbar is about 30 pixels thick, then estimate how many taskbars would fit in the back space. You can drag the taskbar around to do measuring estimates. If the overscan is about half the thickness of the taskbar at the bottom and half at the top, you have about 30 pixels worth of vertical overscan.
3. Next, decide on a new custom vertical resolution to use. Subtract the value you estimated in Step 2 from your current custom resolution. For example, if your vertical computer resolution is 540 pixels and you have estimated overscan of about 30 pixels, use a new custom vertical resolution value of 510.
4. Unplug the computer from the HDTV set and plug it into your computer monitor.
5. Finally, use PowerStrip [Hack #91] to add this new custom resolution.

Now you should have no more overscan.

If you are having underscan problems—resulting in black gaps at the top and bottom of your display—you can increase the vertical resolution instead of decreasing it. Just follow the previous steps, except increase vertical resolution instead of decreasing it.

HACK #93 Play Video Games in Custom Resolutions on HDTVs

If possible, you can have a killer gaming experience by using higher, custom resolutions for playing video games.

This section also applies to HDTV sets that weren't originally designed for computer use **[Hack #90]**.

Although some recent video games such as *DOOM3* support custom computer resolutions, many video games require standard computer resolutions. First, try to figure out if the video game can support custom resolutions. Test with your desktop computer monitor first, trying to get the video game to start in a custom computer resolution.

If not, you need PowerStrip **[Hack #91]** to create custom timings out of standard 800×600, 1024×768, 1280×960, or 1280×1024 resolutions. To fill the whole screen properly, I recommend sticking to 640×480, or going all the way up to 1280×960 or 1280×1024. Once you get that running, here are several tips to further optimize performance and display.

Force the Games to Run at 60 Hz

You need to make sure your video games always run at 60 Hz. Stick to Windows video games that use recent versions of DirectX because PowerStrip has direct control over them. One of the ways to force the video games to run at 60 Hz is to use a monitor *.inf* file of a monitor that doesn't support anything beyond 60 Hz.

As mentioned in the section on resolutions, you might have to specify in PowerStrip to run as 30-Hz vertical refresh if you're using an interlaced format; Windows effectively treats this as 60 Hz.

Underscan Problems

Be warned that you generally can't fix underscan problems with 800×600i and 1024×768i resolutions. If you want to play video games at 800×600i or 1024×768i (centered in the middle of 1920×1080i), you will be stuck with letterbox bars at the top and bottom of your computer image! This is yet another reason to stick with progressive resolutions.

Fill the Whole Screen

If you want video games to fill your full screen at standard computer resolutions, use one of the following resolutions:

- 640×480 (via 480p)
- 1280×960i (via 960i)
- 1280×1024i (via 1080i)

Those with 720p HDTVs can achieve success with two additional resolutions:

- 800×600 (via 720p)
- 960×720 (via 720p)

CHAPTER TWELVE

TiVo

Hacks 94–100

Strictly speaking, TiVo and recording television shows aren't part of a home theater. Realistically, though, I don't know how anyone into watching movies and TV could live without a TiVo, so this chapter belongs in this book as much as any other.

For the most part, TiVo works into the home theater equation as just another device to program into the remote control; there really isn't anything special you'll have to do to ensure TiVo works with your cable or satellite receiver.

In the days of VCR recording, things were much trickier, as you would have to pipe the input audio and video signals through a receiver or processor, out another set of inputs, and into your VCR. With TiVo being self-contained and built into most DirecTV receivers, things are far easier.

Because TiVo is so simple to use, I've selected some of the cooler—and more applicable—hacks from the bestselling *TiVo Hacks* (by Raffi Krikorian, O'Reilly) to whet your appetite. Hopefully, these will drive you into a TiVo frenzy, and you'll purchase the much more comprehensive, complete book.

Must-Skim TV

Slide through boring baseball games, pausing only for crowd-pleasing catches and game-tying errors. Cram that unnecessarily lengthy reality show into a scant 15 minutes of onscreen viewing.

TiVo really is "TV your way." Most people assume it's just about pausing live television to visit the restroom, or watching your shows when you're good and ready to do so. But there are more tidbits to be incorporated into your television-watching modus operandi. Here are a few I find rather useful for reducing downtime in potentially engaging shows.

Sliding Straight to the Instant Replay

While the networks do a fine job of highlighting notable sports plays, it just doesn't help if you're out of the room fetching a cold one or more pretzels for your friends. If you're a sports fan and TiVo fanatic, you've no doubt discovered the ⟲ button. With a flick of the thumb, you've skipped backward a few seconds to catch the splash of that San Francisco homer. Flick it again and you're back at the wind-up.

But did you know that the ⟲ button, when combined with ⏩, can reduce an entire game to just minutes—without missing a single crowd-pleasing catch or game-tying error? You won't even have time for that cold one! Here's what you do…

Select a game from TiVo's Now Playing List and start it playing. Hit the ⏩ button on your TiVo remote three times to zip through the game at high speed. Now keep your eye on the network's overlaid onscreen scoreboard, while keeping your thumb hovering over the ⟲ button on your remote. The second you notice a score change, click ⟲ once or twice and you're right there in the action. Repeat as necessary.

Sure, it takes a trained eye, but you'll get it in no time. Forget Sports Center's take on the best plays of the day—make your own.

The 10-FF40-10 Solution

With project shows like *Trading Spaces* and *Junkyard Wars* and reality shows like *Survivor* all the rage these days, there's no end to the number of hours you can waste watching other people fixing what you should be fixing or doing what you'd never in a million years (or for a million dollars) actually do yourself.

The shows, in case you hadn't noticed, are rather formulaic:

1. Introduce the "problem."
2. Watch teams get frustrated and panic for about 30 to 40 minutes.
3. Applaud as things come together or fall apart, depending on the show.

There's a nice recipe for this kind of "Must-Skim TV":

1. Watch the first 10 minutes at normal speed, skipping commercials, of course.

2. Fast forward at the highest possible speed (hit ⏩ three times) through the next 40 minutes, keeping one eye on the green play-bar at the bottom of the screen, the other on the action, in case there is any.
3. Watch the last 10 minutes—finished rooms, pounded bots, voting off the island, etc.—at normal speed, skipping commercials.

With minor variations on this recipe, you can compress *Antiques Roadshow* to about 15 minutes without missing any of the "action." Or, reduce *Trading Spaces* to just 30 minutes and improve its quality drastically by watching *Changing Rooms*, the original BBC version, but that's more advice than a hack. ;-)

Speed Reading

Local newscasters speak just too slowly for words? Want to skim the cream off that State of the Union Address without waiting for the morning paper to summarize it for you?

Turn on Closed Captioning, and hit ⏩ once. You'll cruise through at twice the speed (speeding up to three ⏩s for commercials), and the Closed Captioning will keep up. Closed Captioning is perfect for skimming the news or cramming two or three of those reality shows into one hour of viewing without missing a single word. Or, if you're embarrassed, you can just claim you're working on your speed reading.

—*Rael Dornfest and Cory Doctorow*

Navigation Shortcuts

Remote control shortcuts mean cruising through the TiVo menu system at high velocity.

Typical TiVo menu navigation is serial, moving step-by-step from one window to another using the Next and Previous buttons. While its menu system is rather well laid out and designed for ease of use, after spending a significant amount of time with your TiVo and its remote control, the travel time and number of button presses can prove rather tedious.

Thankfully, a set of navigation shortcuts are built right in allowing you to leap between major menu items in a single bound, that is, a single button press. Table 12-1 lists known remote control shortcuts and their associated menus.

Table 12-1. TiVo remote control shortcuts

Button sequence	Menu displayed
TiVo → 1 or TiVo → TiVo	Now Playing
TiVo → 2	To Do List
TiVo → 3	WishLists
TiVo → 4	Browse By Name
TiVo → 5	Browse By Channel
TiVo → 6	Browse By Time
TiVo → 7	Record Time/Channel (manual recording)
TiVo → 8	TiVo Suggestions
TiVo → 9	Network Showcases
TiVo → ▶	Messages and Setup

The 30-Second Commercial Skip

Forget about fast forwarding through commercials; blaze through in just three to five clicks of your remote.

One of the religious differences between TiVo and ReplayTV owners is how they fast forward through commercials. While TiVo's ⏩ button will get you through those intrusive breaks soon enough, it requires some trained skill to manipulate those ⏩ and ▶ buttons while keeping a keen eye and trusting your instincts to anticipate the end of the commercials. ReplayTV, on the other hand, has a "30-second skip" button, timed specifically for skipping through commercials. Since television commercials are traditionally a multiple of 30 seconds long, ReplayTV owners just hit the "30-second skip" button three to five times to render commercials only a minor annoyance.

This feature is so effective that it has stirred up quite a bit of controversy with the networks, who are getting their hackles up, labeling commercial skipping as theft and even taking ReplayTV to court.

Don't you wish TiVo had a 30-second skip? It does, thanks to a little Easter egg magic.

The ⏭ button on the TiVo remote will bring you to the end of a program, or if you are at the end, it will bring you to the beginning. If you are fast forwarding, the ⏭ button will skip you to the next tick mark. This hack is all about repurposing that button to act as the 30-second skip.

Bring up any recorded program or Live TV. Then, enter the following sequence on your remote:

Select → ▶ → Select → 3 → 0 → Select

You'll know the combination worked when TiVo rings out three Thumbs Up sounds; that chiming "bling!" sound TiVo makes when you press the 👍 button on your remote control. Your ⏭ button will now skip forward by 30 seconds.

Note that this hack isn't permanent. If at any time your TiVo needs to be rebooted—after becoming unplugged or as a result of a power failure—the hack will go away and you will have to reapply it.

Streaming Internet Audio Broadcasts to TiVo

Stream SHOUTcast Internet audio alongside your own MP3s to your HMO-enabled TiVo.

TiVo's Home Media Option (HMO) allows you to stream MP3 music from your home PC through your TiVo to your television and home audio system speakers. But what of those online music broadcasts, streamed talk shows, and specials? The HMO can handle those too, thanks to Tobias Hoellrich's *m3ugen.pl* (*http://www.kahunaburger.com/blog/archives/000054.html*).

m3ugen.pl is a simple Perl (*http://www.perl.com*) script that generates *.m3u* files from SHOUTcast playlists. These *.m3u* files are what TiVo uses to stream music from PCs and Macs in your house to TiVo.

The Code

Save the following code to a file named *m3ugen.pl* somewhere on your PC or Mac's hard drive:

```
#!c:\perl\bin\perl.exe

use strict;
use HTML::LinkExtor;
use LWP::Simple;
use URI::URL;
use constant PROVIDER => qq{http://www.shoutcast.com/};
use constant DIRECTORY => PROVIDER.qq{directory/};

my $genre=$ARGV[0];
my $results=$ARGV[1];
my $outfile=$ARGV[2];
unless (defined($genre) &&
    defined($results) && $results && $results <= 25 &&
    defined($outfile)) {
  die qq{Usage:\t$0 [genre] [numresults] [outfile]\n}.
    qq{\tgenre=TopTen,House,Blues,Punk,...\n}.
    qq{\tnumresults=1..25\n}.
    qq{\toutfile=m3u output file\n};
}
```

```
my @playlists=getPlaylists(DIRECTORY.
qq{?sgenre=$genre&numresult=$results},PROVIDER);
unless (scalar(@playlists)) {
  die "No results found - unable to create playlist\n";
  exit(0);
}
@playlists=mapForTiVo(@playlists);

open(OUT,">".$outfile) or die "Unable to create output file - $!";
print OUT qq{#EXTM3U\n};

foreach my $entry (@playlists) {
  my($url,$title)=%$entry;
  print OUT qq{#EXTINF:,$title\n$url\n};
}
close(OUT);

sub getPlaylists {
  my($url,$base)=@_;
  my(@results);
  my $content=get($url);

  unless (defined($content) && length($content)) {
    warn qq{Unable to fetch "$url"\n};
    return @results;
  }

  my $parser=HTML::LinkExtor->new(sub {
    my($t,%a)=@_;
    return if $t ne 'a';
    push(@results,$a{href}) if($a{href}=~/filename\.pls$/i);
  });

  $parser->parse($content);
  @results = map {$_=url($_,$base)->abs;} @results;
  return @results;
}

sub mapForTiVo {
  my(@list)=@_;
  my(@results);

  foreach my $url (@list) {
    my $content=get($url);
    next unless(defined($content) && length($content));

    my($file);
    foreach my $line (split(/[\n\r]/,$content)) {
      if ($line =~ /^File\d+=(.*)$/i) {
        my $u=URI::URL->new($1);
        $u->path(""),$file=$u->abs if($u->path eq '/' || $u->path eq '');
      } elsif ($line =~ /^Title\d+\s*=(.*)$/i && defined($file)) {
        push(@results,{$file => $1});
```

```
            last;
          }
        }
      }
      return @results;
    }
```

Mac OS X and Unix/Linux users should alter the first line to point to the proper location of Perl:

```
#!/usr/bin/perl
```

Running the Hack

Let's say I share the path *c:\tivo\mp3* on my Windows PC with my TiVo's HMO Music option. Let's also say that I want the top 15 Punk stations from SHOUTcast (*http://www.shoutcast.com*), so I can listen to the music broadcasts of these stations through my TiVo. I'd invoke the *m3ugen.pl* at my PC's DOS prompt (Start → Run... → *command*), like so:

```
C:\> perl m3ugen.pl Punk 15 c:\tivo\mp3\Punk.m3u
```

If you don't have a copy of the Perl programming language on your system, download and install a copy of ActivePerl from ActiveState (*http://www.activestate.com/Products/ActivePerl/*).

On a Mac, sharing the path *~/tivo/mp3/Punk.m3u*, I'd run *m3ugen.pl* from the Terminal (Applications → Utilities → Terminal), as follows:

```
% perl m3ugen.pl Punk 15 ~/tivo/mp3/Punk.m3u
```

The script will visit *www.shoutcast.com*, look up the Punk category, extract the playlists of the 15 most popular Punk stations, download each individual playlist, find TiVo-compatible server entries, and generate a *Punk.m3u* file.

The HMO only likes streams with URLs of the format:

```
http://hostname:port
```

It doesn't work with more "involved" URLs like:

```
http://hostname:port/dir/dir
```

When asked to play a stream at such a URL, the HMO will simply fail to do so.

You'll then magically find a *Punk.m3u* section under Music and Photos → Music section of your HMO-enabled TiVo. Select it and listen to some of the finest punk music streamed over the Net. See Figure 12-1.

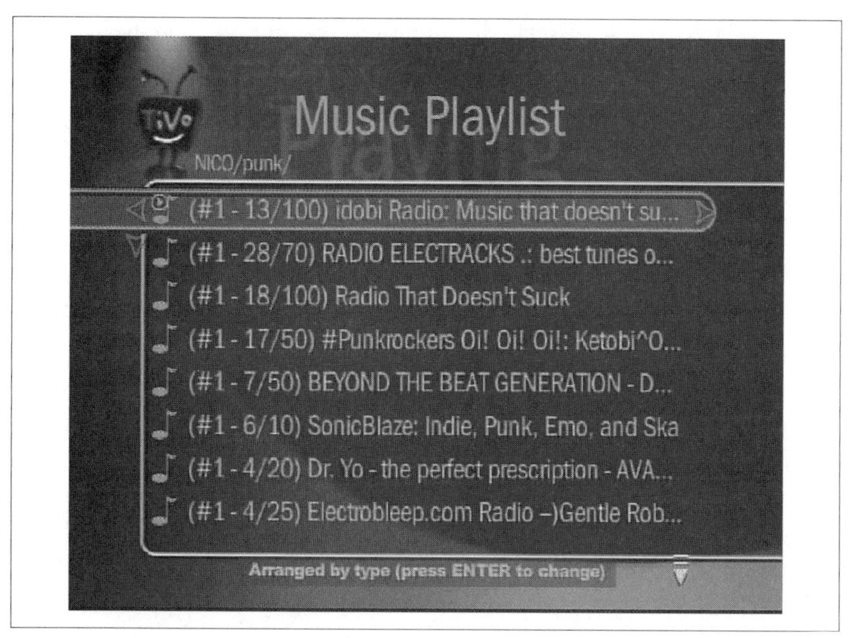

Figure 12-1. The top 15 Punk SHOUTcast stations brought to your TiVo

The HMO doesn't buffer streamed Internet audio. If the stream pauses due to traffic congestion or any other problem, the HMO just punts it and moves on.

If Punk's not quite your cup of tea, visit *http://www.shoutcast.com* and choose a more appropriate genre from the "--Choose a genre --" pull-down list on the right side of the page. The page will refresh, and you'll end up on a page whose URL begins with *http://www.shoutcast.com/directory/?sgenre=*. The word after the = (equals sign) is what you should feed *m3ugen.pl* on the command line to find the most popular streams of that genre.

—*Tobias Hoellrich*

HACK #98 Signing Up for the Home Media Option

Expand your Series 2's capabilities with TiVo, Inc.'s software upgrade.

The TiVo Home Media Option (HMO) is a $99 add-on (available at *http://www.tivo.com/4.9.asp*) that you can purchase for your Standalone Series 2 TiVo. The HMO brings a whole slew of features, including the ability to schedule recordings over the Web, play MP3s through your television and attached stereo system, display digital photos on your TV, and even stream television shows between Series 2 TiVos.

Unfortunately, there is no Home Media Option support for Series 1 TiVos or DirecTiVos (TiVo/DirecTV combination) at the time of this writing.

Signing up for the Home Media Option is pretty simple.

First, get your networked Series 2 TiVo onto your home broadband connection. Any high-speed connection will do just fine.

Point your browser to *http://www.tivo.com* and click the HMO link, or go directly to the HMO page: *http://www.tivo.com/4.9.asp*. You will need the email address and password you provided TiVo, Inc. when you first activated your TiVo. Don't worry if you've forgotten your password; you can either have a new password assigned and sent to your email address, or you can set everything up again by providing TiVo, Inc. with your service number. You can find that number on the New Messages & Setup → System Information screen.

Once you're signed up and have your TiVo connected to your home network, you will have to wait for your TiVo to connect to the service to activate itself. If you're a little impatient, you might try forcing your TiVo to connect and download the HMO option right away by having it make its Daily Call: TiVo Messages & Setup → Settings → Phone & Network Setup → Connect to the TiVo service now. If you notice the addition of Photos and Music to your TiVo Central menu, then the install worked.

Remotely Scheduling a Recording

With the TiVo Home Media Option, remotely scheduling a recording is as easy as going to *http://www.tivo.com* and making your selection.

Scheduling a recording via the Home Media Option (HMO) is supposed to be pretty simple, and indeed it is.

Before you can use TiVo's HMO to administer your TiVo Series 2 remotely, you'll need to log in to TiVo Central Online at *http://www.tivo.com/tco*. You'll be prompted (see Figure 12-2) for the email address and password that you used when you signed up for the HMO.

After successfully logging in, you're presented with a page for scheduling a recording (see Figure 12-3). First, choose which TiVo you want to record on; assuming you're fortunate enough to own more than one, there will be multiple choices.

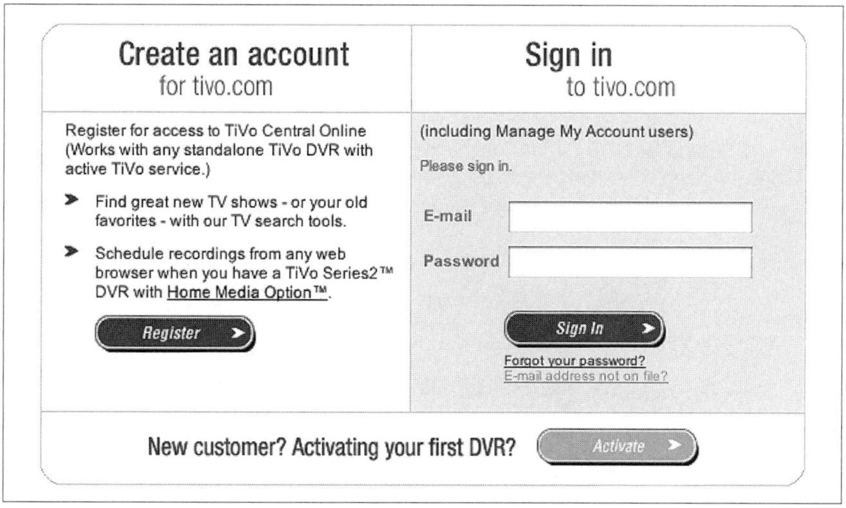

Figure 12-2. Logging in to TiVo Central Online

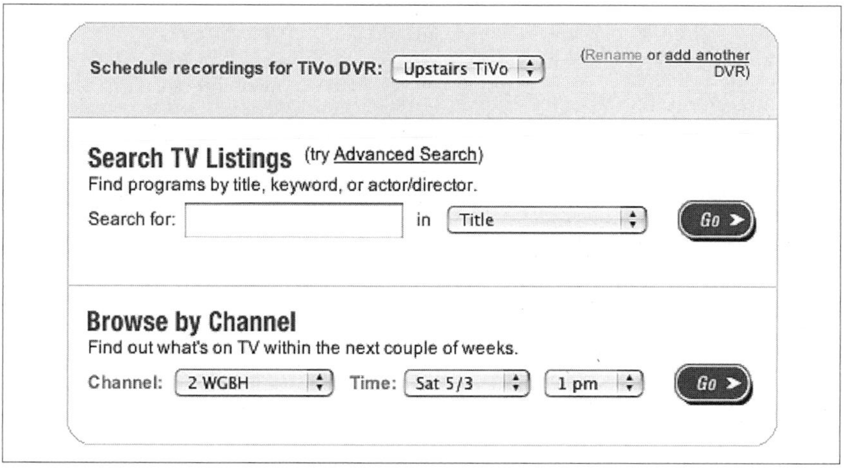

Figure 12-3. Scheduling a recording through TiVo Central Online

In much the same manner as TiVo itself, you can Search TV Listings for a particular show by title, title/description, or actor/director; or, you can Browse by Channel for a gander at what's on TV over the next couple of days.

Advanced Search (shown in Figure 12-4) is reminiscent of the Advanced Wishlists. If you choose that interface, you can set up "and" relationships in the search you want to make. Name that show and the actor/director and the category. TiVo Central Online will try to find it for you.

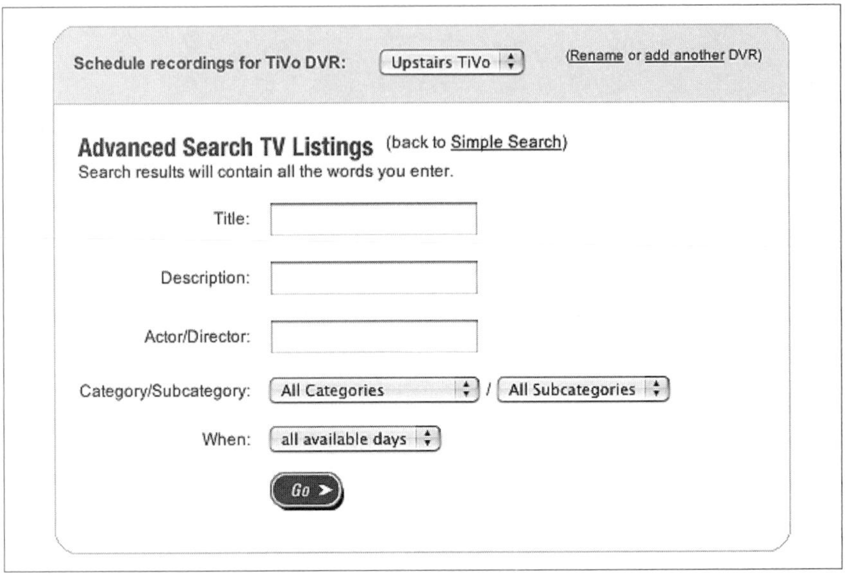

Schedule recordings for TiVo DVR: [Upstairs TiVo] (Rename or add another DVR)

Advanced Search TV Listings (back to Simple Search)

Search results will contain all the words you enter.

Title:

Description:

Actor/Director:

Category/Subcategory: [All Categories] / [All Subcategories]

When: [all available days]

Go

Figure 12-4. Using TiVo Central Online's Advanced Search interface

This site behaves just as you expect your TiVo would. Figure 12-5 shows it listing all the upcoming episodes.

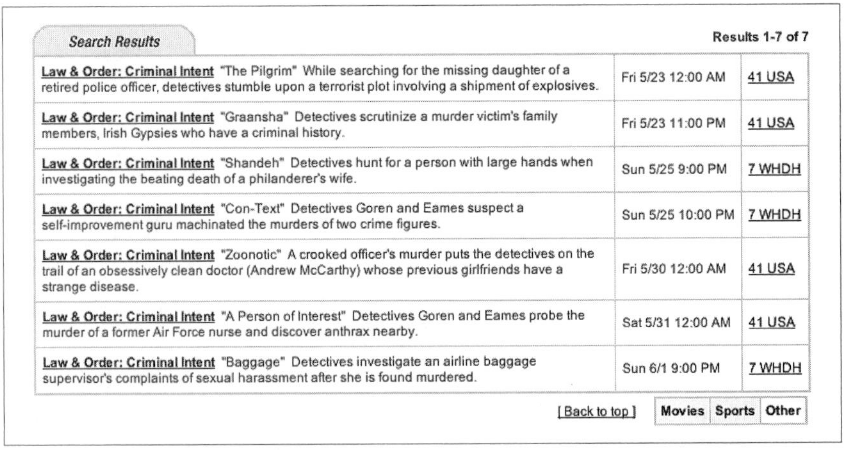

Search Results — Results 1-7 of 7

Program	Time	Channel
Law & Order: Criminal Intent "The Pilgrim" While searching for the missing daughter of a retired police officer, detectives stumble upon a terrorist plot involving a shipment of explosives.	Fri 5/23 12:00 AM	41 USA
Law & Order: Criminal Intent "Graansha" Detectives scrutinize a murder victim's family members, Irish Gypsies who have a criminal history.	Fri 5/23 11:00 PM	41 USA
Law & Order: Criminal Intent "Shandeh" Detectives hunt for a person with large hands when investigating the beating death of a philanderer's wife.	Sun 5/25 9:00 PM	7 WHDH
Law & Order: Criminal Intent "Con-Text" Detectives Goren and Eames suspect a self-improvement guru machinated the murders of two crime figures.	Sun 5/25 10:00 PM	7 WHDH
Law & Order: Criminal Intent "Zoonotic" A crooked officer's murder puts the detectives on the trail of an obsessively clean doctor (Andrew McCarthy) whose previous girlfriends have a strange disease.	Fri 5/30 12:00 AM	41 USA
Law & Order: Criminal Intent "A Person of Interest" Detectives Goren and Eames probe the murder of a former Air Force nurse and discover anthrax nearby.	Sat 5/31 12:00 AM	41 USA
Law & Order: Criminal Intent "Baggage" Detectives investigate an airline baggage supervisor's complaints of sexual harassment after she is found murdered.	Sun 6/1 9:00 PM	7 WHDH

[Back to top] Movies | Sports | Other

Figure 12-5. The results of the program listing search, showing all the shows that match the query

Select which upcoming show you want recorded by clicking on it. You will be given the option of recording just the one episode or getting a Season Pass to it (see Figure 12-6).

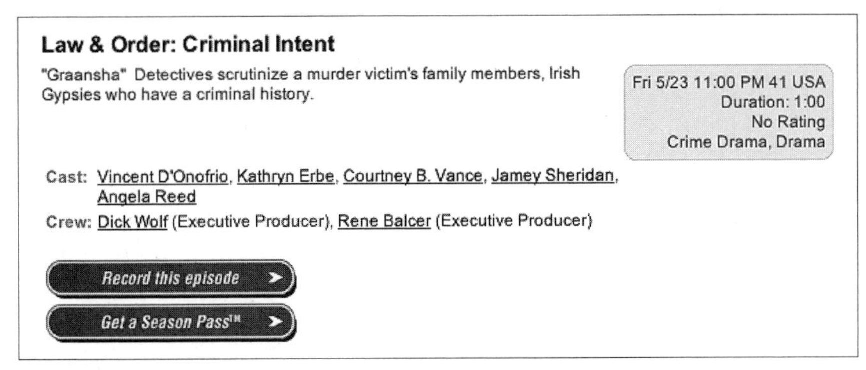

Figure 12-6. Record this episode or get a Season Pass to the show

The recording options (shown in Figure 12-7) are similar to the options you have when you schedule a recording on the TiVo itself: you can have the box keep the recording until space is needed, specify whether to start and stop the episode on time, and pick the quality of the recording. There is one recording option that might not seem that familiar: "What priority?"

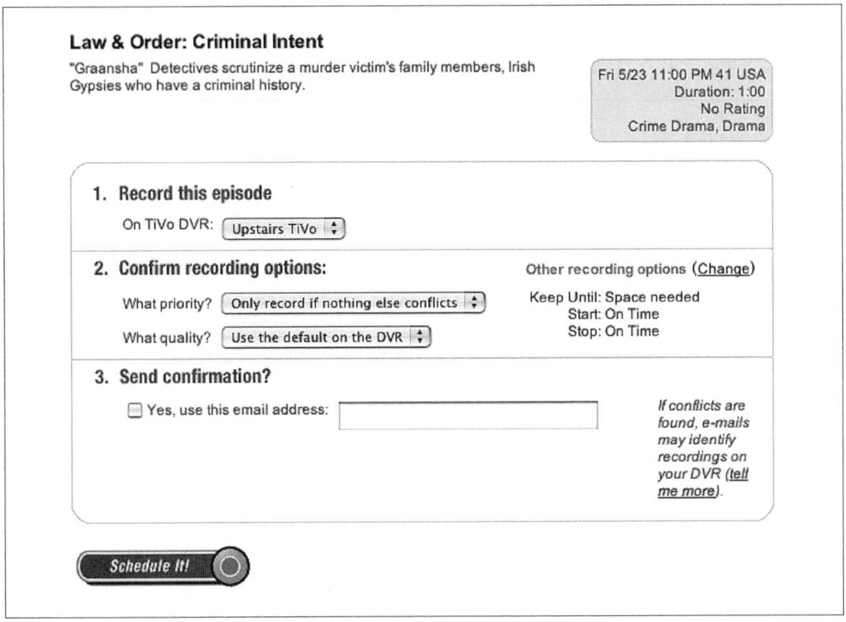

Figure 12-7. Setting all the recording options when scheduling a show to be recorded

Because your TiVo is not continuously connected to TiVo, Inc.'s web site, it doesn't know exactly what shows your TiVo is already scheduled to record. Because your broadband-enabled PVR connects to TiVo, Inc. only a few times a day; the information the HMO has may not be as up to date as it would like to be. So, if you ask the HMO to schedule a recording for you, it can't know if there is a conflict brewing. To work around this problem, there are two options: "Only record if nothing else conflicts" and "Cancel other programs if necessary." These selections do exactly as they say: the first records if there are no other entries on the To Do List, and the second trumps anything else already in the schedule.

If you ask for a Season Pass to be added to your lineup, you're presented with a similar screen, but this time you are able to modify the Season Pass options, as shown in Figure 12-8. The only thing you can't change is the ordering of your passes; to reorder your passes, you're going to have to go to your TiVo.

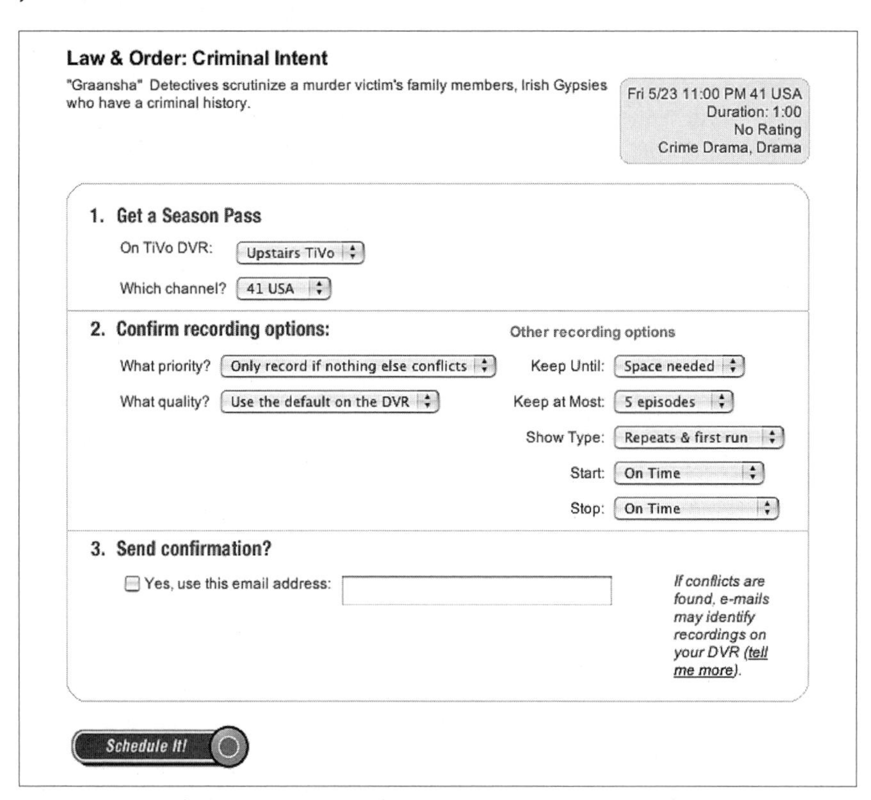

Figure 12-8. Set the Season Pass recording options remotely through the HMO

After you have asked TiVo Central Online to schedule a show for you, TiVo, Inc. just waits for contact from your PVR. And here is the main reason that you're now using your broadband connection for your TiVo: your TiVo doesn't have to use your phone line anymore. Now, it's really simple and basically unobtrusive for your TiVo to make its Daily Call. A few times a day, sometimes as often as once an hour, your broadband-enabled TiVo will connect over the Internet and receive any of these queued-up requests from TiVo Central Online. If it receives any, it will start to incorporate them into its schedule.

If all goes as it should, the next time you turn on your TiVo you should find a message waiting for you, confirming that your request has been added to the To Do List. And if you check out the To Do List, the show will be there, as expected.

If all doesn't go to plan, for example, if there's a scheduling conflict, things get a little more complicated. An hour or so after you ask the HMO to schedule a recording—probably around the time your box decides to connect to TiVo Central Online—you'll receive an email message, informing you of the conflict. A typical email of this sort looks something like this:

```
From: TiVo - Your Upstairs TiVo DVR confirmations@tivo.com
Date: Sat May 3, 2003 15:15:07 America/New_York
To:
Subject: Status of your request for "Law & Order: Criminal Intent"

Your online request for "Law & Order: Criminal Intent : Malignant" has been
received. However, this episode COULD NOT be scheduled to record because it
conflicts with a previously scheduled, higher priority recording.

Will NOT record:
 Law & Order: Criminal Intent 5/4  10:00 pm-11:00 pm   7 WHDH
 overlaps with  Alias  9:00 pm - 11:00 pm

To change which programs will and won't record, go to the
To Do List and Recording History. To get there:
-  press the TiVo button to go to TiVo Central
-  choose "Pick Programs to Record"
-  choose "To Do List"

Best regards,

TiVo
http://www.tivo.com/support
```

This is the TiVo HMO's way of performing conflict resolution. Now it's up to you to go back to TiVo Central Online and reschedule.

Because your TiVo isn't constantly connected, and your web browser is not talking directly to your TiVo, the remote abilities of the HMO are quite limited. There is no way for you to take a look at the To Do List, and there is no way for you to manage your Season Passes. In fact, you are not able to do anything except schedule a recording, and even that may not go to plan, because you have no way of knowing—until an hour has passed—whether there is a conflict. In fact, if you want to record a show that starts in five minutes, there is a very good chance that Remote Scheduling will not do it for you.

Those of you who want it all, take a look at TiVoWeb instead.

Moving Shows Between TiVo Units

Of course you could just extract a television show from one TiVo and insert it into another, but there's an easier way to transfer from one unit to another.

With a growing archive of *tmf* files on your PC, you can easily shuffle shows from one TiVo to another in your home: *The Love Boat* in the bedroom, *Serpico* in the media room, and *Blue's Clues* in the playroom. Think of the collaborative scheduling possibilities: if two television shows are on at the same time, have one TiVo record one, have another TiVo record the other show, extract the *tmf* files, and move them about at will. Stock the family TiVo with only family-friendly programming, keeping those *X-Files* episodes away from your five-year-old.

There's really no need to FTP extract files from one TiVo to your PC, only to insert them into another TiVo—unless, of course, you want to archive the files in the process. Thanks to an extension of FTP called FXP, you can cross-transfer between two FTP servers, using your FTP client only as the middleman for making the introductions, controlling the connection, and deciding what goes where. No data is ever actually sent to or stored on your PC. Your FTP client simply asks one TiVo to send files to the other, and vice versa. The most popular Windows FTP application, WS_FTP (*http://www.ipswitch.com/Products/WS_FTP/index.html*) supports FXP transfers quite nicely. Open up two TiVos in two windows, select the *tmf* file to transfer, drag, and drop. LundFXP (*http://www.lundman.net/unix/lundfxp.html*) is an FXP client (still in alpha testing at the time of this writing) for Mac OS X, Windows, and Unix variants.

Index

We'd like to hear your suggestions for improving our indexes. Send email to *index@oreilly.com*.

B

C

D

I

J

K

L

M

T

U

V

W

X

Z

Colophon

Our look is the result of reader comments, our own experimentation, and feedback from distribution channels. Distinctive covers complement our distinctive approach to technical topics, breathing personality and life into potentially dry subjects.

The tool on the cover of *Home Theater Hacks* is a film loupe. Invented around 1775, the loupe is a small magnifier used by jewelers, watchmakers, and photographers. Loupes come in several varieties. Watchmakers prefer the kind that are held by the eye socket. There are loupes that are worn as eyeglasses; others clip onto eyeglasses. Some are illuminated or have built-in tweezers. Photographers typically use handheld loupes to more closely study negatives and contact sheets.

Mary Anne Weeks Mayo was the production editor, Audrey Doyle was the copyeditor, and Leanne Soylemez was the proofreader for *Home Theater Hacks*. Darren Kelly provided quality control. Mary Agner provided production assistance. Reg Aubrey wrote the index.

Hanna Dyer designed the cover of this book, based on a series design by Edie Freedman. The cover image is an original photograph by Classic PIO Entertainment. Clay Fernald produced the cover layout with QuarkXPress 4.1 using Adobe's Helvetica Neue and ITC Garamond fonts.

Melanie Wang designed the interior layout, based on a series design by David Futato. This book was converted by Julie Hawks to FrameMaker 5.5.6 with a format conversion tool created by Erik Ray, Jason McIntosh, Neil Walls, and Mike Sierra that uses Perl and XML technologies. The text font is Linotype Birka; the heading font is Adobe Helvetica Neue Condensed; and the code font is LucasFont's TheSans Mono Condensed. The illustrations that appear in the book were produced by Robert Romano and Jessamyn Read using Macromedia FreeHand MX and Adobe Photoshop CS. This colophon was written by Mary Anne Weeks Mayo.